SUPERSTRINGS AND PARTICLE THEORY

SUPERSTRINGS and PARTICLE THEORY

Tuscaloosa, Alabama
8–11 November 1989

Editors
L. Clavelli
B. Harms

Published by

World Scientific Publishing Co. Pte. Ltd.

P O Box 128, Farrer Road, Singapore 9128

USA office: 687 Hartwell Street, Teaneck, NJ 07666

UK office: 73 Lynton Mead, Totteridge, London N20 8DH

SUPERSTRINGS AND PARTICLE THEORY

ISBN 981-02-0157-5

Printed in Singapore by Utopia Press.

PREFACE

Six years after the discovery by Green and Schwarz that superstrings are finite and anomaly-free, superstring theory continues to excite the imaginations of particle theorists as a potential theory of everything. There has been much growth in the field since its modern rebirth. On the one hand, there have been attempts to make the theory more realistic by reducing the number of space-time dimensions and by constructing models with more realistic gauge groups. On the other hand, there has been much work of an exploratory nature on the possible mathematical foundations of the theory. Both approaches were strongly represented at the Conference on Superstrings and Particle Theory held at The University of Alabama in Tuscaloosa November 8–11, 1989. Talks involving the first approach included recent work on lower dimensional models, analysis of string loop effects, and thermal effects. Discussions of the mathematical foundations of the theory included conformal field theory, p-adic strings, and W-algebras.

This conference was held in conjunction with the annual meeting of the Southeastern Section of The American Physical Society (SESAPS). It is hoped that joint meetings such as this might help to reduce the polarization of the physics community. An initial grant of seed money from SESAPS was instrumental in bringing about this joint conference and attracting other financial support.

This conference was also supported by generous grants from the Department of Energy, Oak Ridge Associated Universities, and the Alabama Department of Economic and Community Affairs. The University of Alabama provided support through the College of Arts and Sciences and the Capstone International Program.

Many thanks are owed to the staff of the Paul Bryant Conference Center and the College of Continuing Studies, especially to Joe Ann Slact and Joanne Miller. In addition to the editors the local organizing committee consisted of Stanley Jones and Allen Stern of the University of Alabama theory group. The program advisory committee was composed of Paul Frampton, James Gates, Rafael Nepomechie, Joseph Polchinski, and Pierre Ramond. Thanks are also due to the UA Seebeck Computer Center for providing terminals and BITNET accounts for use by the participants.

Tuscaloosa, 1990

Editors
Louis J. Clavelli
Benjamin C. Harms

CONTENTS

Some Topics and Issues in Closed String Field Theory

S.-H. Henry Tye
Newman Laboratory of Nuclear Studies
Cornell University, Ithaca, NY 14853

Abstract

To connect string theory to our four-dimensional world, we must find the string model (or string field vacuum) which describes nature. The status of string models is briefly reviewed to motivate the need for string field theory. Concrete suggestions of a new direction to construct closed string field theory actions are discussed. In particular, the configuration space of the string field is extended to include all classical string vacua. On shell unitarity introduces a new string symmetry beyond the usual BRST symmetry.

I. Introduction

As far as we know, superstring theory is the only theory that offers a realistic hope of unifying all forces and matter in nature. Since the discovery that superstring theories, which naturally incorporate fermionic, gravitational and Yang-Mills interactions, can be finite and anomaly free in ten space-time dimensions[1,2], theorists have focused considerable attention on the possibility of using these models to describe nature, i.e. the four-dimensional world we live in. The most glaring obstacle to overcome in this endeavor (at least at first sight) is transforming an apparently ten-dimensional theory into a four-dimensional one. The initial suggestions for handling this problem centered on compactifying six of the space dimensions to leave an effective four dimensional model.[3,4] This is akin to the original Kaluza-Klein idea. More recently considerable progress has been made on this issue due in part to a philosophical shift away from the viewpoint that superstrings are fundamentally ten dimensional objects, and towards constructing superstring models directly in four dimensions.[5] After all, what singles out the so-called critical dimension of a string theory (e.g. 10 and 26 for the superstring and the bosonic string respectively) is the cancellation of the conformal anomaly[1] which requires only that we choose the appropriate number of world sheet degrees of freedom (for the given local world-sheet (super) symmetry) and not that we must interpret some particular number of these as space-time coordinates. Space-time unitarity and Lorentz invariance impose stringent constraints on the world-sheet theory in particular requiring local reparametrization invariance, world-sheet supersymmetry (for superstrings), and invariance under discrete reparametrizations, referred to as modular transformations.

The simplest way to construct a four-dimensional string model with chiral fermions is to use free world sheet fermions (i.e. Ising fields or chiral bosons) for internal degrees of freedom. Of course, more complicated constructions can involve other conformal field theories. One such construction was discussed in Ref. [6]. The surprise of Gepner's model is its relation to the original Calabi-Yau approach used earlier.[3] This indicates a deep underlying structure of string theory which remains to be understood. Great progress along this direction has been made in Ref. [7].

Let us estimate the number of string models in four dimensions. For heterotic string models the largest gauge group possible has rank 22, where $U(1)$ is counted as rank 1. By counting the number of possible gauge groups and their possible fermion representations, I would guess that the number of four-dimensional string models is of the order of millions. A recent computer generation of four dimensional string models provides a lower bound. Based on spin structure construction using only Ramond Neveu-Schwarz world sheet fermions, Senechal[8] explicitly generated more than a hundred thousand string models with rank 22 gauge groups. Also, the number of Calabi-Yau manifolds is of the order of ten thousand.[9] What are these large sets of string models?

In the language of string field theory, these models correspond to different vacuua (or ground states) of the heterotic string field theory, very much like the many local minima of the potential in a molecule hamiltonian well studied in chemistry. To compare the different vacuua, we must be able to treat the string field dynamics. In usual quantum field theory, this implies a proper treatment of off-shell properties.

There are a number of suggestions of treating off-shell string properties, such as universal moduli space, grassmanian, renormalization flow, integrable systems etc; however, by far the most concrete approach so far is to start from a covariant string field theory action. The covariant string field theory must reproduce the perturbation expansion that is known as dual diagrams. But they should also allow us to analyze off-shell properties as well as non-perturbative properties which are impossible to see from the perturbative expansions. Even if string field theory turns out to be inadequate to treat the many vacua problem, it should reveal the underlying deep structure of string theory and point the correct direction for further progress.

Since the string perturbative series are known in terms of (super) Riemanian surfaces, there must exist field theory actions from which Feynmann rules can be derived. In fact, in the light-cone gauge, they were constructed many years ago.[10] Since the theory is covariant, it is natural to expect that covariant string field actions exist. For open strings, Witten has actually written down such actions.[11] They take the form

$$S = \int \left[\frac{1}{2} \Phi \cdot Q \Phi + \frac{2}{3} \Phi \cdot \Phi \cdot \Phi \right] \tag{1.1}$$

where Φ is a covariant string field, and Q is the BRST operator which obeys $Q^2 = 0$.[12] For open bosonic string, this reproduces the usual dual tree diagrams. However, for the phenomenologically interesting heterotic string theory, the covariant action is not known. In fact, the difficulty partly lies with the closed string field theory in general, where gravity and space-time structure come in. Furthermore, it becomes clear that choosing the string variables X_μ to be the co-ordinates of the string field does not allow us to reach many of the interesting string models that are constructed. Clearly, a more general choice of the configuration space is needed. We shall discuss some of these issues. To simplify the discussions, we shall concentrate mostly on the closed bosonic string theory.

String field theory is an active research topic and there are a number of review articles and many papers on the subject.[13] Here I shall only discuss some thoughts of my own, and these lectures are not meant to be a review of the subject. In Section II, we review briefly conformal field theory and related topics in string theory to set up the notation for later use. Here we shall follow the formulation of Polyakov and LeClair, Peskin and Prietschopf[14], where many earlier references can be found. In Section III, the issue of the configuration space of closed string field theory is discussed. We propose a space for bosonic (and heterotic) string field theories in which all classical vacua are included naturally. We also illustrate how space-time emerges. The existence of a new string symmetry beyond the usual BRST symmetry is pointed out. In Section IV, the structure of the projection operator P introduced earlier is discussed. Remaining problems are outlined in Section V.

II. Preliminaries

Let (τ, σ) describe the two-dimensional string world-sheet. Then, under a Euclidean continuation, the world-sheet can be described by a complex co-ordinate,

$$z = \exp(\tau + i\sigma) \tag{2.1}$$

and its complex conjugate $\overline{z}$.

The flat metric of a plane $g_{\alpha\beta} = \delta_{\alpha\beta}$ becomes, in the complex co-ordinate

$$g_{z\bar{z}} = \frac{1}{2}, \quad g_{zz} = 0 = g_{\bar{z}\bar{z}} \tag{2.2}$$

The energy-momentum tensor of a scale-invariant theory takes the form

$$\begin{aligned} T_{z\bar{z}} &= 0 \\ T_{zz}(z,\bar{z}) &= T(z), \\ T_{\bar{z}\bar{z}}(z,\bar{z}) &= T(\bar{z}) \end{aligned} \tag{2.3}$$

Let us consider only $T(z)$. For a set of free bosons, $X_\mu, \mu = 1, 2, \cdots D$,

$$T(z) = -\frac{1}{2}\sum_\mu :(\partial_z X_\mu)^2: \tag{2.4}$$

where the correlation function

$$< X_\mu(z_1) X_v(z_2) >= -\delta_{\mu v} ln(z_1 - z_2) \tag{2.5}$$

In string theory, X_μ are interpreted as string variables. In a general conformal field theory

$$T(z)T(w) \sim \frac{c/2}{(z-w)^4} + \frac{2}{(z-w)^2}T(w) + \frac{1}{z-w}\partial_w T + \ldots \tag{2.6}$$

where c is the central charge ($c = 1$ for each free boson) and the regular terms as $z \to w$ are neglected in Eq. (2.6). The infinitesimal conformal transformations $\epsilon(z) \propto z^{n+1}$ in $z \to z + \epsilon(z)$ are generated by the Fourier components of $T(z)$ on the circle:

$$L_n = \oint \frac{dz}{2\pi i} z^{n+1} T(z) \tag{2.7}$$

The standard Virasoro algebra follows from Eq. (2.6) and Eq. (2.7):

$$[L_n, L_m] = (n-m)L_{n+m} + \frac{c}{12}n(n^2-1)\delta_{n+m,o} \tag{2.8}$$

Note that $(L_o, L_{\pm 1})$ form a closed algebra. This is the projective algebra which generates the $SL(2,C)$ transformations, i.e. one-to-one mapping on the complex plane. String theory is invariant under such reparametrization on the world sheet. Standard gauge fixing of the bosonic string introduces Fadeev-Popov ghost fields (c^z, b_{zz}). This contributes a term to $T(z)$:

$$T_{\text{ghost}}(z) = -2b\partial_z c - (\partial_z b)c \tag{2.9}$$

where

$$< b_{zz}(z)c^w(w) >= (z-w)^{-1} \tag{2.10}$$

It is straightforward to show that

$$T_{\text{ghost}}(z)T_{\text{ghost}}(w) \sim \frac{-26}{2(z-w)^4} + \cdots \tag{2.11}$$

The BRST charge:

$$Q = \oint \frac{dz}{2\pi i} : c(z)[T(z) + \frac{1}{2}T_{\text{ghost}}(z)] : \tag{2.12}$$

can be shown to be nilpotent if and only if the conformal field theory in $T(z)$ has total central charge $c = 26$:

$$Q^2 = 0 \text{ if } c = 26 \tag{2.13}$$

In preparation for later applications, we shall bosonize the bc ghosts; this can be done by considering a free scalar field η with:

$$T_\eta(z) = \frac{1}{2}(\partial_z \eta)^2 + i\alpha_{gh}\partial^2\eta \tag{2.14}$$

where

$$c = 1 - 12\alpha_{gh}^2 = -26 \tag{2.15}$$

or $\alpha_{gh} = \frac{3}{2}$. Since $< \eta(z)\eta(w) >= -ln(z-w)$ and the conformal dimension of the field $e^{i\rho\eta(z)}$ is $h = \frac{1}{2}\rho(\rho - 2\alpha_{gh})$, we can identify ghosts $b(z)(h=2)$ and $c(z)(h=-1)$ as

$$b(z) \leftrightarrow e^{-i\eta} \qquad c(z) \leftrightarrow e^{i\eta} \tag{2.16}$$

It is convenient to consider the total $T^t(z)$

$$T^t(z) = T(z) + T_\eta(z)$$

and its moments L^t_n whose corresponding Virasoro algebra has central charge $c - 26$. In this bosonized notation,

$$Q = \oint \frac{dz}{2\pi i} : e^{i\eta(z)} T^t(z) : \tag{2.17}$$

$Q^2 = 0$ if the Virasoro algebra for L^t_n has no central charge.

Choosing X_μ as the conformal field theory (CFT) means the bosonic string must live in 26 space-time dimensions.

$$X_\mu(z, \overline{z}) = X_{\mu L}(z) + X_{\mu R}(\overline{z}) \qquad \mu = 0, 1, 2 \cdots, 25 \tag{2.18}$$

We can expand all fields into modes:

$$\begin{aligned} a_{\mu n} &= i \oint \frac{dz}{2\pi i} z^{n+d-1} \partial_z X_\mu \\ b_n &= \oint \frac{dz}{2\pi i} z^{n+d-1} b(z) \\ c_n &= \oint \frac{dz}{2\pi i} z^{n+d-1} c(z) \end{aligned} \tag{2.19}$$

where the conformal dimension d of $\partial_z X_\mu$, $b(z)$, $c(z)$ are 1, 2, –1 respectively. It follows from Eq. (2.5), Eq. (2.10) and Eq. (2.19) that

$$[a_{\mu n}, a_{\nu m}] = n\delta_{\mu\nu}\delta_{n+m,o}$$

$$\{b_n, c_{-m}\} = \delta_{nm} \tag{2.20}$$

Now we are ready to introduce the string field. First we must define the vacuum $| 0 >$. It is natural to demand the vacuum to be $SL(2,C)$ invariant and annihilated by lowering operators, i.e.

$$L_n \mid 0 >= 0 \qquad \forall \ n \geq -1 \tag{2.21}$$

This follows if

$$a_{\mu n} \mid 0 >= 0 \qquad \forall \ n \geq 0 \tag{2.22}$$

Since $b(z)$ has the same dimension as $T(z)$, we require

$$b_n \mid 0 >= 0 \qquad \forall \ n \geq -1 \tag{2.23}$$

Now $c_1 \mid 0 >\neq 0$, since $\{b_{-1}, c_1\} \mid 0 >=\mid 0 >= b_{-1}c_1 \mid 0 >$. Consistency requires

$$c_n \mid 0 >= 0 \qquad \forall \ n \geq 2 \tag{2.24}$$

and

$$< 0 \mid c_{-1}c_0c_1 \mid 0 >= 1 \tag{2.25}$$

where the normalization of the last matrix element is taken to be unity. Since

$$[L_o, f_n] = -nf_n \ ,$$

for $f_n = a_n, b_n$ and c_n, we note that the lowest L_o eigenvalue state is not the $SL(2,C)$ vacuum but instead

$$\mid \Omega >= c_1 \mid 0 > \tag{2.26}$$

where $| \Omega >$ is annihilated by all f_n where $n \geq 1$ (note $c_1^2 = 0$). Now we can define the string field in terms of the 2 dimensional Fock space for both the open and the closed bosonic strings. For open string, we have only one Virasoro algebra. The string field functional can be expanded in the mode basis

$$\Phi[X_\mu, b, c] = \{\phi(x) + A_\mu(x)a^\mu_{-1} + \frac{1}{2}h_{\mu\nu}(x)a^\mu_{-1}a^\nu_{-1} + V_\mu(x)a^\mu_{-2}$$
$$+\eta(x)c_o b_{-1} + f(x)c_{-1}b_{-1} + \cdots\} \mid \Omega > \tag{2.27}$$

where x is the center of mass co-ordinate of the string and become the position of the field $\phi(x), A_\mu(x), h_{\mu\nu}(x)$ etc. Here $\phi(x)$ is the tachyon, $A_\mu(x)$ the massless vector field, $h_{\mu\nu}(x)$ massive tensor field, etc.

For the closed bosonic string, there are left-moving modes $a_{\mu n}$, b_n and c_n and right-moving modes $\tilde{a}_{\mu n}$, $\tilde{b}_n$ and $\tilde{c}_n$. The closed string field

$$\Phi[x_\mu, b, c] = \{\phi(x) + G_{\mu\nu}(x)a^\mu_{-1}\tilde{a}^\nu_{-1} + \cdots\} \mid \Omega >$$

where

$$\mid \Omega >= c_1\bar{c}_1 \mid 0 > \tag{2.28}$$

The open bosonic string theory was written down by Witten and others[11-13]

$$S = \int \frac{1}{2}\Phi * Q\Phi + \frac{2}{3}\Phi * \Phi * \Phi \tag{2.29}$$

where Φ is the string field and the BRST charge operator Q becomes the kinetic operator. The meaning of the multiplication symbol $*$ can be elegantly defined in the language of conformal field theory. Consider the cylinder swept out by the closed string (see Fig. 1). In the complex z variable given in Eq. (2.1), the string propogates from the origin ($z = 0, \tau = -\infty$) to the unit circle ($\mid z \mid= 1, \tau = 0$). Now the unit disk defines the incoming string $\mid \Phi_1 >$.

Take another unit disk which is mapped to the outside via $z \to -\frac{1}{z}$ (see Fig. 2). Call this $< \Phi_2 \mid$. Then

$$\Phi * Q\Phi = < \Phi_2 \mid Q \mid \Phi_1 > \tag{2.30}$$

For the interaction term, we map the unit disk for string 1 to the right half-plane (see Fig. 3). This is then conformally mapped to a third of the complex plane (between $\Theta = \pi/3$ to π). We then take string 2 and map it to another third of the complex plane ($\Theta = -\pi/3$ to $\pi/3$) and string 3 to the remaining plane. For open string field theory, we simply project everything onto the real axis.

To get some feeling of what is going on, let us consider the kinetic energy term of the massless states for the open string field theory. Here, keeping only the relevant terms

$$\begin{aligned}\Phi &= (A_\mu(x)a^\mu_{-1} + \eta(x)c_0 b_{-1} + \cdots) \mid \Omega > \\ Q &= c_0(\frac{p^2}{2} - 1 + a_{-1}\cdot a_1 + b_{-1}c_1 + c_{-1}b_1) \\ &\quad + c_{-1}p\cdot a_1 + c_1 p\cdot a_{-1} + 2c_1c_{-1}b_0 + \cdots\end{aligned} \tag{2.31}$$

we have

$$\Phi * Q\Phi = A^\mu \frac{p^2}{2} A_\mu - (A\cdot p)\eta - \eta(p\cdot A) + 2\eta^2 + \cdots \tag{2.32}$$

Here η is an auxilliary field and the constraint equation from the variation of η implies $2\eta = p\cdot A$ so that

$$S = \int d^{26}p\{A^\mu \frac{p^2}{2} A_\mu - (p\cdot A)^2/2\} = -\frac{1}{4}\int d^{26}x \; F_{\mu\nu}F^{\mu\nu} \tag{2.33}$$

where the last form is obtained after an integration by parts and a Fourier transformation to co-ordinate space.

Recently, major efforts[11-13] have gone to the covariant formulation of string field theory. Covariant formulation of open superstring[15] encountered some difficulty[16]. This problem seems to have been resolved recently[17]. However, the covariant formulation of closed string field theory turns out to be much more difficult than expected[18]. In terms of the operator formalism of string theory, open strings and closed strings are very closely related. Hence many attempts of closed string field theory formulations are various extensions of the open string formulation. So far, none of the covariant formulations proposed can reproduce both the four-point scattering amplitude and the one-loop diagrams, which are necessary requirements for a consistent formulation. Since the heterotic string offers the only hope to describe nature and it is a closed string theory, it is important to continue the search for the covariant formulation of closed string field action. These issues have been discussed extensively in the literature. See e.g. Ref [19-22].

III. Defining the Configuration Space

In terms of closed string field theory, the many string models correspond to the classical vacua of the heterotic string field theory. Dynamics of the string field will hopefully select out the vacuum that describes the world we live in. Hence a proper formulation of string field theories is of great importance in unravelling the underlying string symmetry as well as dynamics.

Unfortunately, closed string field theory as formulated up to now is woefully short of this eventual goal. This statement needs some clarification.

As discussed in the last section, the covariant closed string field theory formulation has encountered difficulties. In particular, the covariant heterotic string field theory is missing. However, even if the covariant closed string theory were written down in the form discussed in the last section, it would still be useless in answering the dynamical questions such as why the theory chooses a particular vacuum. This follows from the choice of space upon which the string field is defined. For simplicity, consider the closed bosonic string field, $\Phi(X_\mu(z,\bar{z}), b(z), c(z), \bar{b}(\bar{z}), \bar{c}(\bar{z}))$ where $X_\mu (\mu = 0,1,2,...,25)$ is the string variable while $c(\bar{c})$ and $b(\bar{b})$ are the ghost and the antighost for left (right) movers. To define the string field, the variables X_μ are chosen to be the coordinates. However this fixes the classical string vacuum to the 26 dimensional string or one of its compactifications. Starting from such a choice, it seems impossible to explore all the other classical vacua (e.g. left-right asymmetric models[5]) that we know exist. To include all classical vacua, we must generalize the choice of coordinates: the configuration space must be enlarged to include all classical vacua; in particular, the choice of X_μ becomes special points in this enlarged configuration space. A proper formulation of a string field theory in general entails an action. To properly define the action, it is necessary to specify the configuration space in which the string fields live, *i.e.*, the coordinates of the string field. It is clearly important, as a necessary first step, to define the string field in a space such that all classical vacua emerge naturally; in fact, these vacua should be on equal footings before dynamics (*i.e.*, interactions) start to play a role.

In this lecture, we propose a configuration space[23] upon which string fields should be defined. We shall explicitly construct this space for the closed bosonic string field. We shall illustrate in general how the classical vacua emerge. In this formulation, space-time is a derived concept. Our proposal clearly shows the existence of a very deep underlying string symmetry. To start, (super)conformal symmetry is taken to be essential to the formulation of string field theory. It is known that a large class of conformal fields can be bosonized[24–30], *e.g.*, (super)conformal ghosts, all (super)conformal models in the minimal series as well as W algebras, which includes parafermions. Our key assumption is that all two-dimensional (super)conformal fields can be bosonized (actually, the assumption needed is somewhat weaker; we require only the bosonization of conformal fields that appear in string theories). We propose that these bosons are the fundamental coordinates of string fields.

Let us concentrate for the moment on the closed bosonic string. We introduce D chiral bosons $(D \gg 26)$:

$$T(z) = \sum_{ij} \eta^{ij} \left\{ -\frac{1}{2}\partial\phi_{Li}\partial\phi_{Lj} + i\alpha_{Li}\partial^2\phi_{Lj} \right\} \tag{3.1}$$

where the diagonal metric is Minkowskian, $\eta=(-1,1,1,....1)$. The central charge of ϕ_{Lj} is $c_{Lj}=1-12\eta_{jj}\alpha_{Lj}{}^2$ (j not summed) where $j=0,1,2,\cdots, D-1$. The total central charge of this set must be 26,

$$\sum_j c_{Lj} = \sum_j (1 - 12\eta_{jj}\alpha_{Lj}^2) = 26 \tag{3.2}$$

or

$$\alpha_L^2 = \eta^{ij}\alpha_{Li}\alpha_{Lj} = \frac{D-26}{12} \tag{3.3}$$

where α_{Lj}^2 are real. The constraint Eq.(3.3) assures the absence of conformal anomaly. Here $\langle\phi_i(z)\phi_j(w)\rangle = -\eta_{ij} ln(z-w)$. Let us introduce a similar set of D right-moving chiral bosons $\phi_{Rj}(\bar{z})$. Denote $\alpha=(\alpha_L, \alpha_R) = (\alpha_{L0}, \alpha_{L1}, \cdots, \alpha_{R0}, \alpha_{R1}, \cdots)$. Now we can define the string field in terms of these chiral bosons and the bc ghosts, $\Phi(\phi_{Li}, \phi_{Ri}, b, c, \bar{b}, \bar{c})$. Of course, the bc ghosts can be bosonized in the same way, if desired (see Eq. (2.16)). These fields can be expanded into modes,

$$
\begin{aligned}
a_{Ljn} &= i\oint \frac{dz}{2\pi i} z^n \partial\phi_{Lj}(z) \qquad \text{where} \qquad [a_{Lin}, a_{Ljm}] = n\delta_{n,-m}\eta_{ij} \\
b_n &= \oint \frac{dz}{2\pi i} z^{n+1} b(z), \quad c_n = \oint \frac{dz}{2\pi i} z^{n-2} c(z)
\end{aligned}
\tag{3.4}
$$

Defining the SL(2,C) invariant world sheet vacuum $|0\rangle$, we have

$$
\Phi = \left\{ s(\mathbf{p}) + g_{ij}(\mathbf{p}) a_{-1}^{Li} a_{-1}^{Rj} + \cdots \right\} c_1 \bar{c}_1 |\mathbf{p}\rangle \tag{3.5}
$$

where $|\mathbf{p}\rangle = e^{i\mathbf{p}\cdot\phi(0)}|0\rangle$. Here $s(\mathbf{p})$ is the tachyon state in the momentum space and $g_{ij}(\mathbf{p})$ contains the "would-be" graviton. It is clear that the space we introduced is much bigger than the Fock space of the physical states in the closed bosonic string. The important point is that by varying the charge vector α (satisfying the constraint Eq.(3.3) for left and right movers separately), the space is big enough to include the Fock spaces corresponding to all the classical vacua of the closed bosonic string. In fact, it is much bigger than that. For any specific choice of the central charges, there are numerous unphysical states. For example, let c_i=1 for $i=0,1,\cdots,25$, and c_i=0 for the rest for both the left and the right movers; this corresponds to the usual bosonic string except that the extra $(D-26)$ (c_i=0 non-unitary) conformal fields introduce numerous unphysical states. In the string field theory, we must project them out. Let us introduce a projection operator, $P(\alpha,\phi)$, to remove these unphysical states when we go on-shell. Although the explicit form of this operator $P(\alpha,\phi)$ is not known, operationally we have a pretty good idea how it functions. (This will be discussed in more detail in the next Section.) For specific choices of α, we can project out all the non-unitary Virasoro representations, leaving behind a unitary conformal family[24–26]. We shall elaborate on this projection in a moment.

The string field action can then be (symbolically) written down as

$$S = \int \mathcal{D}\phi_\alpha \mathcal{D}b\mathcal{D}c \left\{ \frac{1}{2}\langle \Phi K P \Phi \rangle + \frac{1}{3}\langle \Phi\Phi\Phi \rangle \right\} \tag{3.6}$$

Here, the action is essentially the usual string field action, except for the inclusion of P and α; the range of α is constrained by Eq.(3.3). On-shell, the projection operator $P(\alpha, \phi)$ keeps only unitary conformal families for the space-like components. The kinetic operator K includes BRST operators; on-shell, the BRST symmetry removes the remaining unphysical time-like modes. Note that P is required to commute with K. On-shell, the kinetic operator K guarantees left-right level matching and provides the kinetic operator for each physical field. The explicit form of the kinetic operator K is not settled yet. In the absence of the interaction term, we can choose[19,21] either $K = \{\bar{c}_0 Q + c_0 \bar{Q}\}\delta(L_0 - \bar{L}_0)$ or $K = 2\pi Q\bar{Q}/\sin(\pi(L_0 + \bar{L}_0))$ where the left(right) BRST charge operator $Q(\bar{Q})$ obeys $Q^2 = \bar{Q}^2 = 0$, and $L_0(\bar{L}_0)$ is the corresponding Virasoro generator. Actually, the difficulty associated with the covariant formulation of closed string field theory is not essential to our construction. We could easily have applied the idea to the light-cone string field theory where the string field action is known. In that case, we only introduce $(D-2)$ chiral bosons, with c=24 for the transverse modes. All classical vacua that can be recovered from the light-cone formulation will have at least one space, one time dimensions. This is not such a big loss if we can write down the action explicitly.

At this point, it is useful to compare this formulation with the free gauge field theory:

$$S = -\frac{1}{4}\int dx F_{\mu\nu}F^{\mu\nu} = \frac{1}{2}\int dx A_\mu K P^{\mu\nu} A_\nu = \frac{1}{2}\int dx A_\mu \partial^2 \left(\delta^{\mu\nu} - \frac{\partial^\mu \partial^\nu}{\partial^2} \right) A_\nu \tag{3.7}$$

The existence of electric and magnetic field is well established. Special relativity demands a four-vector field, hence the introduction of A_μ. This introduces negative-norm states and we need the projection $P^{\mu\nu}$ to remove them. The gauge symmetry $\delta A_\mu = \partial_\mu \lambda$ follows simply because $P^{\mu\nu}\partial_\nu\lambda = 0$. In string theory, the existence of many classical vacua (i.e. string models) is well established. The preservation of conformal symmetry introduces negative-norm states; hence we need a projection operator. In the free covariant string field theory case, $P(\alpha, \phi)$ clearly allows a very large symmetry, *i.e.*, all $\delta\Phi$ that satisfy $P\delta\Phi = 0$. On-shell, the theory is unitary by construction. All negative-norm states from the $(D-1)$ chiral bosons with Euclidean metric are removed by the projection operator P while the negative norm states from the time-like boson ϕ_0 are removed by the BRST operators in K. Off-shell, negative norm states beyond those from the time-like modes are present in huge numbers in general. We believe this new symmetry, resulting from the projection P, is the true string symmetry. In comparison, the usual BRST symmetry in string field theory is simply a compact way of organizing the gauge symmetries and kinematics that are already present in quantum field theory for massless and massive particles.

This approach is very different from the approach implicit in many recent works, *e.g.*, derivation of the equations of motion via the vanishing of the β functions of non-linear sigma models[31], and the renormalization group flow of two-dimensional non-linear sigma models towards conformal fixed points[32]. In Euclidean space, these approaches demand unitarity off-shell and (super)conformal symmetry is required only on-shell. The approach described here demands (super)conformal symmetry both on and off-shell, while unitarity is required only on-shell. This new approach embraces the same philosophy of gauge theories.

The operational meaning of the projection operator $P(\alpha,\phi)$ needs some clarification. Consider any boson with $c<1$. For any given p in a set of discrete momenta (discrete because chiral boson with anomaly takes value on a circle), the Fock space F^c_p corresponding to any $e^{ip\phi}$ is *apriori* very large,

$$F^c_p = \left\{ |n,p\rangle = a^{n_1}_{-1} a^{n_2}_{-2} \cdots |p\rangle \right\} \tag{3.8}$$

Furthermore, the Fock space has a natural Virasoro structure from Eq.(3.1). Let us consider any model in the $c<1$ unitary series, which can be bosonized. In general F^c_p as Virasoro module contains negative-norm states. For specific choices of the highest weight $h=p(p-2\alpha)/2$ permitted in the unitary series, the Fock space F^c_p is still much bigger than the irreducible representation of Virasoro algebra V_h, and in general contains negative-norm states. Feigin and Fuchs[24] developed a way to project out all states except those in the Virasoro representation: $P_{FF} : F^c_p \rightarrow V_h$ where h is an allowed highest weight.

Going through all allowed choices of **p**, we have $P(\alpha,\phi) : \{F^c_p\} \rightarrow \oplus_i V_{h_i}$. For $c<1$, the final result is well-known and the set of representations $\oplus V_{h_i}$ is automatically guaranteed to form a unitary conformal family C_α, *i.e.*, the operator product expansion of any two fields in C_α closes in C_α. For $c=26$, we further require the closure property on the set of irreducible representations. Otherwise it is projected out. For a given choice of α, there should be a unique unitary conformal family C_α (up to discrete symmetries) and $P(\alpha,\phi)$ projects out everything else. For some choices of α, C_α can be trivial, *i.e.*, it contains only the identity element. For a fixed α, the equation of motion for the free string field, $KP\Phi = 0$, may have more than one solution. Given the classical solution, we may still have to check if it is consistent with quantum mechanics, *i.e.*, check for anomalies as in the case in quantum field theory. This means checking the consistency of the one loop diagrams. In particular, modular invariance at the one loop level must be maintained.

Let us illustrate the above points with the case of two bosons (c_1, c_2) whose total central charge is $c_1+c_2=1$. We discuss how the projection P operates on the Fock space as we move from the $(\frac{1}{2}, \frac{1}{2})$ case to the $(1,0)$ case. This example also illustrates how a space dimension can arise. Of course we can connect the two cases by going along the Z_2 orbifold and then along the torus by changing radius, so that the theory remains unitary throughout[33]; but instead we shall vary *away* from unitary conformal theory by following $c_1=\frac{1}{2}+y$ and $c_2=\frac{1}{2}-y$ where y varies smoothly from 0 to $\frac{1}{2}$.

At the point $(\frac{1}{2}, \frac{1}{2})$ the projection operator P essentially eliminates all the Fock space except the three unitary representations of the Ising model, *i.e.*,

$$P(\alpha, \phi) : F_p^{\frac{1}{2}} \otimes F_p^{\frac{1}{2}} \rightarrow (V_0 \oplus V_{\frac{1}{2}} \oplus V_{\frac{1}{16}}) \otimes (V_0 \oplus V_{\frac{1}{2}} \oplus V_{\frac{1}{16}}) \tag{3.9}$$

Therefore the theory describes *either* two copies of Ising models *or* a complex Dirac fermion. For $y \neq 0, \frac{1}{2}$, the theory is not unitary and hence is projected out. At the end point $(1,0)$, $P(\alpha, \phi)$ acts on the c=0 boson alone. Since the only unitary representation of the c=0 Virasoro algebra is the identity representation, *i.e.*, $P(\alpha, \phi) : \oplus_p F_p^0 \rightarrow I$; the projection essentially eliminates this boson. For the c=1 boson, the projection operator acts like the identity operator. The closure condition for unitary conformal family (coming from interaction vertex) implies the single-valuedness on the world sheet. Consider the case of a $c = 1$ left moving boson with a $c = 1$ right moving boson,

$$e^{ip_L\phi(z)+ip_R\phi(\bar{z})} e^{iq_L\phi(w)+iq_R\phi(\bar{w})}$$
$$= (z-w)^{-p_Lq_L}(\bar{z}-\bar{w})^{-p_Rq_R} e^{i(p_L+q_L)\phi(w)+i(p_R+q_R)\phi(\bar{w})} + \dots \tag{3.10}$$

the single-valuedness condition is satisfied if $p_Lq_L - p_Rq_R \in \mathbf{Z}$. The level-matching condition from K implies $\frac{p_L^2}{2}+N = \frac{p_R^2}{2}+\bar{N}$. Therefore the momenta are

$$p_L = (nR + \frac{m}{2R}), \quad p_R = (nR - \frac{m}{2R}) \qquad n,\ m \in \mathbf{Z} \tag{3.11}$$

where the radius R is arbitrary. In the limit $R \to \infty$, the momentum $p_L{=}p_R$ takes continuous values, and a_{L0} can be identified with a_{R0} . If we choose 26 bosons, including ϕ_0, to have c=1 and the rest to have c=0, we can repeat the above procedure for each c=1 boson. The $g_{ij}(\mathrm{p})$ in Eq.(3.5) become $g_{\mu\nu}(\mathbf{p})$ as functions of continuous momenta $\mathbf{p}$, where $\mu,\ \nu$=0,1,···,25. The space-time then emerges from the Fourier transformation of $\mathbf{p}$. Clearly this is merely one particular solution out of many possibilities.

Note that the requirement that P and K commute imposes a strong constraint on P. In particular, P can only decouple c=0 conformal fields. Suppose, for fixed α, there are central charges not in the unitary series, say some $c_i<0$; then *either* P projects out everything leaving behind a trivial solution, *or* the corresponding $c_i<0$ fields combine with some $c>0$ fields to yield non-trivial unitary conformal families. That is to say, P cannot completely project out any boson with non-zero central charge; otherwise, the effective central charge in the BRST operator will not be 26, *i.e.*, P would not commute with K.

In summary, we introduce a configuration space upon which string fields are defined and demand (super)conformal symmetry both on and off-shell while unitarity only on-shell. The enlarged space includes all classical vacua and hopefully the true vacuum of nature as well. We point out that the projection on an enlarged space introduced here also suggests a way to classify conformal field theories. This point will be clear in the next section.

IV. Discussions on the Projection Operator

If string field theory with the new enlarged configuration space is of any use, we must know the action explicitly in a compact form. A preliminary step is to know the free string case. From the previous discussions, it is clear we must know the projection operator P explicitly. The construction of P is a challenging problem which remains to be solved. Here I shall briefly discuss some of its properties which, in my opinion, provide the first steps in its construction. At isolated points in the configuration space, a compact form for the projection operator is suggested. These points correspond to conformal field theories of the unitary series with $c \leq 1$, parafermion and/or W algebra. What we need is a unique projection operator at every point in the configuration space. This is a problem which remains to be solved.

It is generally believed that all conformal field theories can be bosonized. We shall further assume that all unitary conformal field theories that appear in the string field theory besides the time-like fields can be represented by space-like bosons with appropriate background charges. For a given CFT with central charge c, n number of space-like bosons (where $n \geq c$) are needed to represent it. For $n > c$, the bosonized version has a Fock space larger than that of the corresponding CFT and so we need a projection P to project out all the spurious states to obtain a proper bosonized representation.

At each point α in the configuration space, we must associate a unique projection operator P which projects in a unique unitary CFT plus the ϕ_0. It is possible that the resulting CFT at some points in the configuration space is trivial, i.e., all states except the identity state have been removed. At more interesting points where the resulting CFT is non-trivial, typically it is a direct product of a set of holomorphic conformal fields and a set of anti-holomorphic conformal fields where the central charges add up to the critical value. All other points in the configuration space are off-shell points.

In string field theory, CFT naturally fall into two categories, which shall be denoted as reducible CFT and irreducible CFT. A reducible CFT with central charge c can be built from other reducible and/or irreducible CFT whose central charges add up to c. An irreducible CFT with central charge c cannot be built from other unitary CFT with smaller central charges which add up to c. Examples of irreducible CFT are, as we know today, the unitary series with $c < 1$, c=1 boson with arbitrary radius, parafermions and W-algebras. Examples of reducible CFT are Kac-Moody algebras and minimal N=2 superCFT. Kac-Moody algebras at level one can be built from c=1 bosons, while minimal N=2 superCFT and SU(2) Kac-Moody algebras can be built from a c=1 boson and a parafermion. In string field theory, the projection operator P yields a direct product of irreducible unitary CFT plus the time component. The kinetic operator K combines the irreducible unitary CFT with the time-like bosons to form string solutions, which typically contain reducible CFT.

Let me first briefly summarize the results of the bosonization of the unitary series due to Feigin and Fuchs and others.[24–25] Let me start with the $c \leq 1$ case first. Here I shall follow Felder's elegant presentation.[26] I shall assume some familiarity with the work and be brief. The unitary series[30] can be described by a single boson with background charge α_0:[25]

$$S = \frac{1}{2\pi}\int d^2z\sqrt{g}\{\partial_z\phi\partial_{\bar{z}}\phi + \frac{i}{2}R\alpha_0\phi\} \tag{4.1}$$

where R is the world-sheet curvature scalar, and

$$\frac{1}{2\pi}\int d^2z\sqrt{g}R = 4(1-h) \tag{4.2}$$

where h is the number of handles of the Riemann surface under consideration. Now single-valuedness under $\phi \to \phi + n\pi L$

$$e^{iS} \to e^{iS} e^{2\pi i \alpha_o L n}$$

demands

$$\alpha_0 L = 1 \tag{4.3}$$

Let us consider only the holomorphic part (setting the antiholomorphic part to zero), then $\phi = \phi(z)$ and

$$T(z) = -\frac{1}{2}\partial\phi\partial\phi + i\alpha_o\partial^2\phi \tag{4.4}$$

with central charge

$$c = 1 - 12\alpha_o^2 = 1 - \frac{6}{p(p+1)} \tag{4.5}$$

or

$$\alpha_o^2 = \frac{1}{2p(p+1)} \tag{4.6}$$

where the integer $p \geq 3$ for the unitary series. Eq. (4.3) and Eq. (4.6) fix the compactification radius L. Given any momentum k, the weight is given by

$$h = \frac{1}{2}k^2 - k\alpha_o \tag{4.7}$$

A general vertex operator $\exp(2ik\phi)$ has dimension h. Let us define screening charges to have dimension 1:

$$\frac{1}{2}\alpha^2 - \alpha_o\alpha = 1 \tag{4.8}$$

or

$$\alpha_+ = \sqrt{\frac{2(p+1)}{p}}$$

$$\alpha_- = -\sqrt{\frac{2p}{p+1}} \tag{4.9}$$

Then we can construct operators which do not change the weight of a given state:

$$Q_m = \frac{1}{m}\int dv_o dv_1 \cdots dv_{m-1} e^{i\alpha_+\phi(v_o)} e^{i\alpha_+\phi(v_1)} \cdots e^{i\alpha_+\phi(v_{m-1})} \tag{4.10}$$

Here the integration of v_i is a closed loop beginning and ending at v_o and outside the v_{i+1} loop; the last integration of v_o is a closed circle. Next a general vertex operator with momentum k obeys

$$exp\{2ik\phi(z)\} = exp\{2ik(\phi + \pi L)\} \tag{4.11}$$

or $kL\epsilon Z$. Here we can write k in terms of $\alpha_\pm$, for some integers l, l' or m, m',

$$k = \frac{n}{L} = \frac{n}{\sqrt{2p(p+1)}}$$

$$= \frac{1}{\sqrt{2p(p+1)}}(lp + l'(p+1))$$

$$= \frac{1}{2}\alpha_+(1-m) + \frac{1}{2}\alpha_-(1-m') \tag{4.12}$$

where we note that p and $(p+1)$ are co-prime.

The Fock space is

$$F_c' = \{F_{m'm}^c\}$$

$$F_{m\prime m}^c = \{a_{-l_1}a_{-l_2}\cdots a_{-l_n} \mid m\prime m >\} \tag{4.13}$$

where $| m'm >= e^{2ik\phi(0)} | 0 >$ with k given by Eq. (4.12). Now the closure condition from the fusion rule shows the set of momenta $\{k\} = \{1 \leq m < p-1, 1 \leq m' \leq p\}$ close and they correspond to the primary states. Their weights are the same as that given in Eq. (4.7). The Fock space with this set of momenta is much bigger than that of the CFT with central charge c.

$$F^c = \{F^c_{m'm}, \quad 1 \leq m < p-1, 1 \leq m' \leq p\}$$
$$\supset F_{CFT} \tag{4.14}$$

It is clear that $F_{m'm} = F_{m'-p',m-p}$. It can be shown that $(p' = p+1)$

$$\begin{array}{lccc}
(1) & F_{m',2p-m} & & F_{m'm} \xrightarrow{Q_m} F_{m',-m} \\
& \| & & \| \\
& F_{m'-p',p-m} & \xrightarrow{Q_{p-m}} & F_{m'-p',m-p}
\end{array}$$

$$(2) \quad Q_m Q_{p-m} = 0 \quad \text{or simply} \quad Q^2 = 0 \tag{4.15}$$

Now $F_{m'm}$ is isomorphic to its dual $F_{-m',-m}$. Let us define the dual BRST operator by $F_{m'm} \xrightarrow{Q^d} F_{m',2p-m}$ or, in more detail

$$\begin{array}{ccc}
F_{m'm} & & F_{m',2p-m} \\
d \downarrow & & \uparrow d \\
F_{-m',-m} & & F_{-m',m-2p} \\
\| & & \| \\
F_{p'-m,p-m} & \xrightarrow{Q_{p-m}} & F_{p'-m',m-p}
\end{array} \tag{4.16}$$

Then it can be shown that the projection operator can be written as

$$P = \lim_{\beta \to \infty} \exp\{-\beta(QQ^d + Q^dQ)\} \tag{4.17}$$
$$P : F^c \longrightarrow F^c_{CFT}$$

Eq. (4.17) implies that only states that are annihilated by both Q and Q^d are projected in. This is equivalent to the BRST cohomology discussed by Felder. For example, for $c = \frac{1}{2}$, P projects F^c to the Ising model.

The above projection operator can be easily generalized to W-algebra[28] where

$$c = (n-1)(1 - \frac{n(n+1)}{p(p+1)}) \tag{4.18}$$

This includes the unitary series ($n = 2$) and parafermion ($p = n+1$). Our discussion is quite similar to that of Caselle and Narain.[34] Let us generalize the case of one boson to $d = (n-1)$ bosons

$$S = \frac{1}{2\pi}\int d^2z\sqrt{g}\{\partial\vec{\phi}\bar{\partial}\vec{\phi} + \frac{iR}{2}\vec{\alpha}_o \cdot \vec{\phi}\} \tag{4.19}$$

where $\vec{\phi} = (\phi_1, \phi_2, \cdots, \phi_d)$ and $\vec{\alpha}_o = (\alpha_{01}, \alpha_{02}, \cdots, \alpha_{0d})$ and

$$T(z) = -\frac{1}{2}\partial\vec{\phi}\partial\vec{\phi} + i\vec{\alpha}_o \cdot \partial^2\vec{\phi}$$

where

$$c = d - 12\vec{\alpha}_o^2 \tag{4.20}$$

Single-valuedness of e^{iS} under

$$\vec{\phi} \longrightarrow \vec{\phi} + \pi\sum n_i\vec{u}_i \qquad n_i \epsilon Z \tag{4.21}$$

implies

$$\vec{\alpha}_o \cdot \vec{u}_i \ \epsilon\ Z \tag{4.22}$$

for each i. Single-valuedness of the chiral vertex operator $e^{2i\vec{k}\vec{\phi}(z)}$ implies $\vec{k}\cdot\vec{u}_i\epsilon Z$ or

$$\vec{k} = \sum m_i \vec{u}_i^* \tag{4.23}$$

where the set of vectors u_i^* obey $\vec{u}_i^* \cdot \vec{u}_j \epsilon Z$. In general $\{u_i\}$ form an integer lattice. For W-algebra $\{u_i\}$ are the root lattices of $SU(n)$ and form an integer lattice. $\{u_i^*\}$ is the dual lattice and $\{u_i^*\} \supset \{u_i\}$. Then

$$g_{ij} = \vec{u}_i \cdot \vec{u}_j \epsilon Z \tag{4.24}$$

and

$$\vec{u}_i = \sum_j g_{ij} \vec{u}_j^* \tag{4.25}$$

At the point $\vec{\alpha}_o = \sum_j \vec{u}_j^*$ in the configuration space, we can define the screening charges $\vec{\alpha}_i$ which must satisfy the condition

$$\frac{1}{2}\vec{\alpha}_i^2 - \vec{\alpha}_o \cdot \vec{\alpha}_i = 1 \tag{4.26}$$

in the following way

$$\vec{\alpha}_{i+} = \frac{1}{p}\vec{u}_i \tag{4.27}$$

It is easy to check that they satisfy Eq. (4.26) if

$$g_{ii} = 2p(p+1) \tag{4.28}$$

From the α_{i+} we can construct the BRST charge operators

$$Q_m^{(j)} = \frac{1}{m}\int dv_o dv_i \cdots dv_m e^{i\alpha_{j+}\phi(v_o)} \cdots e^{i\alpha_{j+}\phi(v_m)} \tag{4.29}$$

where the contour integrations are defined as in Eq.(4.10). Again, from the unitarity condition and the closure condition, we restrict the momenta from Eq.(4.23) to

$$\vec{k} = \frac{1}{2}\sum\{(1-m_i)\vec{\alpha}_{i+} + (1-m\prime_i)\vec{\alpha}_{i-}\} \tag{4.30}$$

where $\vec{\alpha}_{i-} = -\vec{u}_i/(p+1)$. Then the projection operator P

$$P = \lim_{\beta \to \infty} \exp\{-\beta \sum_i Q^{(i)} Q^{(i)d} + Q^{(i)d} Q^{(i)}\} \tag{4.31}$$

acting on

$$F = \{F_{\vec{m}'\vec{m}} \quad \text{where} \quad 1 \leq m_i \leq p-1 \quad , \quad 1 \leq m_i' \leq p\}$$

$$F_{\vec{m}'\vec{m}} = \{a_{-l_1}^{(i)} a_{-l_2}^{(j)} \cdots a_{-l_t}^{(n)} \mid \vec{m}', \vec{m} >\} \tag{4.32}$$

where $|\vec{m}', m >= e^{2i\vec{k}\cdot\vec{\phi}(0)} \mid 0 >$ gives the Virasoro representation for W algebra, i.e.

$$P : F \longrightarrow F_W \tag{4.33}$$

To define $Q^{(i)d}$, we note that that $F_{\vec{m}'\vec{m}}$ is isomorphic to its dual $F_{-\vec{m}', -\vec{m}}$. Note that in general $\{Q^{(i)}, Q^{(j)}\} \neq 0$. However, in the projection operator, we only need the zero eigenstates. For this sub-space, individual terms commute with each other's. It is likely that this approach can be generalized to cover all points in the configuration space.

V. Discussions

Let us conclude with a few remarks. In string field theory, what we need is a unique projection operator at every point in the configuration space. This remains to be constructed. As we mentioned earlier, this entails the classification of irreducible CFT.

One crucial question is how does the string field at one point in the configuration space communicate with string fields at other points. (Otherwise string fields at different points simply decouple and dynamics cannot move it from point to point.) One possibility is through the interaction term, where string fields at different points in the configuration space couple. However, it is more likely that the interaction term is local in the configuration space. Then the kinetic term must provide the coupling. This can be achieved in a number of ways. For example, the β parameter in the projection operator P actually is an operator involving $L_o, \overline{L}_o$ and ∇_α, the angular part of the quadratic derivative operator which leaves Eq. (3.3) invariant. Now $\beta \longrightarrow \infty$ as $L_o, \overline{L}_o, \nabla_\alpha$ take zero eigenvalues. The inclusion of derivatives with respect to the background charges α_i allows the string field to probe different points in the configuration space.

The physical meaning of arbitrary central charge is not clear. It is interesting to note that at least for the time component, c away from 1 implies the presence of a cosmological term in the gravity sector.[35] Further investigation hopefully will clarify the structure of the configuration space.

It is likely that the string field action is more compact than that shown in Eq. (3.6). In particular, there may be only a ϕ^3 term as proposed in Ref. [36], or the kinetic operator actually does not separate like KP as in Eq. (3.6). To achieve the more compact form, we require some higher principle, such as the equivalence principle in general relativity or the gauge symmetry. However, at the moment, we do not know what is the corresponding guiding principle for string symmetry. The bottom line is "What is the equivalence of the equivalence principle for string theory?" Obviously this is a very deep question.

Acknowledgement

I thank Zongan Qiu and Ed Lyman for collaborations. Discussions with Jacques Distler, Zvonko Hlousek, Hikaru Kawai, Andre LeClair, David Lewellen, David Sénéchal, and Barton Zwiebach have been most useful. This work is supported in part by the National Science Foundation.

REFERENCES

1. See, e.g. M. B. Green, J. H. Schwarz, and E. Witten, "Superstring Theory", Vol. I and II (Cambridge University Press, 1986); J. H. Schwarz, CALT-68-1432 (1987).

2. D. J. Gross, J. A. Harvey, E. Martinec, and R. Rohm, Phys. Rev. Lett. **54**, 502 (1985); Nucl. Phys. **B256**, 253 (1985) and **B267**, 75 (1986).

3. P. Candelas, G. Horowitz, A. Strominger, and E. Witten, Nucl. Phys. **B258**, 46 (1985).

4. L. Dixon, J. Harvey, C. Vafa, and E. Witten, Nucl. Phys. **B261**, 651 (1985); Nucl. Phys. **B274**, 285 (1986).

5. K. S. Narain, Phys. Lett. **169B**, 41 (1986); K. S. Narain, M. H. Sarmadi, and E. Witten, Nucl. Phys. **B279**, 369 (1987); H. Kawai, D. C. Lewellen, and S.-H. H. Tye, Phys. Rev. Lett. **57**, 1832 (1986) **58**, 429E (1987), Nucl. Phys. **B288**, 1 (1987); I. Antoniadis, C. Bachas, and C. Kounnas, Nucl. Phys. **B289**, 87 (1987); M. Mueller and E. Witten, Phys. Lett. **182B**, 28 (1986); R. W. Lerche, D. Lust, and A. N. Schellekens, Nucl. Phys. **B287**, 477 (1987); K. Narain, M. H. Sarmadi and C. Vafa, Nucl. Phys. **B288**, 551 (1987).

6. D. Gepner, Nucl. Phys. **B296**, 757 (1988); Princeton University preprint (1988).

7. B. R. Greene, C. Vafa and N. P. Warner, Nucl. Phys. **B324**, 371 (1989); E. Martinec, Phys. Lett. **217B**, 431 (1989).

8. D. Sénéchal, Phys. Rev. **D39**, 3717 (1989).

9. P. Candelas, A. Dale, C. A. Lutken and R. Schimmrigk, Nucl. Phys. **B298**, 493 (1988).

10. M. Kaku and K. Kikkawa, Phys. Rev. **D10**, 1110, 1823 (1974).

11. E. Witten, Nucl. Phys. **B268**, 253 (1986); **B276**, 291 (1986).

12. W. Siegel, Phys. Lett. **151B**, 391, 396 (1985); M. Kato and K. Ogawa, Nucl. Phys. **B212**, 443 (1983).

13. See e.g. W. Siegel, "Introduction to String Field Theory", (World Scientific 1988); S. Giddings, Nucl. Phys. **B278**, 242 (1986); S. Giddings and E. Martinec, Nucl. Phys. **B278**, 91 (1986); S. Giddings, E. Martinec and E. Witten, Phys. Lett. **176B**, 362 (1986); D. Gross and A. Jevicki, Nucl. Phys. **B282**, 1 (1987); **B287**, 362 (1987); E. Cremmer, C. B. Thorn and A. Schwimmer, Phys. Lett. **179B**, 57 (1986); K. Itoh, T. Kugo, H. Kunitomo and H. Ooguri, Prog. Theor. Phys. **75**, 162 (1986); H. Hata, K. Itoh, T. Kugo, H. Kunitomo and K. Ogawa, Phys. Rev. **D34**, 2360 (1986), **D35**, 1318 (1987); W. Siegel and B. Zweibach, Nucl. Phys. **B263**, 105 (1985); T. Banks and M. E. Peskin, Nucl. Phys. **B264**, 513 (1986).

14. A. M. Polyakov, Phys. Lett. **103B**, 207 (1981); D. Friedan, in 1982 Les Houches Summer School "Recent Advances in Field Theory and Statistical Mechanics" edited by J. B. Zuber and R. Stora, North-Holland (1984); A. LeClair, M. E. Peskin and C. R. Prietschopf, Nucl. Phys. **B317**, 411, 464 (1989). See also M. E. Peskin, Proceedings of the TASI

in Elementary Particle Physics, 1986, edited by H. E. Haber (World Scientific, Singapore, 1987).

15. E. Witten, Nucl. Phys. **B276**, 291 (1986); D. Gross and A. Jevicki, Nucl. Phys. **B243**, 29 (1987); A. LeClair and J. Distler, Nucl. Phys. **B273**, 552 (1986); J. P. Yamron, Phys. Lett. **187B** 67 (1987).
16. C. Wendt, Nucl. Phys. **B314**, 209 (1989).
17. C. R. Prietschopf, C. Thorn and S. Yost, University of Florida preprint (1989). See Talk by C. Thorn in this proceeding.
18. See e.g. B. Zwiebach, Annals of Phys. **186**, 111 (1988).
19. A. Neveu, H. Nicolai and P. C. West, Nucl. Phys. **B264**, 173 (1986); J. Lykken and S. Raby, Nucl. Phys. **B278**, 256 (1986).
20. R. P. Woodard, Phys. Lett. **213B**, 144 (1988).
21. S.-H. H. Tye, Phys. Rev. Lett. **63**, 1046 (1989).
22. See e.g. H. Sonoda and B. Zwiebach, MIT preprint (1989).
23. Z. Qiu and S.-H. H. Tye, Phys. Rev. Lett. **63**, 1203 (1989).
24. A. Chodos and C. Thorn, Nucl. Phys. **B72**, 509 (1974); B. L. Feigin and D. B. Fuchs, Func. Anal. Appl. **16**, 114 (1982); **17**, 241 (1983).
25. V. I. Dotsenko and V. A. Fateev, Nucl. Phys. **B240**, 312 (1984), **B251**, 691 (1985); C. Thorn, Nucl. Phys. **B248**, 551 (1984).
26. G. Felder, Nucl. Phys. **B317**, 215 (1989).
27. D. Friedan, E. Martinec and S. Shenker, Nucl. Phys. **B271**, 93 (1986).

28. A. B. Zamolodchikov, Theor. Mat. Fiz. **59**, 10 (1985); V. A. Fateev and S. L. Lykyanov, Int. Journ. of Mod. Phys. **A3**, 507 (1988).

29. J. Distler and Z. Qiu, Cornell Preprint, CLNS 89/911 (1989). D. Nemeshansky, Phys. Lett. **224B**, 121 (1989).

30. A. A. Belavin, A. M. Polyakov and A. B. Zamolodchikov, Nucl. Phys. **B241**, 333 (1984); D. Friedan, Z. Qiu and S. Shenker, Phys. Rev. Lett. **52**, 1575 (1984).

31. D. Friedan, Phys. Rev. Lett. **45**, 1057 (1980); E. S. Fradkin and A. A. Tseytlin, Phys. Lett. **158B**, 316 (1985); C. G. Callan, D. Friedan, E. Martinec and M. J. Perry, Nucl. Phys. **B262**, 593 (1985); A. Sen, Phys. Rev. Lett. **55**, 1864 (1985).

32. See e.g. T. Banks and E. Martinec, Nucl. Phys. **B294**, 733 (1987).

33. P. Ginsparg, Nucl. Phys. **B295**, [FS21], 153 (1988).

34. M. Caselle and K. S. Narain, Nucl. Phys. **B323**, 673 (1989).

35. R. C. Myers, Phys. Lett. **199B**, 371 (1989). S. P. DeAlwis, J. Polchinski and R. Schimmrigk, Phys. Lett. **B218**, 449 (1989). I. Antoniadis, C. Bachas, J. Ellis and D. Nanopoulos, CERN-TH 5231/89 (1989).

36. G. Horowitz, J. Lykken, R. Rohm and A. Strominger, Phys. Rev. Lett. **57**, 283 (1989).

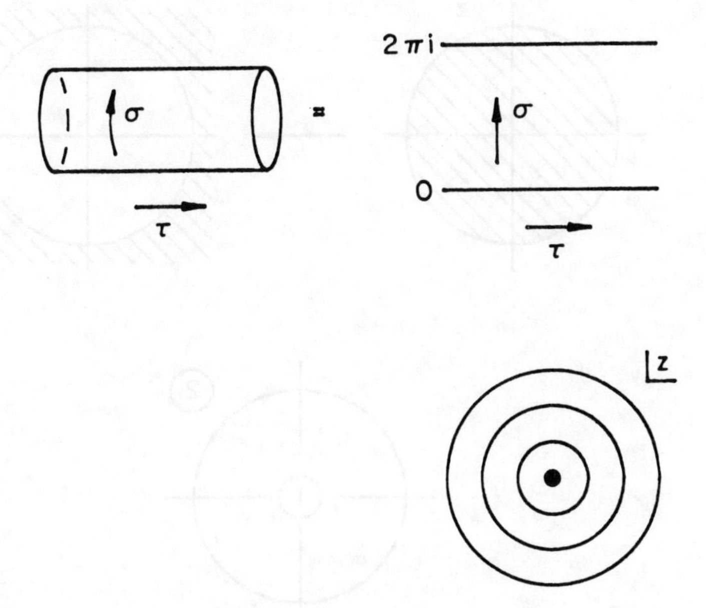

Fig. 1 The closed string world sheet propagating from $\tau = -\infty$ to $\tau = 0$ can be mapped to the unit disk in $z = \exp(\tau + i\sigma)$ complex plane.

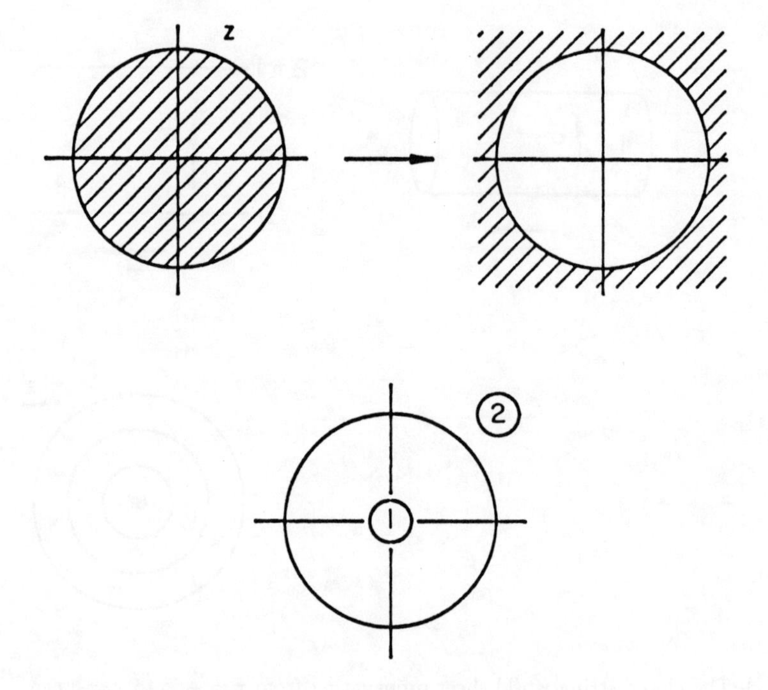

Fig. 2 The construction of the kinetic term for string field theory by mapping two canonically defined string states into the conformal plane.

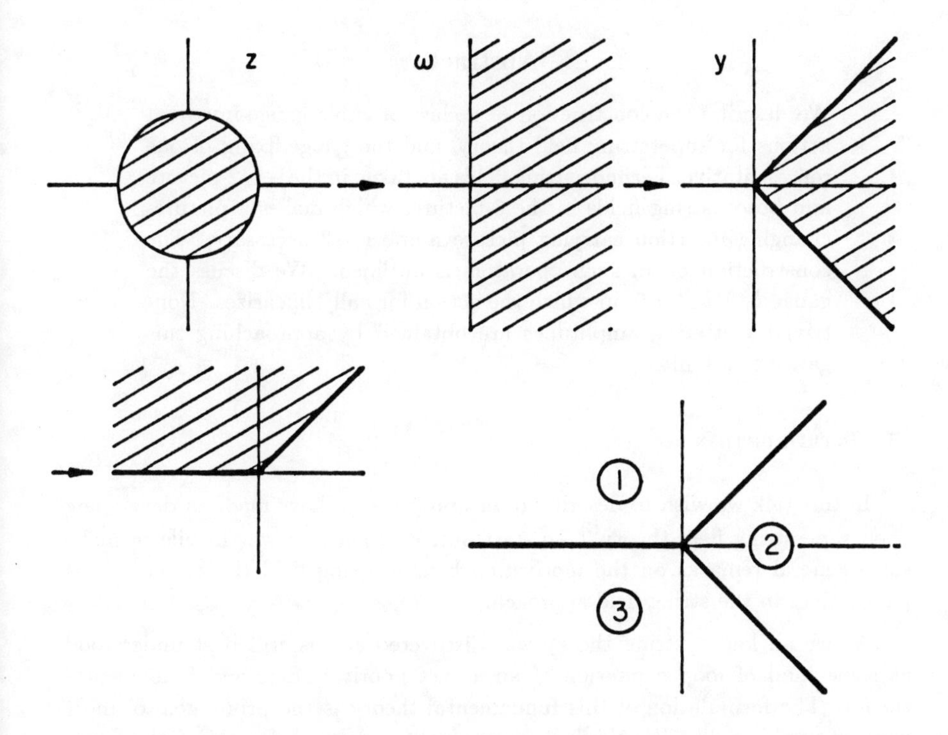

Fig. 3 The construction of the interaction term by conformally mapping three string states into the complex plane.

Superstring Field Theory*

CHRISTIAN R. PREITSCHOPF,
CHARLES B. THORN AND SCOTT A. YOST
Department of Physics
University of Florida, Gainesville, Florida 32611

Abstract

We describe the construction of a class of cubic gauge-invariant actions for superstring field theory, and the gauge-fixing of one representative. Fermion string fields are taken in the $-1/2$-picture and boson string fields in the 0-picture, which makes a picture-changing insertion carrying picture number -2 necessary. The construction of all such operators is outlined. We discuss the gauge $b_1 + b_{-1} = 0$, in which the action formally linearizes. Non-trivial scattering amplitudes are obtained by approaching this gauge as a limit.

1. INTRODUCTION

In this talk we wish to describe recent progress we have made in developing open superstring field theory.[1] To start out, it is perhaps worthwhile to make some general remarks on the motivation behind string field theory as well as alternatives to the string field approach.

As we all know, string theory was discovered and is still best understood as some kind of loop expansion of an as yet poorly understood fundamental theory. The formulation of this fundamental theory is the prime goal of most workers in this field. We think it is reasonable to search for this formulation in the framework of an action principle, but it is not clear from the beginning even what the fundamental dynamical variables should be. String field theory is the most straightforward way to discover this action principle, because it takes as dynamical variables fields associated with the particle states present in the theory at zero loops.

---

* Work supported in part by the Department of Energy, contract DE-FG05-86ER-40272

But it is by no means necessary that the zero loop particle states be associated in such a direct way with the fundamental dynamical variables. A popular analogy to contemplate is the relation between the fundamental variables of QCD and the hadrons. This analogy is most persuasive in the context of 't Hooft's $1/N_{colors}$ expansion of QCD which resembles very closely the dual loop expansion.[2] However, if something like this is at work, a simple flat space quantum field theory like Yang-Mills theory can't possibly be the answer. For one thing, string theory contains gravity, so the putative fundamental theory must be generally covariant. For another thing, the short distance structure of string theory is ultrasoft, unlike the hard parton structure associated with the asymptotic freedom of nonabelian gauge theory. One might speculate that the answer is some generally covariant quantum field theory, but in how many dimensions? In fact, this kind of picture of string theory was tried over a decade ago in the context of a "wee parton" approximation to QCD.[3] Perhaps the "wee parton" assumption is linked to the requirement of general covariance; it certainly accounts for the softness of string theory.

Yet another possibility is a fundamental formulation in terms of the two dimensional world sheet in which topology change is taken as a dynamical variable. A version of this idea was proposed over a decade ago in the context of the light cone gauge.[4] Or maybe the framework for this idea should be universal moduli space.[5,6]

In any case, even if string field theory is not the ultimate formulation of string theory, it should be possible to develop an effective action principle which would at least be valid at the classical level. Wendt[7] discovered a flaw in Witten's initial extension of his action principle for the interacting open bosonic string to the superstring. It would be unsettling if this flaw were fatal. Our work, we believe, provides a new formulation of open superstring field theory which surmounts this difficulty. It does not touch on efforts to develop a superstring field theory based on the manifestly supersymmetric formalism,[8] nor does it deal with the problem of developing a string field theory which does not contain open strings.

2. String and Superstring Field Theory

We turn now to an introductory description of what string field theory is.[9] Ordinary field theory assigns a number (or finite collection of numbers) to each point in space-time. String field theory generalizes this concept by replacing the space of points in space-time by the space of paths in space-time. Thus the string field is a functional of curves: $A[x^\mu(\sigma), c(\sigma)]$. Because paths in space-time are not restricted to lie in equal-time hyperplanes, the theory must be a very special field theory for which it is possible to recover our customary notion of a unitary time evolution. We can implement this special feature by requiring the dynamics to be such that the domain of the string field can be consistently restricted to those paths which do lie in equal time hyperplanes. That is, it should be possible to formulate the theory on a submanifold of the full space of paths. In this sense we might describe such a theory as "topological" on path space. From this point of view, Witten's proposal that the string field action be a Chern-Simons three form on path space is very natural.

The basic ingredients of Witten's version of string field theory[10] are: string fields A; a derivation acting on string fields, Q, which is taken to be the first-quantized $BRST$ operator; an associative exterior product, $*$, for which Q is a derivation:

$$Q(A * B) = QA * B + (-)^A A * QB;$$

and a volume form on path space, $\int$, which assigns a number, $\int A$, to each string field. The ghost numbers assigned to these objects are $-1/2$ for A, $+1$ for Q, $+3/2$ for $*$, and $-3/2$ for $\int$. In terms of these quantities Witten's action for open bosonic string field theory takes the form

$$S = \frac{1}{2}\int A * QA + \frac{1}{3}\int A * A * A,$$

and its gauge invariance is just

$$\Delta A = Q\Lambda + A * \Lambda - \Lambda * A.$$

In order to extend these ideas to the spinning string, Witten introduced a fermionic partner Ψ for the bosonic string field A and took the natural generalizations to the spinning string of Q, $*$, and $\int$. These objects have total ghost number 1, 1/2, and $-1/2$, respectively. Unfortunately, there is no assignment of ghost number to A and Ψ which allows the construction of a Chern-Simons three form action without the use of picture-changing operators. In his initial formulation Witten assigned A ghost and picture number $-1/2$ and -1, respectively.

These assignments for Ψ were 0 and $-1/2$ respectively. Then his proposed action takes the form

$$S = \frac{1}{2}\int A * QA + \frac{1}{2}\int Y(\sigma = \frac{\pi}{2})\Psi * Q\Psi + \frac{1}{3}\int X(\sigma = \frac{\pi}{2})A * A * A + \int A * \Psi * \Psi.$$

In this expression X is the local operator introduced by Friedan, Martinec, and Shenker[11] which changes picture number by $+1$ unit. Y is the inverse picture changing operator which changes picture number by -1. Both are inserted into the action at the midpoint of the string singled out by the definition of $*$ and $\int$ and denoted by $\sigma = \frac{\pi}{2}$. These operators have the explicit representations[12,13]

$$\begin{aligned} Y(z) =& c(z)\delta'(\gamma(z)) \\ X(z) =& \{Q, \Theta(\beta(z))\} \end{aligned}$$

where δ is the Dirac delta function and Θ is the Heaviside step function satisfying $\Theta' = \delta$. Here c is the reparametrization ghost, γ the superghost, and β the superantighost. The argument of the fields is just $z = e^{i\sigma+\tau}$. X and Y satisfy the short distance product relation

$$X(z)Y(z') \underset{z' \to z}{\sim} I.$$

Now we can describe the flaw that Wendt[7] discovered in this version of superstring field theory. The source of the difficulty is that the picture changing operator $X(z)$ has a singular operator product with itself:

$$X(z)X(z') \underset{z' \to z}{\sim} \frac{\Omega}{(z-z')^2}.$$

This causes a difficulty with the proof of gauge invariance because the bosonic gauge transformation of A is

$$\Delta A = Q\Lambda + X(\sigma = \frac{\pi}{2})(A * \Lambda - \Lambda * A).$$

When one tries to check the nonlinear gauge invariance of the action, one finds two X's colliding at the same point. The hope that somehow this singularity is cancelled is dashed by Wendt's explicit calculation of the four string function, which gives the wrong result. One can try to fix the problem by adding terms to the action with higher powers of A with coefficients designed to cancel the discrepancy. This is less than satisfactory because (1) the coefficient of the A^4 term is infinite and this presumably is true also of higher terms and (2) it would destroy the very attractive Chern-Simons form of the action.

Our solution to the difficulty is based on the idea that the classical string field should be described in the 0-picture.[14] There are several motivations for this idea:

(1) The $SL(2)$ invariant string state is in the 0-picture sector.

(2) Working in this picture corresponds to the manifestly dual form of the Neveu-Schwarz dual resonance amplitudes for spinning strings.

(3) Most importantly, picture changing operators are in less dangerous configurations.

This last point is easily understood. In the 0-picture, A carries total ghost number $+1/2$. Thus QA and $A * A$ both carry ghost number $3/2$, so there is no need for a picture changing operator to balance ghost number. For example the field equations in this picture read:

$$
\begin{aligned}
QA + A * A + X[\Psi * \Psi] &= 0 \\
Q\Psi + A * \Psi + \Psi * A &= 0
\end{aligned}
$$

and X appears only in the coupling of Ψ to A. Moreover the bosonic gauge transformation of A is now simply

$$\Delta A = Q\Lambda + A * \Lambda - \Lambda * A$$

so there is no collision of insertions in proving the gauge covariance of the field equations. One does need a new inverse picture changing operator Y_{-2} in order to balance ghost number in the action, however:

$$
\begin{aligned}
S = &\int Y_{-2}(\sigma = \frac{\pi}{2})[\frac{1}{2} A * QA + \frac{1}{3} A * A * A] \\
&+ \int Y(\sigma = \frac{\pi}{2})[\Psi * Q\Psi + A * \Psi * \Psi].
\end{aligned}
$$

The operator Y_{-2} had not previously appeared in the literature so our first task was to construct it. It must be a world sheet operator attached to the midpoint of the string and so must be local or bilocal in the complex plane: $z = \pm i$ are both associated with the midpoint of the string. It must also change picture number P and ghost number each by -2 units. We actually employ the bilocal choice

$$Y_{-2} = Y(i)Y(-i) \equiv Y\bar{Y}$$

for our construction because it leads to the simplest and most manageable formulation. In the -1-picture bilocal insertions had been previously proposed by

Lechtenfeld and Samuel.[15] Since $\partial_z Y$ is BRST trivial, one can say that such a choice is in some sense equivalent to a completely local choice. However, there is a short distance singularity in $Y(i)Y(z)$ as $z \to i$, so there is a divergence in the trivial part.

We systematically searched for all local candidates for Y_{-2}. Since there is a 1-1 correspondence between local operators and Fock states $|Y_{-2}\rangle = Y(0)\,|0\rangle$ with $|0\rangle$ the $SL(2)$ invariant Fock state, we searched for a Fock state with picture number -2 and ghost number $-5/2$ (since $|0\rangle$ has ghost number $-1/2$) satisfying

a) Lorentz and scale invariance (in particular $\alpha_0 = 0$ and $L_0 = 0$);

b) *BRST* invariance (*i.e.* it is annihilated by Q);

c) It is not *BRST* trivial;

d) It is normalized so that $X(z)Y_{-2}(z) = Y(z)$.

Condition a) plus ghost and picture number constraints limit the choice to 15 candidates. Of these 10 turn out to be primary and among the primary ones 7 are *BRST* invariant. However, the space of trivial states satisfying these conditions is found to be 10 dimensional so the coset space of candidates is only 5 dimensional. Only a 1 dimensional subspace of the coset space of nontrivial states is *BRST* invariant, so finally Y_{-2} is unique up to *BRST* equivalence. Although different choices are equivalent on-shell, they do lead to different off-shell actions. We work with the simple bilocal choice $Y\bar{Y}$.

3. New Gauge Invariances

For the fermionic kinetic term of the string field action, it has been pointed out[16,17] that the necessary inverse picture changing insertion gives rise to new gauge invariances in addition to those associated with BRST invariance. A similar feature applies to our new bosonic kinetic term. To understand this, notice that $Y(z) = c(z)\delta'(\gamma(z))$ is annihilated by $c(z)$ and by $\gamma^2(z)$. In fact, these new gauge invariances are needed to choose the gauge $\beta_0\Psi = 0$. Kugo and Terao achieve this gauge by first constructing a BRST invariant projection operator that annihilates the redundancies due to the new gauge invariances. They start by introducing a nonlocal version of X,

$$X_0 \equiv \{Q, \Theta(\beta_0)\}.$$

Then one can easily show that $YX_0Y = Y$ and $X_0YX_0 = X_0$, so that $\mathcal{P} = X_0Y$ is a projection operator. By virtue of these properties one can first restrict Ψ to satisfy $\Psi = \mathcal{P}\Psi$ and then use the Q gauge invariance to set $\beta_0\Psi = 0$. The resulting propagator is then $b_0X_0/L_0 = b_0\delta(\beta_0)/F_0$.

It is clear that identical considerations apply to our form of the bosonic kinetic term, only now all of the four operators $c(i), c(-i), \gamma^2(i), \gamma^2(-i)$ annihilate Y_{-2}. Thus we introduce two nonlocal X's

$$X_\pm = \{Q, \Theta(\beta_\pm)\},$$

where

$$\beta_\pm = \frac{1}{2}(e^{\pm\pi i/4}\beta_{-1/2} + e^{\mp\pi i/4}\beta_{1/2}).$$

Then $\mathcal{P} \equiv X_+ X_- Y(i)Y(-i)$ is a BRST invariant projection operator which kills the new redundancies. One can then first restrict A to satisfy $A = \mathcal{P}A$, and then fix the Q invariance in a convenient way.

Before turning to gauge-fixing, we note that the interaction terms are separately invariant under the new gauge invariances because each of the four operators $c(i), c(-i), \gamma^2(i), \gamma^2(-i)$ have negative conformal weight. Insertion of a negative weight local primary field at the midpoint of a Witten vertex always vanishes because of the curvature singularity there.

4. Gauge-Fixing

We next consider a general class of Siegel-like gauges for the fixing of the Q gauge invariances. For the bosonic string this class of gauges is simply $v \cdot b\, A = 0$ where $v \cdot b = \sum_n v_n b_n$, and the v_n are any set of numbers. For general v the propagator following from this condition is

$$\Pi_v = \frac{b \cdot v}{L \cdot v} Q \frac{b \cdot I(v)}{L \cdot I(v)}$$

where $I(v)_{-n} = (-)^{n+1} v_n$. If $I(v) \propto v$, the propagator simplifies to $\frac{b \cdot v}{L \cdot v}$. This is true of the standard Siegel gauge $v \cdot b = b_0$, for example.

For the spinning string, it is awkward to attempt general gauges of exactly this form. Instead, we use v and the projection operator $\mathcal{P}$ to construct a new projection operator

$$\mathcal{P}_v \equiv \mathcal{P} \frac{b \cdot v}{L \cdot v} Q \mathcal{P}.$$

This construction works equally well in the fermionic and bosonic sectors, with $\mathcal{P}$ being the appropriate projector. The general class of gauges we consider then are specified by

$$A = \mathcal{P}A = \mathcal{P}_v A,$$

and the corresponding propagators are just

$$\Pi_v^B = \mathcal{P}_v \frac{b \cdot v}{L \cdot v} X_+ X_- \mathcal{P}_v^T$$
$$\Pi_v^F = \mathcal{P}_v \frac{b \cdot v}{L \cdot v} X_0 \mathcal{P}_v^T$$

for the bosonic and fermionic components respectively, where the transpose is taken with respect to the string inner product. In general these propagators are prohibitively complicated. However, for special choices of v they become manageable. The fermionic propagator simplifies for $v \cdot b = b_0$ or $b_+ = \frac{1}{2}(b_1 + b_{-1})$. But the bosonic propagator only simplifies significantly for the latter choice, which is the one we studied most intensively.

5. b_+ GAUGE

The gauge $v \cdot b = b_+$ is particularly simple because b_+ commutes with local operators located at $z = \pm i$. In fact, the simplification is spectacular: the gauge choice formally causes the cubic term in the action to vanish! The reason is that b_+ is a derivation of the $*$ algebra. For example, applying b_+ to the l.h.s. of the field equation for fields satisfying $b_+ A = 0$ gives

$$0 = L_+ A + b_+(A * A) = L_+ A + (b_+ A) * A - A * b_+ A = L_+ A,$$

linearizing the field equation. Similarly the cubic term in the action can be shown to formally vanish. One uses the fact that $b_+ A = 0$ implies $A = b_+ c_+ A$ to replace one of the A's by $b_+ c_+ A$. Then one integrates by parts to throw the b_+ onto the remaining two A's, which then vanish. This gauge would have the same dramatic consequences for the bosonic string so as preparation we studied this case very carefully.

Of course, we know the interactions can't really be a gauge artifact. The resolution presumably lies in the fact that one can not reach this gauge for all field configurations. In fact, at the linearized level one can prove that for field configurations in the nonvanishing eigenspace of L_+, such a gauge can be reached. But this proof breaks down for fields in the kernel of L_+. This same caveat applies also to Siegel gauge $v \cdot b = b_0$, but there it is less problematic because a nontrivial kernel of L_0 exists only for special on-shell values of the space-time momentum p. In contrast, the kernel of L_+ is nontrivial for all p. Thus one can't regulate the problem by taking p slightly off-shell. Instead, we regulate our calculations by employing a nearby gauge $v \cdot b = b_+ + i\epsilon b_0$, letting $\epsilon \to 0$ at the end of the calculation. In this way we explicitly confirm that general tree amplitudes for the bosonic string and selected ones for the superstring come out correctly. The results of these calculations are independent of ϵ, as they should be. Since the vertex function is proportional to ϵ, this implies that there are compensating singularities in that amplitude.

The manner in which moduli space is covered in the limit $\epsilon \to 0$ is interesting. In Siegel gauge, all the diagrams of a cubic field theory provide an essential contribution. In particular, the six string tree diagram with three internal lines meeting at a cubic vertex is nonzero. We find that in the limit $\epsilon \to 0$ this diagram vanishes! The multiperipheral diagrams in which every vertex has at least one external line suffice to cover moduli space.

6. Conclusions

Our work strongly indicates that the difficulty discovered by Wendt is absent in our formulation of superstring field theory. The only qualification is that we haven't done a careful study of loop diagrams. But since the problem was initially present at tree level, it is significant that we have removed the difficulty at that level. Since the product of two Y's is just as singular as the product of two X's, the potential for a problem is present. However, in our scheme such singularities are prevented because of the nonlocal $X_\pm$'s that each propagator inserts between each pair of potentially colliding Y's.

In our work, we were led to consider a fascinating gauge choice which formally linearizes the theory. This gauge is somewhat analogous to the temporal or axial gauge choice in Chern-Simons gauge theories in $2+1$ dimensions. In that case one also linearizes the field equations, but nonlinearities do remain in the constraints due to gauge invariance. One might expect similar features in the case of superstring field theory. Linearization of the action does not guarantee linearization of constraints. The problem with checking this idea is that the canonical phase space of string field theory is not well understood. If one considers the path integral approach to Chern-Simons theories, one can see the analogy with our situation more clearly. In that case it is not really allowed to set A_0 to zero for all field configurations: one can only set the nonzero frequency modes to zero. The zero mode of A_0 then participates in the interactions, and in fact integration over it precisely imposes the constraints of the phase space approach. In the case of string field theory, the analogue of the zero frequency mode of A_0 would be the kernel of L_+.

To conclude, let us review the status of string field theory. Our work has provided us with a satisfactory formulation of spinning string field theory in the NSR formalism for *open* strings. Such a formulation does not implement space-time supersymmetry manifestly. For that one would have to employ the Green-Schwarz formalism, which is not yet developed sufficiently to apply to string field theory. There has been much recent activity in this direction along the lines of seeking linearizing gauge choices.[18] In spite of dramatic claims of progress, this approach is still problematic.[19] Another gap in our understanding is the absence of a satisfactory string field theory involving *only* closed strings. In this area there has been recent progress in finding at least a gauge fixed action which produces correct tree amplitudes.[20]

With our work, we now have an action principle for all the open string theories. Since it uses string fields as dynamical variables, it is an admittedly cumbersome formulation. Because of this, one hopes that a simpler action principle might be possible based on a different choice of variables. We have suggested other approaches in the introduction, and it is clearly important to explore these and other possibilities.

References

1. C.R. Preitschopf, C.B. Thorn and S.A. Yost, University of Florida preprint UFTP-HEP-89-19, to be published in Nucl. Phys. B.
2. G. 't Hooft, *Nucl. Phys.* **B72** (1974) 461.
3. C.B. Thorn, *Phys. Rev.* **D19** (1979) 639.
4. R. Giles and C. B. Thorn, *Phys. Rev.* **D16** (1977) 366.
5. T. Banks and E. Martinec, *Nucl. Phys.* **B294** (1987) 733.
6. D. Friedan and S. Shenker, *Phys. Lett.* **175B** (1986) 287.
7. Wendt, C., *Nucl. Phys.* **B314** (1989) 209.
8. M. B. Green and J. H. Schwarz, *Phys. Lett.* **136B** (1984) 367; *Nucl. Phys.* **B243** (1984) 285.
9. C. B. Thorn, *Physics Reports* **175** (1989) 1-101.
10. E. Witten, *Nucl. Phys.* **B268** (1986) 253.
11. D. Friedan, E. Martinec, and S. Shenker, *Phys. Lett.* **160B** (1985) 55; *Nucl. Phys.* **B271** (1986) 93.
12. E. Verlinde and H. Verlinde, *Phys. Lett.* **192B** (1987) 99; *Nucl. Phys.* **B288** (1987) 357.
13. J. P. Yamron, *Phys. Lett.* **174B** (1986) 69.

14. C. B. Thorn, "Lectures on String Theory", in *Superstrings '88*, Proc. of the Trieste Spring School, M. Green, M. Grisaru R. Iengo, and A. Strominger, World Scientific Publishing Co. (1988).

15. O. Lechtenfeld and S. Samuel, *Phys. Lett.* **B213** (1988) 431.

16. T. Kugo and H. Terao, *Phys. Lett.* **208B** (1988) 416.

17. I. Ya. Arefeva and P. B. Medvedev, *Phys. Lett.* **202B** (1988) 510; *Phys. Lett.* **212B** (1988) 299.

18. R.E. Kallosh, Phys. Lett. **B224** (1989) 273;
S.J. Gates, M.T. Grisaru, U. Lindström, M. Roček, W. Siegel, P. van Nieuwenhuizen and A.E. van de Ven, Phys. Lett. **B225** (1989) 44;
R.E. Kallosh, Phys. Lett. **B225** (1989) 49;
M.B. Green and C.M. Hull, Phys. Lett. **B225** (1989) 57;
U. Lindström, M. Roček, W. Siegel, P. van Nieuwenhuizen and A.E. van de Ven, Stony Brook preprint ITP-SB-89-38.

19. J.M.L. Fisch and M. Henneaux, Université Libre de Bruxelles preprint ULB TH2/89-04;
F. Bastianelli, G.W. Delius and E. Laenen, Stony Brook preprint ITP-SB-89-51;
U. Lindström, M. Roček, W. Siegel, P. van Nieuwenhuizen and A.E. van de Ven, Stony Brook preprint ITP-SB-89-76.

20. M. Saadi and B. Zwiebach, *Ann. Phys.* **192** (1989) 213;
T. Kugo, H. Kunitomo and K. Suehiro, *Phys. Lett.* **B226** (1989) 48;
H. Sonoda and B. Zwiebach, MIT preprints MIT-CTP-1774, 1758;
B. Zwiebach, MIT preprint MIT-CTP-1787;
T. Kugo and K. Suehiro, Kyoto University preprint KUNS-988.

CONFORMAL METHODS AND STRING FIELD THEORY AMPLITUDES

Robert Bluhm*
Physics Department, Indiana University
Bloomington, IN 47405, U.S.A.

and

Stuart Samuel
Physics Department, The City College of New York,
New York, NY 10031, U.S.A.

The method of applying conformal field theoretic techniques to the study of string field theory is discussed. In particular, the tree-level and one-loop off-shell N-point tachyon amplitudes are obtained for Witten's covariant string field theory. The derivation of these results involves proving an identity between the Jacobian, which relates the string field theory variables to the standard variables, and the ghost correlation functions. The covering of moduli space by the field theory Feynman diagrams and the off-shell singularity structure of the amplitudes are discussed as well.

1. Introduction

String field theory [1] is a second-quantized field theory of strings. The string field operators $A(X^\mu(\sigma))$ are functionals of the string coordinates $X^\mu(\sigma)$, and the theory describes a many-body system in which entire strings are created or destroyed. This is to be contrasted with a first-quantized theory of strings, which is described by a worldsheet that is swept out as a string evolves through time, that is, $X^\mu(\sigma, \tau)$ is a function of the worldsheet parameters σ and τ. The first-quantized theory of strings is an inherently perturbative theory. To calculate a scattering amplitude to a given order, one has to specify the topology of the worldsheet, and the total amplitude is given as the sum over all such topologies. There

---

* Speaker

are problems, however, with defining string theory in this way. For one thing, it has been shown that string perturbation theory diverges [2]. Furthermore, it is believed that there ought to be some fundamental principle underlying string theory analogous to the way differential geometry is the basis for general relativity, and it is difficult to imagine how such a fundamental principle could be discovered in a perturbative framework. Lastly, first-quantized string theory describes the scattering of on-shell ($p^2 = -m^2$), physical states, and thus the types of nonperturbative techniques that were developed for particle field theories (such as (off-shell) high-energy behavior of scattering amplitudes, renormalization group, and static effective potentials) are not applicable to the case of first-quantized strings. It is toward remedying this latter problem—i.e., of going off shell—that most of the work described here has been directed. To address these sorts of questions, we must consider a second-quantized field theory of strings.

There are many formulations and types of string field theories: light-cone gauge or covariant approaches for bosonic, supersymmetric, or heterotic strings. For nonperturbative questions, a covariant string theory is necessary; and the one that is best understood is Witten's covariant string field theory for the open bosonic string [3]. It is for this theory that we have developed techniques for calculating off-shell N-point amplitudes at the tree [4] and one-loop levels [5].

To calculate a string scattering amplitude from string field theory, and to show that it agrees with the results of first-quantized string theory, there are two ingredients that must match up—namely, the integrands of the expressions and the regions of integration. Having the correct region of integration means that moduli space is correctly covered. This problem has been analyzed in ref. [6] where it was shown that covariant string field theory provides an appropriate triangulation of moduli space for vacuum bubble diagrams having exactly one boundary and no external states. When there are such states, the dimension of moduli space is increased and additional analysis is necessary. Ref. [7] established that the correct Koba-Nielsen integration range is achieved for the 4-point tree amplitude, and ref. [4] extended the result to the N-point case. In ref. [5], we showed that the Witten string field theory correctly produces the N-point one-loop integration region.

The integrand involves two factors: a momentum dependent term and a measure term. The latter is generated by the bc-ghost correlation function, $\langle\,\rangle_{bc}$, and the integration over the lengths, T_i, of intermediate worldsheet strips. In changing variables from the T_i to Koba-Nielsen variables, x_i, a Jacobian, J, is generated. For the $N = 4$ tree amplitudes, Giddings showed [7] that $\langle\,\rangle_{bc}$ cancels most of J thereby leaving a relatively simple expression. This cancellation was the key algebraic step in proving that the four-tachyon amplitude in covariant string field theory agrees on-shell with that of the dual-resonance model. For the N-point case, general arguments were given by Giddings and Martinec [8] that this cancellation should again occur. This was then proved rigorously as an identity between the ghost correlation function and the Jacobian for the N-point tree-level case in ref. [4], and in this way the full Koba-Nielsen amplitude was obtained [4]. A similar proof was then given for the one-loop level in ref. [5].

The off-shell N-point amplitudes were obtained using the method of off-shell conformal field theory [9]. This is the generalization of using conformal techniques for calculating string amplitudes to the case where the external states are allowed to go off the mass shell. This method has the advantage that it is much easier to use to obtain off-shell amplitudes [11,12] than by working directly with string field theory vertex functions. This method will be reviewed in the next section; then in section 3, we will show how it can be used to obtain the off-shell Koba-Nielsen amplitude [4]. In section 4, we present the off-shell N-point one-loop amplitudes [5]. The $N = 2$ off-shell one-loop amplitude was computed in ref. [10]. In ref. [5] we verified that one-loop off-shell conformal field theory gives the same result. The nonplanar contribution has an unusual off-shell singularity structure [10]: new poles appear. For $N = 2$, we found the same result, and for $N > 2$ it seems that more unusual behavior is present [5]. Lastly, in section 4, conclusions and subsequent results are described.

2. Off-Shell Conformal Field Theory

In Witten's string field theory, the action is given as a Chern-Simons form:

$$S = \int \left(A * QA + \frac{2}{3} A * A * A \right) , \tag{1}$$

where A is the string field and Q is the BRST charge. In quantizing the theory, a set of Feynman rules is obtained that describes how the worldsheets of the first-quantized theory are produced. By expanding the string fields in the Fock space of the first-quantized states and using the string field theory Feynman rules, we are able to reproduce the scattering amplitudes of the first-quantized string modes, such as the tachyon or massless vectors.

The Feynman rules are the following. In the Siegel gauge, the propagator is given by $b_0 L_0^{-1}$, which can be put in integral form as:

$$\frac{b_0}{L_0} = b_0 \int_0^\infty dT\, e^{-TL_0}\,. \tag{2}$$

The operator e^{-TL_0} acts to evolve a string through a time T, creating a rectangular strip of worldsheet. These rectangular strips are then joined together in the Witten manner, in which the widths of the strings are all equal, and three strings join forming a singular point in the middle [3]. At the positions of the external states, vertex operators, $V(p)$, are inserted, which describe the the emission of a particular string mode with momentum p. Lastly, we have to sum over all possible Feynman diagrams connecting the incoming and outgoing states.

The final expression for tachyon scattering at tree level is the following:

$$A = \sum_{\substack{\text{Feynman}\\ \text{diagrams}}} \int_0^\infty dT_1 \ldots \int_0^\infty dT_{N-3}\, \langle V(p_1) b_0 V(p_2) b_0 \ldots b_0 V(p_{N-1}) V(p_N) \rangle\,, \tag{3}$$

where the correlation function of the vertex operators and ghost zero modes is evaluated on the Witten configuration. For an N-point tree amplitude, there are $(N-3)$ integration parameters T_i and ghost operators b_0. The b_0's can be represented as contour integrals across the propagator strips.

The problem then is how to evaluate the correlation function in eq. (3) on the Witten configuration, at the same time allowing the external states to go off the mass shell (i.e., not requiring that $p^2 = -m^2$). Off-shell conformal field theory provides a technique for doing this [9].

In conformal field theory in the usual on-shell case, the correlation functions of vertex operators are invariant under conformal mappings because, when they satisfy the mass-shell condition, the vertex operators have weight zero. This permits a transformation from one configuration to another without picking up any additional factors. If, however, the mass-shell condition is relaxed, the vertex operators will no longer transform simply. In ref. [9] it was determined how these and the other primary fields and their correlation functions transform. It was proved [9] that the nontrivial factors that are obtained off shell lead to an answer for the off-shell scattering amplitudes that is identical to that which would be obtained working directly with the string field theory vertex functionals. The advantage is that the off-shell conformal field theory approach is much easier to use—one can write down right away what the off-shell expressions for the amplitudes are. The only remaining problem for N-point amplitudes is to show that one gets the correct measure factors in the z-plane—i.e., that the ghost-Jacobian identity holds and that moduli space is covered.

3. The Off-Shell Koba-Nielsen Formula

The usual on-shell string scattering amplitude can be written formally as

$$A = \int_{\substack{\text{moduli}\\ \text{space}}} dm_i \left\langle V(p_1) \ldots V(p_N) \right\rangle , \tag{4}$$

where the moduli, m_i, are a set of variables that describe the positions of the vertex operators as well as the shape of the worldsheet. The range of moduli space is picked so that each inequivalent class of surfaces is integrated over only once.

We show that the expression in eq. (3), coming from string field theory, correctly reproduces the on-shell expression (4), which was derived from first-quantized string theory. Then we find the off-shell extension of eq. (4).

If we apply the techniques of off-shell conformal field theory to eq. (3) and use the mapping given in ref. [8] that goes from the Witten configuration to the upper-half of the complex plane, we get the following expression (ignoring some irrelevant factors):

$$A = e^{\sum_{r=1}^{N} N_{00}^{rr} \left(\frac{p_r^2}{2} - 1\right)} \sum_{\substack{\text{Feynman}\\ \text{diagrams}}} \int_{R_i} dm_i \, J \left\langle \right\rangle_{bc} \left\langle V(p_1) \ldots V(p_N) \right\rangle , \tag{5}$$

where we have set the slope parameter $\alpha\prime$ equal to $\frac{1}{2}$. The regions R_i are the subregions of moduli space corresponding to the $i^{\underline{th}}$ Feynman diagram. The exponential factor is the off-shell contribution; for on-shell tachyons, $p^2 = 2$, and this factor goes away. The zero-zero Neumann functions, N_{00}^{rr}, are constants that arise as conformal factors in going from the Witten string configuration to the complex plane [9]. The Jacobian, J, is acquired when the change of variables is made from the propagator strip parameters, T_i, to the moduli m_i.

It can be seen that eq. (5) agrees on shell with the Koba-Nielsen result (4) if both of the following are true:

$$J\langle\,\rangle_{bc} = 1\,, \tag{6}$$

which is the ghost-Jacobian identity, and

$$\sum_{\substack{\text{Feynman}\\ \text{diagrams}}} \int_{R_i} = \int_{\substack{\text{moduli}\\ \text{space}}}\,, \tag{7}$$

which is the requirement that moduli space be properly covered. Both of these were proved for the N-point case in ref. [4], and the final result for the off-shell Koba-Nielsen amplitude is then given by [4]:

$$\begin{aligned} A = \prod_{i=1}^{N-3} \int_{-\infty}^{+\infty} & dx_i\,(x_{N-2} - x_{N-1})(x_{N-2} - x_N)(x_{N-1} - x_N) \\ & \prod_{1\leq r<s\leq N} (x_r - x_s)^{p^r\cdot p^s} \left(\prod_{r=1}^{N} e^{N_{00}^{rr}\left(\frac{p_r^2}{2}-1\right)} \right)\,, \end{aligned} \tag{8}$$

where the x_i's are the usual Koba-Nielsen variables.

4. Off-Shell One-Loop Amplitudes

In ref. [5], it was shown how off-shell conformal field theory may be applied to the case of one-loop amplitudes. The mapping that takes the one-loop Witten configuration to a half-annulus in the z-plane was obtained. The ghost-Jacobian identity as well as the covering of moduli space were proved at the one-loop level, and agreement was found with the on-shell one-loop result [13]. Finally, the off-shell contributions were evaluated, and in this way the N-point one-loop off-shell amplitude was obtained.

For the $N = 2$ case, we compared our results to those found by Freedman, Giddings, Shapiro, and Thorn [10], who studied the nonplanar one-loop case. We find the same singularity structure as they had found—namely, that off shell there are unphysical poles in addition to the usual closed-string poles that are expected to appear in a nonplanar loop amplitude. We showed that this is a generic feature of the one-loop amplitudes—that for $N > 2$, unphysical poles will occur as well [5].

We also found expressions for the case of Mobius-strip loop diagrams and for the case of amplitudes involving Chan-Paton factors. All of the explicit calculations were done for tachyon amplitudes, but it would be straightforward to adapt our methods to compute amplitudes involving fields other than tachyons, such as massless vectors, excited states, auxiliary fields, and second-quantized Faddeev-Popov fields.

5. Conclusions

We hope our work illustrates that concrete results are obtainable from string field theory. Our endeavor is to obtain a deeper mathematical understanding of strings. Furthermore, there are several potential physical applications of our results—among them the understanding of vacuum structure [14].

These techniques have also been applied to a modified version [15] of Witten's superstring field theory [16]. In ref. [17], we showed that the modified superstring field theory [15] gives the correct N-point tree-level amplitudes on shell, but that the off-shell amplitudes are infinite. More recently, some new proposals for modifying Witten's superstring field theory have been made [18,19] that appear promising. Perhaps our techniques can be applied to one of these theories to obtain higher-point off-shell superstring amplitudes.

Lastly, it has recently been shown in ref. [20] that off-shell conformal field theory can be applied to the case of genus $g \geq 2$ scattering. In principle, any open bosonic string amplitude can now be obtained [20].

Acknowledgments

This work was supported in part by the United States Department of Energy under contracts DE-AC02-84ER40125 and DE-AC02-83ER40107.

References

1. J. H. Schwarz, ed., Superstrings, Vols. I and II, (World Scientific, Singapore, 1985); M. B. Green and D. J. Gross, eds., Unified String Theories, (World Scientific, Singapore, 1986); C. Thorn, Physics Reports 175 (1989) 1; G. Horowitz in The Santa Fe TASI-87, (World Scientific, Singapore, 1988); R. Slansky and G. West, eds.
2. D. J. Gross and V. Periwal, Phys. Rev. Letts. 60 (1988) 2105.
3. E. Witten, Nucl. Phys. B268 (1986) 253.
4. R. Bluhm and S. Samuel, Nucl. Phys. B323 (1989) 337.
5. R. Bluhm and S. Samuel, Nucl. Phys. B325 (1989) 275.

6. S. Giddings, E. Martinec and E. Witten, Phys. Lett. 176B (1986) 362.
7. S. Giddings, Nucl. Phys. B278 (1986) 242.
8. S. Giddings and E. Martinec, Nucl. Phys. B278 (1986) 91.
9. S. Samuel, Nucl. Phys. B308 (1988) 317.
10. D. Freedman, S. Giddings, J. Shapiro and C. Thorn, Nucl. Phys. B298 (1988) 253.
11. J. H. Sloan, Nucl. Phys. B302 (1988) 349.
12. S. Samuel, Nucl. Phys. B308 (1988) 285.
13. J. H. Schwarz, Physics Report 89 (1982) 223.
14. V.A. Kostelecký and S. Samuel, Indiana preprint IUHET 173 (September, 1989).
15. O. Lechtenfeld and S. Samuel, Phys. Letts. 213B (1988) 431.
16. E. Witten, Nucl. Phys. B276 (1986) 291.
17. R. Bluhm and S. Samuel, Indiana preprint IUHET 170 (August, 1989).
18. C. R. Preitschopf, C. B. Thorn, and S. A. Yost, Florida preprint UFIFT-HEP-89-19
19. I. Ya. Aref'eva, P. B. Medvedev, and A. P. Zubarev, Steklov preprint SMI-10-1989
20. S. Samuel, Indiana preprint IUHET 171 (August 1989)

Progress Toward Covariant Formulation of All D = 4 GS-type σ-model Actions

S. James Gates, Jr.[1]

Department of Physics and Astronomy
University of Maryland, College Park, MD 20742

Abstract

The initial steps toward the development of the general formalism for the description of four dimensional Green-Schwarz superstrings coupled to massless background fields are discussed. A number of open problems are described.

I. Introduction

In order to use string and superstring theories [1] for four dimensional physics, the concept of compactification developed along diverse lines [2]. However, the net result of all of these seemingly different compactification techniques is the specification of a four dimensional internal symmetry group and a spectrum of ordinary fields which form representations of the internal symmetry group. Thus, a conceptually simpler way to formulate these string and superstring theories is as *ab initio* four dimensional theories but in which the internal space arises in the manner of the conventional fiber bundle viewpoint [3].

In this way, the symmetries (both spacetime and internal) represented by fundamental particles all have their origins in the Kac-Moody algebras of the string theories. String theories treat configuration space and 'isotopic' space in almost the same

[1]Work supported by National Science Foundation Grant PHY 88-16001.

manner. The only differences between configuration space and 'isotopic' space arise because of the different treatment of zero modes. With this view in mind, I will discuss progress toward the formulation of four dimensional superstring theories wherein all symmetries are manifest at the level of the corresponding two-dimensional actions. In particular, for spacetime supersymmetry this implies that a Green-Schwarz type description is our ultimate goal.

II. The Ultimate D = 4 Superstring σ-model Problem

We seek to formulate all D = 4 superstring nonlinear σ-models describing the propagation of D = 4 superstrings in a complete background of massless modes. Such σ-models should possess an explicit "β-function coupling constant" (equivalent to vertex operators) for every massless mode of the superstring theory. Thus we are attempting to extend our previous works[4] in NSR σ-models which cover the bosonic fields only. In the present problem we want to introduce the *superfield* coupling constants listed below in our first table. The coupling constants should appear in

σ-model coupling tensor	string massless supermultiplet
$E_{\underline{A}}{}^{\underline{M}}(\Theta, X)$	*superfield supergravity multiplet*
$B_{\underline{AB}}(\Theta, X)$	*axion multiplet*
$\Gamma_{\underline{M}}{}^{\hat{\alpha}}(\Theta, X)$	*gravi-photon multiplets*
$\Gamma_{\underline{M}}{}^{\hat{I}}(\Theta, X)$	*right gauge-group*
	vector multiplets
$\Phi_{\hat{\alpha}\hat{I}}(\Theta, X)$	*scalar multiplets*

Table 1: Background Superfields of the GS σ-model

a two dimensional world-sheet action $\mathcal{S}(E_{\underline{A}}{}^{\underline{M}}, B_{\underline{AB}}, \Gamma_{\underline{A}}{}^{\hat{\alpha}}, \Gamma_{\underline{A}}{}^{\hat{I}}, \Phi_{\hat{\alpha}\hat{I}})$ where $\mathcal{S}$ should satisfy the condition that for $E_{\underline{A}}{}^{\underline{M}} = B_{\underline{AB}} = \Gamma_{\underline{A}}{}^{\hat{\alpha}} = \Gamma_{\underline{A}}{}^{\hat{I}} = \Phi_{\hat{\alpha}\hat{I}} = 0$ it describes the action for a free four dimensional heterotic string. This is an ambitious undertaking of which we are not yet completely capable. Having defined the problem this way, we will in the remainder of this talk describe the present state of the art in reaching this goal.

III. The D = 10 Paradigm

As a preliminary illustration of the use of manifest realization of internal and spacetime symmetries in a string theory, we may use the D = 10 heterotic string[5] as an example. A proposed action for $\mathcal{S}(E_{\underline{A}}{}^{\underline{M}}, B_{\underline{AB}}, \Gamma_{\underline{A}}{}^{\hat{I}})$ is given by $\mathcal{S} = S_{GS} + S_R + S_N$, where the terms refer to a Green-Schwarz type action, a non-abelian righton action[4] and a noton action[6], respectively. The Green-Schwarz action has the familiar form

$$S_{GS} = \int d^2\sigma V^{-1}\Big[-\tfrac{1}{2}e^{-\Phi}\Pi_{++}{}^{\underline{a}}\Pi_{--\,\underline{a}} + \int_0^1 dy \hat{\Pi}_y{}^{\underline{C}}\hat{\Pi}_{++}{}^{\underline{B}}\hat{\Pi}_{--}{}^{\underline{A}}\hat{G}_{\underline{ABC}}\Big],$$

$$\Pi_{++}{}^{\underline{A}} = V_{++}{}^{m}(\partial_m Z^{\underline{M}})E_{\underline{M}}{}^{\underline{A}} \ , \ \Pi_{--}{}^{\underline{A}} = V_{--}{}^{m}(\partial_m Z^{\underline{M}})E_{\underline{M}}{}^{\underline{A}} \ ,$$

$$\hat{Z}^{\underline{M}} = Z^{\underline{M}}(\sigma,\tau,y) \ , \ \hat{\Pi}_y{}^{\underline{A}} = (\partial_y \hat{Z}^{\underline{M}})E_{\underline{M}}{}^{\underline{A}}(\hat{Z}) \ , \ \hat{G}_{\underline{ABC}} = G_{\underline{ABC}}(\hat{Z}) \ . \quad (3.1)$$

where $Z^{\underline{M}}(\tau,\sigma)$ is the superstring coordinate $(Z^{\underline{M}}(\tau,\sigma) \equiv (\Theta^{\mu}(\tau,\sigma), X^{\underline{m}}(\tau,\sigma))$, $G_{\underline{ABC}}$ is the field strength supertensor for a super 2-form $B_{\underline{AB}}(Z)$. The "hatted" coordinates are the usual extensions used in the Vainberg construction[7].

The next term in $\mathcal{S}$ introduces the internal degrees of freedom in a manifest way. For this purpose non abelian rightons $\phi_R^{\hat{I}}(\tau,\sigma)$ are used

$$S_R = -\frac{1}{4\pi}\int d^2\sigma V^{-1} Tr\{ (\mathcal{D}_{++}U^{-1})(\mathcal{D}_{--}U) - \lambda_{--}{}^{++}(U^{-1}D_{++}U)^2$$
$$+ \int_0^1 dy(\tilde{U}^{-1}\frac{d}{dy}\tilde{U})[(\mathcal{D}_{++}\tilde{U}^{-1})(\mathcal{D}_{--}\tilde{U}) - (\mathcal{D}_{--}\tilde{U}^{-1})(\mathcal{D}_{++}\tilde{U})]$$
$$-2\Pi_{++}{}^{\underline{B}}\Gamma_{\underline{B}}{}^{\hat{I}} t_{\hat{I}}(U^{-1}\mathcal{D}_{--}U) \} \ , \quad (3.2)$$

with $D_{++}U \equiv \mathcal{D}_{++}U - i\Pi_{++}{}^{\underline{B}}\Gamma_{\underline{B}}{}^{\hat{I}}Ut_{\hat{I}}$. The quantity $U \equiv exp[i\phi_R^{\hat{I}}(\tau,\sigma)t_{\hat{I}}]$ is an element of an arbitrary group. The matrices $t_{\hat{I}}$ generate a compact Lie algebra for the right-gauge group $\mathcal{G}_R$ where $\hat{I} = 1,\ldots,d_G$, $[t_{\hat{I}},\ t_{\hat{J}}] = i f_{\hat{I}\hat{J}}{}^{\hat{K}} t_{\hat{K}}$, $f_{\hat{I}\hat{J}\hat{K}} f^{\hat{I}\hat{J}}{}_{\hat{L}} = c_2\delta_{\hat{K}\hat{L}}$, and $Tr(t_{\hat{I}}t_{\hat{J}}) = 2k\delta_{\hat{I}\hat{J}}$. Above and in the following discussion, we use the notation $\mathcal{D}_{\pm\pm}$ to denote the world-sheet two-dimensional gravitationally covariant derivative.

The final term of $\mathcal{S}$ is a noton action which is required for the covariant removal of the Siegel anomaly by use of the noton fermions $\rho_+^{\hat{\imath}}$ (with $\hat{\imath} = 1,...,20$)

$$S_N = -i\tfrac{1}{2}\int d^2\sigma V^{-1}\delta_{\hat{\imath}\hat{\jmath}}[\ \rho_+^{\hat{\imath}}\mathcal{D}_{--}\rho_+^{\hat{\jmath}} + \lambda_{--}{}^{++}\rho_+^{\hat{\imath}}\mathcal{D}_{++}\rho_+^{\hat{\jmath}}\ ] \ . \quad (3.3)$$

These notons do not introduce any physical degrees of freedom into the model. The actions in (3.2) and (3.3) possess the local gauge invariance known as Siegel symmetry[8]. The would-be anomaly in this symmetry is precisely cancelled between the contributions coming from (3.2) and (3.3). This is the Hull mechanism[6]. Although it is sufficient for the genus zero theory, it has not been studied for $g > 0$ two-surfaces. Thus, our present approach is well suited to only calculate string tree-level contributions to the effective action.

A conceptual advantage of such a σ-model action is that it clearly demonstrates the relation of the closed GS string theory and particle field theory approaches to the introduction of internal symmetries. In the particle theory approach, we would simply specify that the Yang-Mills matter multiplet, $\Gamma_{\underline{B}}{}^{\hat{I}}(Z)$, should transform as the adjoint representation of the right-gauge group $\mathcal{G}_R$, and thus construct a principal fiber bundle over D $= 10$ superspace. The string takes this process one step further by actually "coordinate-izing" the Lie algebra of the fiber with $\phi_R^{\hat{I}}$. Thus, the non-abelian rightons may be regarded as providing maps from the world-sheet into isotopic charge space.

The σ-model we have described so far is actually inconsistent for arbitrary choices of the right-gauge group. In fact, if we demand the absence of anomalies, only the well-known $E_8 \otimes E_8$ or $SO(32)/Z_2$ groups are found to lead to consistent theories. This illustrates how the kinematical description of a candidate string theory is divorced from questions of its anomaly-freedom.

An important feature of the action $\mathcal{S}$ is that it possesses κ-symmetry[9]. This is most easily seen if we set the background fields to zero. A direct calculation shows $\delta_\kappa \mathcal{S} = 0$ under the variations

$$\begin{aligned}
\delta_\kappa Z^{\underline{M}} &= i(\sigma_{\underline{c}})^{\alpha\beta}(\kappa_{--\alpha}\Pi_{++}{}^{\underline{c}})E_\beta{}^{\underline{M}} \quad , \\
\delta_\kappa h_{--}{}^{++} &= \kappa_{--\alpha}\Pi_{--}{}^{\alpha}e^{2l}[\,1 - h_{--}{}^{++}h_{++}{}^{--}\,] \quad , \\
\delta_\kappa \lambda_{--}{}^{++} &= -\kappa_{--\alpha}\Pi_{--}{}^{\alpha} \quad , \quad \delta_\kappa h_{++}{}^{--} = \delta_\kappa \phi_R^{\hat{I}} = \delta_\kappa \rho_+^{\hat{\imath}} = 0 \quad , \\
\delta_\kappa \psi^{-1} &= -\tfrac{1}{2}\kappa_{--\alpha}\Pi_{--}{}^{\alpha}e^{2l}h_{++}{}^{--} \quad , \quad \delta_\kappa l = \tfrac{1}{2}\kappa_{--\alpha}\Pi_{--}{}^{\alpha}e^{2l}h_{++}{}^{--} \quad ,
\end{aligned} \tag{3.4}$$

where we have used the Beltrami decomposition of the zweibein $V_a{}^m$

$$\begin{aligned}
V_{++}{}^m\partial_m &\equiv \psi^{-1}e^{l}(\partial_{++} + h_{++}{}^{--}\partial_{--}) \quad , \\
V_{--}{}^m\partial_m &\equiv \psi^{-1}e^{-l}(\partial_{--} + h_{--}{}^{++}\partial_{++}) \quad ,
\end{aligned} \tag{3.5}$$

into its fundamental parts. In particular, global information about the 2-surface is carried by the left Beltrami field, $h_{--}{}^{++}$, and the right Beltrami field, $h_{++}{}^{--}$ along with the transition functions required to patch the surface together. (Taken together (3.4) and (3.5) imply $\delta_\kappa V_{--}{}^m = \kappa_{--\alpha} \Pi_{--}{}^{\alpha} V_{++}{}^m \;\; , \;\; \delta_\kappa V_{++}{}^m = 0$.)

It is clear from (3.5) that the left and right Beltrami fields are treated non-symmetrically. The κ-symmetry transformation is only nontrivial on the left Beltrami field. If we let the ordered pair (N_L, N_R) denote the number of nontrivial κ-symmetry transformations realized on the left and right Beltrami fields respectively, then the heterotic theory is a (1,0) theory. The fact that the κ-symmetry is nontrivial in the (1,0) theory only on the left Beltrami field means that there must be a deep connection between the "left" topology of the 2-surface and κ-symmetry.

The intrinsic advantage of the action $\mathcal{S}$ is that it permits all of the massless states of the D $= 10$ heterotic string to be represented by the superfield coupling functions $E_{\underline{M}}{}^{\underline{A}}$, $B_{\underline{AB}}$ and $\Gamma_{\underline{A}}{}^{\hat{I}}$ that explicitly appear in a world-sheet action. In other words, every massless state has an explicit representation among the component field expansions of $E_{\underline{M}}{}^{\underline{A}}$, $B_{\underline{AB}}$ and $\Gamma_{\underline{A}}{}^{\hat{I}}$. The superfield "equations of motion" for these quantities, when considered to be fields in super-spacetime, are determined from the principle that the β-functions calculated from $\mathcal{S}$ must vanish [10] as was first noted by Friedan. Alternately, these "equations of motion" should be derivable from an action

Vanishing β-function	Geometrically Constrained Superfield
$\beta(E_{\underline{A}}{}^{\underline{M}})$	$R_{\underline{AB}de} \;\; , \;\; T_{\underline{AB}}{}^{\underline{C}}$
$\beta(B_{\underline{AB}})$	$G_{\underline{ABC}}$
$\beta(\Gamma_{\underline{A}}{}^{\hat{I}})$	$F_{\underline{AB}}{}^{\hat{I}}$

Table 2: Beta-functions & Superfield Equations of Motion

principle. It is possible to directly derive this action by use of the "c-theorem" [11] combined with the calculation of the "averaged" anomaly [12]. Thus, the σ-model approach is seen to be sufficient to describe the complete massless string effective action in D $= 10$ superspace. This discussion also shows why the GS σ-model approach must ultimately prove to be superior to the NSR σ-model approach. In the latter, we can represent the bosonic states but not the fermionic states as "coupling constants" of a d = 2 σ-model. The "coupling constants" of a d = 2 σ-model are equivalent to the construction of vertex operators for the emission or absorption of massless states from the the superstring.

IV. The "Kernel" of D = 4 GS Actions

In the last section, we saw that the notion of a manifest and covariant action which represented all of the symmetries of D = 10 heterotic superstrings has an explicit realization. This suggests a general philosophy for treating all superstring theories, especially those in four dimensions. Namely, it should be possible to construct four dimensional GS-type actions which necessarily include some degrees of freedom associated with the presence of internal symmetries. Furthermore, our "discovery" that the D = 10 heterotic theory is a (1,0) theory implies that it should be possible to describe four dimensional theories which have an arbitrary (N_L, N_R) set of κ-symmetries. Such theories would be the GS analogs of the type (p,q) theories [13] known in the NSR formalism. To this end we first introduce the notion of the "kernel" of all D = 4 GS-type actions.

Let $Z^{\underline{M}} \equiv (\Theta^{\mu\ i}, \Theta^{\mu\ i'}, \bar{\Theta}^{\dot{\mu}}{}_i, \bar{\Theta}^{\dot{\mu}}{}_{i'}, X^{\mu\dot{\mu}})$ define the supercoordinate of the string. We introduce the four dimensional bosonic string coordinates in the form of a two by two hermitian matrix,

$$X^{\mu\dot{\mu}}(\tau,\sigma) = \begin{pmatrix} X^0 + X^3 & X^1 - iX^2 \\ X^1 + iX^2 & X^0 - X^3 \end{pmatrix} \quad , \tag{4.1}$$

where the μ and $\dot{\mu}$ indices each take on two values. This will ultimately facilitate the derivation of low energy four dimensional results expressed in the notation of two component Weyl spinors. Our previous experience in D = 4, N = 1 superspace has amply demonstrated the convenience of such a notation.

The fermionic coordinates are also introduced in the form of two component spinors which carry additional "isospin" indices i and i',

$$\begin{aligned} &\Theta^{\mu i}(\tau,\sigma) \qquad i = 1, ..., N_L \ (\text{4D} - \text{Weyl spinor}) \ , \\ &\Theta^{\mu i'}(\tau,\sigma) \qquad i' = 1, ..., N_R \ (\text{4D} - \text{Weyl spinor}) \ . \end{aligned} \tag{4.2}$$

We also introduce the symbol $\hat{G}_{\underline{ABC}}$ defined by

$$\hat{G}_{\underline{ABC}} = i\tfrac{1}{2}C_{\alpha\gamma}C_{\dot{\beta}\dot{\gamma}} \left\{ \begin{array}{ll} \delta_i{}^j & : if\ \underline{A} = \alpha\ i\ ,\ \underline{B} = \dot{\beta}\ j\ ,\ \underline{C} = \gamma\dot{\gamma} \\ & \text{or any even permutation,} \\ -\delta_i{}^j & : \text{for any odd permutation,} \\ -\delta_{i'}{}^{j'} & : if\ \underline{A} = \alpha\ i'\ ,\ \underline{B} = \dot{\beta}\ j'\ ,\ \underline{C} = \gamma\dot{\gamma} \\ & \text{or any even permutation,} \\ \delta_{i'}{}^{j'} & : \text{for any odd permutation,} \\ 0 & : \text{ otherwise.} \end{array} \right\} . \quad (4.3)$$

Next we note the same action given in (3.1) (except with $\Phi = 0$) can be defined here for a D = 4 theory. We define κ-symmetry variations by

$$\begin{aligned} \delta_\kappa Z^{\underline{N}} = &\ i\Pi_{++}{}^{\alpha\dot{\alpha}}(\bar{\kappa}_{--\ \dot{\alpha}}{}^{i} E_{\alpha\ i}{}^{\underline{N}} + \kappa_{--\ \alpha\ i} E^{i}_{\dot{\alpha}}{}^{\underline{N}}) \\ & + i\Pi_{--}{}^{\alpha\dot{\alpha}}(\bar{\kappa}_{++\ \dot{\alpha}}{}^{i'} E_{\alpha i'}{}^{\underline{N}} + \kappa_{++\ \alpha\ i'} E^{i'}_{\dot{\alpha}}{}^{\underline{N}}) \quad , \\ \delta_\kappa h_{--}{}^{++} = &\ \kappa_{--\ i\alpha}\Pi_{--}{}^{\alpha i} e^{2l}[\,1 - h_{++}{}^{--}h_{--}{}^{++}\,] + \text{h.c.} \quad , \\ \delta_\kappa h_{++}{}^{--} = &\ \kappa_{++\ i'\alpha}\Pi_{++}{}^{\alpha i'} e^{-2l}[\,1 - h_{++}{}^{--}h_{--}{}^{++}\,] + \text{h.c.} \quad , \\ \delta_\kappa \psi = &\ -\tfrac{1}{2}[\,\kappa_{++\ i'\alpha}\Pi_{++}{}^{\alpha i'} e^{-2l} h_{--}{}^{++} + \kappa_{--\ i\alpha}\Pi_{--}{}^{\alpha i} e^{2l} h_{++}{}^{--}\,] + \text{h.c.} \quad , \\ \delta_\kappa l = &\ \tfrac{1}{2}[\,\kappa_{++\ i'\alpha}\Pi_{++}{}^{\alpha i'} e^{-2l} h_{--}{}^{++} - \kappa_{--\ i\alpha}\Pi_{--}{}^{\alpha i} e^{2l} h_{++}{}^{--}\,] + \text{h.c.} \quad , \end{aligned} \quad (4.4)$$

and remarkably enough the action is invariant for arbitrary integers N_L and N_R! The case of N = 1 spacetime supersymmetry is clearly the special case of $N_L = 1$ and $N_R = 0$ in our conventions.

This kernel can be used in several different ways. First we can impose the light-cone gauge condition on it. Then additional conformal field theories may be added to it so as to insure anomaly cancellation. In this way one would arrive at the light-cone formulation of (presumably all) D = 4 superstring theories! It seems likely that the condition of anomaly cancellation will imply that only theories with $1 \le N_L + N_R \le 8$ will be free of anomalies. This would provide a stringy reason why the maximally supersymmetric four dimensional theory would be one with eight spacetime supersymmetries. Although only the case of $N_L = 1,\ N_R = 0$ is phenemologically interesting, presumably the other theories provide all possible stringy extensions of D = 4 supergravity theories.

V. SUSY Augmentation of the Kernel

Our eventual aims are more lofty than the construction of light-cone gauge four dimensional superstring theories. We are interested in the covariant realization of both super-spacetime and internal symmetries for four dimensions along the lines demonstrated for the ten dimensional theories. This is presently beyond our reach. But it seems to require that we augment the kernel in two ways:

(a.) supersymmetry augmentation,

(b.) internal symmetry augmentation.

The reason such augmentation is required is that the action given purely by the kernel $S_{GS}^{(K)}$ is anomalous for all values of N_L and N_R. This brings us to the notoriously difficult problem of the covariant quantization of the GS action.

Despite the recent efforts [14] at covariant quantization of the GS action, it remains an unsolved problem! We are in a similar position with the covariant quantization of the GS formulation of superstring theories as was the case with QED between the time it was first formulated in the 1920's and its successful quantization <u>and</u> renormalization in the late 1940's. This problem poses quite a challenge to the progress of really constructing a <u>theory</u> of superstrings as opposed to the collection of facts that passes under the name of superstring theory for now.

Presently, the best hope to meet this challenge seems to lie in the idea of augmentation of the kernel. The only principle which presently seems to be available is that all augmentations of the kernel must be consistent with κ-symmetry. An example, of augmentation in four dimensions can be constructed by a slight modification of the model proposed by Siegel [15]. Consider the action given by

$$\begin{aligned} S_{GSS} = \int d^2\sigma V^{-1}\Big[& -\tfrac{1}{2}\Pi_{++}{}^{\underline{a}}\Pi_{--\,\underline{a}} + \int_0^1 dy \hat{\Pi}_y{}^{\underline{C}}\hat{\Pi}_{++}{}^{\underline{B}}\hat{\Pi}_{--}{}^{\underline{A}}\hat{G}_{\underline{ABC}} \\ & + \Pi_{--}{}^{\alpha i}\mathsf{d}_{++\alpha i} + \Pi_{--}{}^{\dot{\alpha}}{}_i\bar{\mathsf{d}}_{++\dot{\alpha}}{}^i + \Pi_{++}{}^{\alpha i'}\mathsf{d}_{--\alpha i'} + \Pi_{++}{}^{\dot{\alpha}}{}_{i'}\bar{\mathsf{d}}_{--\dot{\alpha}}{}^{i'} \\ & + \lambda_{[-4]}{}^{\alpha\dot{\alpha}}\mathsf{d}_{++\alpha i}\bar{\mathsf{d}}_{++\dot{\alpha}}{}^i + \lambda_{[4]}{}^{\alpha\dot{\alpha}}\mathsf{d}_{--\alpha i'}\bar{\mathsf{d}}_{--\dot{\alpha}}{}^{i'} \\ & + i\Pi_{++}{}^{\alpha\dot{\alpha}}[\,\bar{\mathsf{d}}_{++\dot{\alpha}}{}^i\psi_{[-4]\alpha i} + \mathsf{d}_{++\alpha i}\bar{\psi}_{[-4]\dot{\alpha}}{}^i\,] \\ & + i\Pi_{--}{}^{\alpha\dot{\alpha}}[\,\bar{\mathsf{d}}_{--\dot{\alpha}}{}^{i'}\psi_{[4]\alpha i'} + \mathsf{d}_{--\alpha i'}\bar{\psi}_{[4]\dot{\alpha}}{}^{i'}\,] \\ & + \phi_{[-6]}{}^{\alpha\dot{\alpha}}[\,\mathsf{d}_{++\alpha i}\mathcal{D}_{++}\bar{\mathsf{d}}_{++\dot{\alpha}}{}^i - \bar{\mathsf{d}}_{++\dot{\alpha}}{}^i\mathcal{D}_{++}\mathsf{d}_{++\dot{\alpha} i}\,] \\ & + \phi_{[6]}{}^{\alpha\dot{\alpha}}[\,\mathsf{d}_{--\alpha i'}\mathcal{D}_{--}\bar{\mathsf{d}}_{--\dot{\alpha}}{}^{i'} - \bar{\mathsf{d}}_{--\dot{\alpha}}{}^{i'}\mathcal{D}_{--}\mathsf{d}_{--\dot{\alpha} i'}\,]\,\Big] \ . \end{aligned} \tag{5.1}$$

This action includes several new (two-dimensional) fields that are not present in the original GS action. These new fields are $d_{++\alpha i}$, $d_{--\alpha i'}$, $\psi_{[-4]\alpha i}$, $\psi_{[4]\alpha i'}$, $\lambda_{[4]}{}^{\alpha\dot{\alpha}}$, $\lambda_{[-4]}{}^{\alpha\dot{\alpha}}$, $\phi_{[6]}{}^{\alpha\dot{\alpha}}$ and $\phi_{[-6]}{}^{\alpha\dot{\alpha}}$ where I have used the notation device $\psi_{[-4]\alpha i} \equiv \psi_{----\alpha i}$ etc. to simplify the appearance of these fields.

The astute reader may at this point wonder why this augmentation process is necessary. A simple answer to this question is that in covariant gauges, the GS action suffers from a serious drawback; it does <u>not</u> define a propagator for the Θ-variables! This is a different problem from that usually encountered in a gauge theory. For example, the QED action, before gauge-fixing, contains terms quadratic in the photon field. The operator between these terms, however, is not invertible. Thus, the need for gauge-fixing. In the GS action there are simply no such terms at all for the Grassmann coordinates of the superstring. In the light-cone gauge quantization of the GS theory, this difficulty is overcome by performing a certain redefinition which, in the language of two dimensional field theory, turns a would-be three point function into a two-point function. There are a number of indications [16] that performing a similar such redefinition in covariant gauges leads to a new type of anomaly. The way the augmentation would solve this problem is seen by noting that the action S_{GSS} contains terms of the form $\Pi_{--}{}^{\alpha} d_{++\alpha}$. It is envisioned that these lead to a new two-point function involving Θ after gauge fixing $< 0|\Theta^{\alpha} d_{++\beta}|0 > \sim \delta_{\beta}{}^{\alpha} p_{++}/p^2$. It has been suggested that such propagators are absolutely essential in type-II GS theories. Suggestive evidence has been found [17] that such propagators are required for renormalizability of the type-II σ-model. However, adding only new $\Pi_{--}{}^{\alpha} d_{++\alpha}$-terms to the old GS action is not sufficient because κ-symmetry invariance is broken. Since κ-symmetry is to strings in the GS formulation as world-sheet supersymmetry is to the NSR formulations, one must find a way to restore κ-symmetry in the presence of the new terms. This is precisely the role of the new fields which occur in the modification of the GS action suggested by Siegel. Such extensions of the GS action are not unique [2]

An extended and modified version of the κ-symmetry transformations may be

[2] Other extensions have also been suggested (see *Introduction to String Field Theory* by W. Siegel (World Scientific,1988) p. 100).

defined with the goal of yielding an invariant action under the κ-symmetry,

$$
\begin{aligned}
\delta_\kappa Z^{\underline{N}} = & \; i\Pi_{++}{}^{\alpha\dot{\alpha}}\bar{\kappa}_{--\,\dot{\alpha}}{}^{i}E_{\alpha\,i}{}^{\underline{N}} + i\tfrac{1}{2}\kappa_{--\,\alpha i}\bar{\mathsf{d}}_{++\dot{\alpha}}{}^{i}E^{\alpha\dot{\alpha}\,\underline{N}} + \text{h.c.} \\
& + i\Pi_{--}{}^{\alpha\dot{\alpha}}\bar{\kappa}_{++\,\dot{\alpha}}{}^{i'}E_{\alpha\,i'}{}^{\underline{N}} + i\tfrac{1}{2}\kappa_{++\,\alpha i'}\bar{\mathsf{d}}_{--\dot{\alpha}}{}^{i'}E^{\alpha\dot{\alpha}\,\underline{N}} + \text{h.c.} \; , \\
\delta_\kappa \mathsf{d}_{++\alpha i} = & -2\kappa_{--\alpha i}[\; \Pi_{++}{}^{\beta j}\mathsf{d}_{++\beta j} + \Pi_{++}{}^{\dot{\beta}}{}_{j}\bar{\mathsf{d}}_{++\dot{\beta}}{}^{j}\;] \\
& + \Pi_{++}{}^{\dot{\beta}}{}_{i}[\; \kappa_{--\alpha j}\bar{\mathsf{d}}_{++\dot{\beta}}{}^{j} - \bar{\kappa}_{--\dot{\beta}}{}^{j}\mathsf{d}_{++\alpha j}\;] \; , \\
\delta_\kappa \mathsf{d}_{--\alpha i'} = & -2\kappa_{++\alpha i'}[\; \Pi_{--}{}^{\beta j'}\mathsf{d}_{--\beta j'} + \Pi_{--}{}^{\dot{\beta}}{}_{j'}\bar{\mathsf{d}}_{--\dot{\beta}}{}^{j'}\;] \\
& + \Pi_{--}{}^{\dot{\beta}}{}_{i'}[\; \kappa_{++\alpha j'}\bar{\mathsf{d}}_{--\dot{\beta}}{}^{j'} - \bar{\kappa}_{++\dot{\beta}}{}^{j'}\mathsf{d}_{--\alpha j'}\;] \; , \\
\delta_\kappa \psi_{[-4]\alpha i} = & -\mathcal{D}_{--}\kappa_{--\alpha i} + 2\psi_{[-4]\alpha i}[\; \Pi_{++}{}^{\beta j}\kappa_{--\beta j} + \Pi_{++}{}^{\dot{\beta}}{}_{j}\bar{\kappa}_{--\dot{\beta}}{}^{j}\;] \\
& - 2\kappa_{--\alpha i}[\; \Pi_{++}{}^{\beta j}\psi_{[-4]\beta j} + \Pi_{++}{}^{\dot{\beta}}{}_{j}\bar{\psi}_{[-4]\dot{\beta}}{}^{j}\;] \\
& + \Pi_{++}{}^{\dot{\beta}}{}_{i}[\; \kappa_{--\alpha j}\bar{\psi}_{[-4]\dot{\beta}}{}^{j} + \bar{\kappa}_{--\dot{\beta}}{}^{j}\psi_{[-4]\alpha j}\;] \; , \\
\delta_\kappa \psi_{[4]\alpha i'} = & -\mathcal{D}_{++}\kappa_{++\alpha i'} + 2\psi_{[4]\alpha i'}[\; \Pi_{--}{}^{\beta j'}\kappa_{++\beta j'} + \Pi_{--}{}^{\dot{\beta}}{}_{j'}\bar{\kappa}_{++\dot{\beta}}{}^{j'}\;] \\
& - 2\kappa_{++\alpha i'}[\; \Pi_{--}{}^{\beta j'}\psi_{[4]\beta j'} + \Pi_{--}{}^{\dot{\beta}}{}_{j'}\bar{\psi}_{[4]\dot{\beta}}{}^{j'}\;] \\
& + \Pi_{--}{}^{\dot{\beta}}{}_{i'}[\; \kappa_{++\alpha}{}^{j'}\bar{\psi}_{[4]\dot{\beta}}{}^{j'} + \bar{\kappa}_{++\dot{\beta}}{}^{j'}\psi_{[4]\alpha j'}\;] \; , \\
\delta_\kappa \lambda_{[-4]}{}^{\alpha\dot{\alpha}} = & \tfrac{1}{96}[\; \kappa_{--}{}^{\alpha}{}_{i}\mathcal{D}_{++}\bar{\psi}_{[-4]}{}^{\dot{\alpha}i} - \bar{\kappa}_{--}{}^{\dot{\alpha}i}\mathcal{D}_{++}\psi_{[-4]}{}^{\alpha}{}_{i}\;] \\
& - \tfrac{1}{96}[\; (\mathcal{D}_{++}\kappa_{--}{}^{\alpha}{}_{i})\bar{\psi}_{[-4]}{}^{\dot{\alpha}i} - (\mathcal{D}_{++}\bar{\kappa}_{--}{}^{\dot{\alpha}i})\psi_{[-4]}{}^{\alpha}{}_{i}\;] \\
& - \tfrac{1}{24}\phi_{[-6]}{}^{\alpha\dot{\beta}}[\; (\mathcal{D}_{++}\bar{\kappa}_{--(\dot{\beta}}{}^{i})\Pi_{++}{}^{\dot{\alpha})}{}_{i} + \text{h.c.}\;] \\
& - \lambda_{[-4]}{}^{\alpha\dot{\alpha}}[\; \kappa_{--\beta i}\Pi_{++}{}^{\beta i} + \bar{\kappa}_{--\dot{\beta}}{}^{i}\Pi_{++}{}^{\dot{\beta}}{}_{i}\;] \\
& - i\tfrac{1}{2}\lambda_{[-4]}{}^{\alpha\dot{\beta}}\bar{\kappa}_{--(\dot{\beta}}{}^{i}\Pi_{++}{}^{\dot{\alpha})i} + \text{h.c.} \; , \\
\delta_\kappa \lambda_{[4]}{}^{\alpha\dot{\alpha}} = & \tfrac{1}{96}[\; \kappa_{++}{}^{\alpha}{}_{i'}\mathcal{D}_{--}\bar{\psi}_{[4]}{}^{\dot{\alpha}i'} - \bar{\kappa}_{++}{}^{\dot{\alpha}i'}\mathcal{D}_{--}\psi_{[4]}{}^{\alpha}{}_{i'}\;] \\
& - \tfrac{1}{96}[\; (\mathcal{D}_{--}\kappa_{++}{}^{\alpha}{}_{i'})\bar{\psi}_{[4]}{}^{\dot{\alpha}i'} - (\mathcal{D}_{--}\bar{\kappa}_{++}{}^{\dot{\alpha}i'})\psi_{[4]}{}^{\alpha}{}_{i'}\;] \\
& - \tfrac{1}{24}\phi_{[6]}{}^{\alpha\dot{\beta}}[\; (\mathcal{D}_{--}\bar{\kappa}_{++(\dot{\beta}}{}^{i'})\Pi_{--}{}^{\dot{\alpha})}{}_{i'} + \text{h.c.}\;] \\
& - \lambda_{[4]}{}^{\alpha\dot{\alpha}}[\; \kappa_{++\beta i'}\Pi_{--}{}^{\beta i'} + \bar{\kappa}_{++\dot{\beta}}{}^{i'}\Pi_{--}{}^{\dot{\beta}}{}_{i'}\;] \\
& - i'\tfrac{1}{2}\lambda_{[4]}{}^{\alpha\dot{\beta}}\bar{\kappa}_{++(\dot{\beta}}{}^{i'}\Pi_{--}{}^{\dot{\alpha})i'} + \text{h.c.} \; ,
\end{aligned}
$$

$$
\begin{aligned}
\delta_\kappa \phi_{[-6]}{}^{\alpha\dot\alpha} &= i\tfrac{1}{4}\kappa_{--}{}^{\alpha}{}_{i}\bar\psi_{[-4]}{}^{\dot\alpha i} + \text{h.c.} \ , \quad \delta_\kappa \phi_{[6]}{}^{\alpha\dot\alpha} = i\tfrac{1}{4}\kappa_{++}{}^{\alpha}{}_{i'}\bar\psi_{[4]}{}^{\dot\alpha i'} + \text{h.c.} \ , \\
\delta_\kappa V_{--}{}^{m} &= -2[\ \kappa_{--\alpha i}\Pi_{--}{}^{\alpha i} + \bar\kappa_{--\dot\alpha}{}^{i}\Pi_{--}{}^{\dot\alpha}{}_{i}\ ]V_{++}{}^{m} \ , \\
\delta_\kappa V_{++}{}^{m} &= -2[\ \kappa_{++\alpha i'}\Pi_{++}{}^{\alpha i'} + \bar\kappa_{++\dot\alpha}{}^{i'}\Pi_{++}{}^{\dot\alpha}{}_{i'}\ ]V_{--}{}^{m} \ .
\end{aligned} \tag{5.2}
$$

But even with the extension in (5.1) it turns out [18] that still we cannot covariantly quantize the GS action! So it remains an unsolved question how to proceed.

VI. Internal Augmentation of the Kernel

Leaving aside the difficulties with the covariant realization of spacetime supersymmetry in D = 4 σ-model actions, we can also spend some deliberations on the situation for the covariant and manifest realization of internal symmetries. Here we have a simple answer to the question of the need for such augmentation. All D = 4 supergravity theories that arise from the zero-slope limit of a superstring theory possess spin-one gauge fields. Following our philosophy that every massless state should be associated with an explicit renormalizable coupling constant in a σ-model[3] implies that such constants should occur as

$$
S_{Spin\ 1} \sim \int d^2\sigma V^{-1}\Pi_{\pm\pm}{}^{\underline{B}}\Gamma_{\underline{B}}{}^{\tilde a}(Z) J^{\tilde a}_{\mp\mp} \tag{6.1}
$$

where $J^{\tilde a}_{\mp\mp}$ is some explicit current which is realized in the action. The kernel possesses no fields from which these internal symmetries currents can be constructed.

But here too κ-symmetry is expected to play a role. Note in our D = 10 paradigm, even in the presence of the internal symmetry currents, the total action possessed an invariance under a κ-transformation. This suggests that the same should be true for any D = 4 theory also. We have been able to find thus far only one example in which this is true, so as an end to this talk, this model will be presented. Consider an action obtained by taking the GS kernel action in the case of $N_L = 4,\ N_R = 0$ and add three terms to it, i.e. $S = S^{(K)}_{GS} + S_1 + S_2 + S_3$. The first of these three is simply the action of (3.2) for the case of $\mathcal{G}_R = U(1)^{16}$. The second is given by the action in

[3] This supposition is certainly supported by our experience in both the NSR formulation of superstring theories and the GS formulation of the D = 10 heterotic theory.

(3.3). The final term has the explicit form

$$S_3 = \int d^2\sigma V^{-1}\big[-\tfrac{1}{4}\mathcal{P}_{++}{}^{[ij]}\mathcal{P}_{--}{}^{[ij]} + \int_0^1 dy \hat{\mathcal{P}}_y{}^{[ij]}\hat{\Pi}_{++}{}^{\underline{B}}\hat{\Pi}_{--}{}^{\underline{A}}\hat{\mathcal{B}}_{\underline{AB}[ij]} \\ + \tfrac{1}{2}\int_0^1 dy \hat{\Pi}_y{}^{\underline{B}}(\hat{\Pi}_{++}{}^{\underline{A}}\hat{\mathcal{P}}_{--}{}^{[ij]} - \hat{\Pi}_{--}{}^{\underline{A}}\hat{\mathcal{P}}_{++}{}^{[ij]})\hat{\mathcal{B}}_{\underline{AB}[ij]} \big] \quad , \tag{6.2}$$

$$\mathcal{P}_{\pm\pm}{}^{[ij]} \equiv (\mathcal{D}_{\pm\pm}\chi^{[ij]}) - \Pi_{\pm\pm}{}^{\underline{A}}\Gamma_{\underline{A}}{}^{[ij]} \quad , \quad \hat{\mathcal{B}}_{\underline{AB}[kl]} \equiv (\, C_{\alpha\beta}C_{ijkl} \,,\, C_{\dot{\alpha}\dot{\beta}}\delta_k{}^{[i}\delta_l{}^{j]} \,) \quad , \tag{6.3}$$

where $\chi^{[ij]}$ are six $(i,j = 1,...,4)$ world-sheet scalar fields. The quantities $\Gamma_{\underline{A}}{}^{[ij]}$ are N = 4 superspace gauge connections whose field strengths satisfy the conditions,

$$F_{\alpha i \beta j}{}^{[kl]} \equiv C_{\alpha\beta}\delta_i{}^{[k}\delta_j{}^{l]} \quad , \quad F^{ij}_{\dot{\alpha}\dot{\beta}}{}^{[kl]} \equiv C_{\dot{\alpha}\dot{\beta}}C^{ijkl} \quad . \tag{6.4}$$

These conditions are known to be satisfied by the gravi-photon supergauge connections [19] in the D = 4, N = 4 superspace supergravity theory.

Now the encouraging feature of this model is that it is a D = 4 theory which introduces internal degrees of freedom via S_2 and S_3 and simultaneously possesses an invariance with respect to κ-symmetry. For the fields $\chi^{[ij]}$, we take their variations to be given by

$$\delta_\kappa \chi^{[kl]} = i\Pi_{++}{}^{\alpha\dot{\alpha}}[\kappa_{--\alpha i}\Gamma^{i}_{\dot{\alpha}}{}^{[kl]} + \bar{\kappa}_{--\dot{\alpha}}{}^{i}\Gamma_{\alpha i}{}^{[kl]}] \quad . \tag{6.5}$$

It is then a simple but tedious task to show that the entire action for this D = 4 model is invariant. We have at least shown that the suggestion of internal symmetry augmentation of the kernel action is realizable in an explicit construction.

We end our talk here. As we have shown there are many, many unsolved or incomplete problems that remain. However, we are optimistic that progress in our understanding of superstring theories will shed further light on these.

Acknowledgement

I wish to acknowledge the collaborative efforts of D. A. Depireux, J. W. Durachta and B. Radak in the research into the topics discussed in this presentation.

REFERENCES

[1] M. B. Green, J. H. Schwarz, and E. Witten, *Superstring Theory*, Camb. Press. (1987) NY, NY; M. Kaku, *Introduction to Superstrings* Springer-Verlag (1988) NY, NY.

[2] P. Candelas, G. Horowitz, A. Strominger, and E. Witten; Nucl. Phys. **B258** (1985) 46, L. Dixon, J. Harvey, C. Vafa and E. Witten, Nucl. Phys. **B261** (1985) 678; K. S. Narain, Phys. Lett. **169B** (1986) 41; I. Antoniadis, C. Bachas, C. Kounnas and P. Windey, Phys. Lett. **171B** (1986) 51; H. Kawai, D. C. Lewellyn, and S.-H.-H. Tye, Phys. Rev. Lett. **57** (1986) 1832; W. Lerche, D. Lust and A. N. Schellekens, Nucl. Phys. **B287** (1987) 477.

[3] R. Kerner, Ann. Inst. Henri Poincaré **A9** (1968) 143; A. Trautman, Rep. Math. Phys. (1975) 1; Y. M. Cho, J. Math. Phys. **16** (1975) 2029; Z. Ezawa and H. Tze, J. Math. Phys. **17** (1976) 2228; M. E. Mayer in *"Fiber Bundle Techniques in Gauge Theories" Lectures in Mathematical Physics at the Univ. of Texas at Austin* ed. A. Bohm and J. D. Dollard (Springer,Berlin (1977) p. 1; M. Daniel and C. M. Viallet, Rev. Mod. Phys. **52** (1980) 175.

[4] S. J. Gates, Jr. and W. Siegel, Phys. Lett. **206B** (1988) 631; D. A. Depireux, S. J. Gates, Jr. and Q-Han Park, Phys. Lett. **224B** (1989) 364; S. Bellucci, D. A. Depireux and S. J. Gates, Jr., Phys. Lett. **232B** (1989) 67.

[5] D. J.Gross, J. A. Harvey, E. Martinec and R. Rohm, Nucl. Phys. **B256** (1986) 253.

[6] C. M. Hull, Phys. Lett. **206B** (1988) 234; **212B** (1988) 437; D. A. Depireux, S. J. Gates, Jr. and B. Radak, Univ. of Md. preprint # UMDEPP 90-036, Sept., 1989 (see also third work of ref. [4]).

[7] M. M. Vainberg, *Variational Methods for the Study of Nonlinear Operators* (Holden Day, San Francisco, 1964) p.135; E. Witten, Nucl. Phys. **B223** (1983) 422; M. Henneaux and L. Mezincescu, Phys. Lett. **152B** (1985) 340.

[8] W. Siegel, Nucl. Phys. **B238** (1984) 307.

[9] W. Siegel, Phys. Lett. **128B** (1983) 397; M. B. Green and J. H. Schwarz, Phys. Lett. **136B** (1984) 367; Nucl. Phys. **B198** (1984) 252, 441.

[10] D. Friedan, Phys. Rev. Lett. **45** (1980) 1057; T. Curtright and C. Zachos, Phys. Rev. Lett. **53** (1984) 1799; E. S. Fradkin and A. A. Tseytlin, Phys. Lett. **158B** (1985) 316; A. Sen, Phys. Lett. **166B** (1986) 300.

[11] A. B. Zamolodchikov, JETP Lett. **43** (1986) 731; Sov. J. Nucl. Phys. **46** (1987) 1090; A. A. Tseytlin, Phys. Lett. **194B** (1987) 63.

[12] S. V. Ketov, Class. Quantum Grav. **4** (1987) 1163; A. A. Tseytlin, Nucl. Phys. **B294** (1987) 383; S. V. Ketov, and O. A. Soloviev, Phys. Lett. **232B** (1989) 75.

[13] C. M. Hull and E. Witten, Phys. Lett. **160B** (1985) 398.

[14] S. J. Gates, Jr., M. T. Grisaru, U. Lindstrom, M. Roček, W. Siegel, P. van Nieuwenhuizen and A. E. van de Ven, Phys. Lett. **225B** (1989) 44; M. B. Green and C. M. Hull, Phys. Lett. **225B** (1989) 49; R. Kallosh, Phys. Lett. **225B** (1989) 57.

[15] W. Siegel, Phys. Lett. **203B** (1988) 79; Nucl. Phys. **B263** (1986) 93; L. Romans, Nucl. Phys. **B281** (1987) 639.

[16] S. Bellucci, Phys. Lett. **227B** (1989) 61; S. J. Gates, Jr., P. Majumdar, B. Radak and S. Vashakidze, Phys. Lett. **226B** (1989) 237; U. Kraemmer and A. Rebhan, "Anomalous Anomalies in the Carlip-Kallosh Quantization of the Green-Schwarz Superstring," Tech. Univ. of Wien preprint, to appear in Phys. Lett.; S. Bellucci and R. Oerter, INFN-LNF preprint # LNF-89/089 PT, Dec.,1989.

[17] D. A. Depireux, S. J. Gates, Jr., P. Majumdar, B. Radak and S. Vashakidze, Univ. of Md. preprint # UMDEPP 89-169, to appear in Nucl. Phys. B.

[18] J. M. L. Fisch and M. Henneaux, Univ. Libre de Bruxelles preprint # ULB TH2/89-04; U. Lindstrom M. Roček, W. Siegel, P. van Nieuwenhuizen and A. E. van de Ven, Stony Brook preprint # ITP-SB-89-76; A. Miković, M. Roček, W. Siegel, P. van Nieuwenhuizen and A. E. van de Ven, Stony Brook preprint # ITP-SB-89-77.

[19] S. J. Gates, Jr., Nucl. Phys. **B213** (1983) 409; J. W. Durachta and S. J. Gates, Jr., Mod. Phys. Lett. **A21** (1989) 2007.

The Green-Schwarz String in Curved Superspace

Joel A. Shapiro
Department of Physics and Astronomy
Rutgers, the State University of New Jersey
Piscataway, NJ 08855-0849, USA

Abstract

Siegel invariance, an essential local fermionic symmetry of superparticles and superstrings, is present only if the background superfields in which they propagate satisfy certain constraints. For super Yang-Mills theory and for curved superspace, we show that these constraints are essentially the Yang-Mills or supergravity equations of motion. In the supergravity case, this is modulo an equivalence relation which relates different formulations of supergravity, and the full set of equations comes only from the superstring, with only the supertorsion constraints arising from the superparticle.

1. Introduction

In ordinary field theories, the dynamical degrees of freedom are maps from a specified base manifold into a specified target space. Gravity theories are peculiar in that the dynamical degrees of freedom determine the geometry of their own domain. In the σ model formulation of string theories of gravity, this situation is changed. Here the dynamical fields are maps from two dimensional manifolds into spacetime, and the underlying topology of the two dimensional manifolds is not determined by the dynamical fields. Perturbation theory about a given background spacetime gives the dynamics of the σ model fields, which do not directly affect the geometry of the background. Nonetheless, this dynamics includes fluctuations which we regard as gravitons and gauge bosons, so it must in some way determine the background spacetime. We are not able to demonstrate this method of determining the background dynamically, but we can ask for consistency conditions, and indeed it has been found that, in order to consistently define string dynamics, there are constraints on the background space.

One way such constraints have arisen occurs even for the bosonic string, in which quantization of the Polyakov action destroys conformal invariance unless certain conditions are imposed on the background fields. For the genus 0 world-sheet (0 loops in the string expansion), this gives gravity equations of motion. The same conclusions apply to the heterotic string in its Neveu-Schwarz-Ramond formulation, in which the domain is a supermanifold of (2,1) dimensions, and the fields map into ordinary, albeit 10 dimensional, spacetime. We will be concerned, however, with a classical effect in the Green-Schwarz[1] formulation, which is manifestly space-time supersymmetric, involving maps from ordinary 2-d manifolds into a chiral superspace of dimension (10,16). Here the constraints on the background fields are not quantum effects, but arise from requiring a local gauge invariance in motions through superspace. This Siegel invariance has been very widely discussed, as it makes extremely difficult the covariant quantization.

This symmetry, and many of the constraints it imposes, applies in the simpler case of the Brink-Schwarz[2] superparticle as well. In the hamiltonian formulation, there are constraints on the momenta, basically the momenta corresponding to motion through the θ's. This gives rise to a spinor of constraints we call u_α. These may be divided into two sets by the operation of the Dirac operator for the particle, and half are first class constraints and the other half second class. The first class constraints are associated, as always, with a gauge invariance, here the Siegel invariance, while the second class constraints correspond to degrees of freedom which must be solved for before quantization, or treated in some other way which effectively prevents integrating over them in a functional integral formulation of quantum mechanics.

This local fermionic symmetry was discovered by Siegel[3] for the free superparticle, in "flat" superspace. It was extended by Green and Schwarz[4] to the free superstring in "flat" space. This gauge invariance is necessary for the string theory to make sense — for example, in the light cone gauge, it is the fixing of this gauge which halves the number of fermionic degrees of freedom to 8, so that it can be equal to the number of transverse degrees of freedom of x. When the particle or string experiences a gauge field, the preservation of this invariance imposes a differential condition on these fields, which may be thought of as an integrability condition on a generalization of the "null-line" to superspace. This consists of points whose bosonic coordinates are separated by a null-vector n^a and θ coordinates separated by $n^a\Gamma_a\kappa$. It is thus a 1+8 dimensional space. While any gauge field-strength is integrable on a one-dimensional line, to be integrable in the 8 fermionic directions imposes a constraint on the field-strength supertensor. Requiring this integrability for all null lines requires $F_{\alpha\beta} = 0$, so that the spinor-spinor component of the field strength vanishes. Due to Bianchi identities, this also constrains the spinor-vector piece, and eventually requires the Yang-Mills superfields to satisfy their equations of motion. This connection of super-null-lines, $F_{\alpha\beta} = 0$, and the super-Yang-Mills equations of motion was developed by

Witten[5]. This connection is further elaborated in a recent paper[6].

An extension of this formulation to the string in curved superspace was given by Witten[5], who showed that if a set of supergravity equations were satisfied by the background fields, Siegel's symmetry extends to curved space. Similar results were also shown by others. These equations, for $N = 1$, 10 dimensional supergravity, are formulated in terms of a dilaton superfield, vielbein superfields, and an antisymmetric tensor superfield B_{MN}. Taylor and I decided to explore[7–9] in what sense the supergravity equations are necessary conditions for the Siegel symmetry. In the process we cast some light on the relations between the various different forms of supergravity equations in superspace which have been presented by different (an sometimes the same) authors. Formulations of supergravity in superspace generally impose constraints on various components of the torsion and the H fields. Which components are constrained appears arbitrary. I will consider the first set of Nilsson[10] and the set Witten gave in his twistor paper[5].

	Nilsson	Witten
${T_{\alpha\beta}}^{\gamma}$	$\Gamma^{abcde}_{\alpha\beta}\Gamma^{\gamma\delta}_{abcde}T_\delta$	0
${T_{\alpha\beta}}^{c}$	$2\Gamma^{c}_{\alpha\beta}$	$2\Gamma^{c}_{\alpha\beta}$
${T_{a\beta}}^{\gamma}$	—	$\Gamma_{a\beta\delta}\phi^{\delta\gamma}$
${T_{\alpha b}}^{c}$	$\delta^c_b V_\alpha$	0
${T_{ab}}^{\gamma}$	—	—
${T_{ab}}^{c}$	0	—
$H_{\alpha\beta\gamma}$	0	0
$H_{a\beta\gamma}$	$2\Gamma_{a\beta\gamma}$	$2e^{2\phi}\Gamma_{a\beta\gamma}$
$H_{ab\gamma}$	${\Gamma_{ab\gamma}}^{\delta}\lambda_\delta$	${\Gamma_{ab\gamma}}^{\delta}\phi_{,\delta}$
H_{abc}	—	—

Table 1. The supergravity constraints given by Nilsson[10] and Witten[5]. Note Greek indices are spinorial and Latin are vector indices. Components without constraints beyond those implied by Bianchi identities are marked by —.

These constraints on the torsion do lead to additional constraints due to Bianchi identities. In the end the dynamical degrees of freedom can be thought of as lying in one scalar superfield. We will see how viewing the string in a background of these superfields gives a perspective to these differing formulations. As I will discuss, we have shown that the existence of Siegel symmetry for the superparticle imposes these torsion constraints, while symmetry of the string imposes both these and the constraints on the H_{ABC} field strength.

2. The Superparticle and Superstring Actions

The massless particle action in flat superspace was described by Brink and Schwarz[2]. In the language of vielbeins, we can write

$$I = \int d\tau \, \frac{1}{2} \, e(\tau) \eta_{ab} \dot{Z}^M E_M{}^a \dot{Z}^N E_N{}^b \tag{1}$$

where

$$Z^M = (x^m, \theta^\mu) \tag{2}$$

is the position in a (10,16) dimensional superspace, and

$$E^A = (E^a, E^\alpha) = dZ^M E_M{}^A \tag{3}$$

are the supervielbeins, which determine the metric structure. τ is the timelike parameter describing the world line and $e(\tau)$ is the einbein for the one dimensional manifold. Brink, Schwarz and Siegel's work concerned "flat" superspace, for which

$$E^a = dx^a + d\theta^\alpha \Gamma^a_{\alpha\beta} \theta^\beta, \qquad E^\alpha = d\theta^\alpha . \tag{4}$$

The Siegel symmetry involves a local fermionic parameter κ with

$$\delta\theta = \not{p}\kappa, \qquad \delta x^a = \theta \Gamma^a \not{p}\kappa \ . \tag{5}$$

For null p, $\not{p}$ has rank 8, so that κ transformations permit the gauging away of 8 of the θ degrees of freedom, as required by supersymmetry.

We will discuss this transformation in the language of Dirac constraints. Before I do that, I want to introduce the heterotic string Lagrangian as well:

$$\mathcal{L} = -\frac{1}{2}\sqrt{-g} g^{rs} e^{2\phi} \partial_r Z^M E_M{}^a \partial_s Z^N E_N{}^b \eta_{ab} + \frac{1}{2} \epsilon^{rs} \partial_r Z^M \partial_s Z^N B_{NM}. \tag{6}$$

I have included the dilaton field ϕ included by some authors, although it is not necessary, as it can be absorbed into the vielbein. The antisymmetric tensor field B is essential to maintain the Siegel supersymmetry, as well as playing a crucial rôle in the coupling to gauge fields when they are included. g_{rs} is the 2 dimensional metric which we consider as determined by the zweibein $v_i{}^r v_j{}^s \eta^{ij}_{(2)} = g^{rs}$.

Before we begin our analysis of the string dynamics, with an expectation of finding constraints on the background fields, it is appropriate to point out that, in fact, the particle and the string do not feel all the background fields. Most obviously, E^α, the spinorial vielbeins, do not enter the action at all. Thus string dynamics can place no constraints on E^α itself. Thus there are different background configurations which are equivalent as far as the string is concerned, and we have an equivalence relation defined by the string. Arbitrary changes

$$\delta E^\alpha = E^A \lambda_A{}^\alpha \tag{7}$$

are equivalence relations, as are Lorentz transformations

$$\delta E^a = E^b \lambda_b{}^a, \qquad \text{with} \qquad \lambda_c{}^a \lambda_d{}^b \eta_{ab} = \eta_{cd}. \tag{8}$$

When a dilaton field is introduced, there is also clearly an equivalence

$$\phi \to \phi + u, \qquad E^a \to e^{-u} E^a. \tag{9}$$

No consistency conditions on string dynamics can distinguish representatives within an equivalence class. If background field dynamics is determined by string dynamics, these equivalence relations must be a more general form of gauge transformations on the background fields. Current superspace supergravity equations build in the Lorentz gauge invariance but impose constraints, often very arbitrary looking, to fix these other gauges transformations.

What the string dynamics can do is restrict the allowed equivalence classes in which the background superfields can lie. We will show that, for each allowed class, there is a representative which satisfies the set of superspace supergravity equations proposed by Nilsson[10], and another representative satisfying the equations proposed by Witten[5]. We will also have demonstrated the transformations which are necessary to go from one description to the other.

3. Siegel Invariance and Constraints from the Superparticle

Let us now proceed to analyzing the particle using the Dirac Hamiltonian approach. The Lagrangian is

$$\mathcal{L} = \frac{1}{2} e(\tau) \eta_{ab} \dot{Z}^M E_M{}^a \dot{Z}^N E_N{}^b. \tag{10}$$

The canonical momenta are

$$P_M = \frac{\partial L}{\partial \dot{Z}^M} = e(\tau)\eta_{ab} E_M{}^a \dot{Z}^N E_N{}^b$$
$$\Pi = \frac{\partial L}{\partial \dot{e}} = 0. \tag{11}$$

Π is a primary constraint in Dirac's language. So are the u_α, where

$$u_A = V_A{}^M P_M, \qquad \text{with } V_A{}^M E_M{}^B = \delta_A^B \tag{12}$$

defining the inverse vielbein V. The canonical Hamiltonian is given by

$$H_C = \dot{Z}^M P_M - L \approx \frac{1}{2} e^{-1} u_a u^a \tag{13}$$

but more generally terms proportional to the constraints may be added. One must check for additional constraints coming from the time-variation of the primary ones. In particular,

$$\{H, \Pi\} \approx 0 \Rightarrow \chi := \frac{1}{2} u^a u_a \approx 0 \tag{14}$$

χ, Π, and u_α are the full set of constraints. We must divide these into first and second class constraints. The first correspond to local gauge invariances and have Poisson brackets with constraints which are themselves constraints. We must be sure to find Siegel's fermionic symmetry among the first class constraints. The constraints satisfy the algebra

$$\{\chi, \chi\} = 0, \quad \{u_\alpha, u_\beta\} = \Omega_{\alpha\beta}{}^A u_A, \quad \{u_\alpha, \chi\} = \eta^{ab} u_b \Omega_{\alpha a}{}^B u_B \tag{15}$$

where we have introduced the object of anholonomicity,

$$\Omega^A = dE^A = \frac{1}{2} E^C \wedge E^B \, \Omega_{BC}{}^A \, . \tag{16}$$

In flat superspace, $\Omega_{\alpha\beta}{}^a = 2\Gamma^a_{\alpha\beta}$, with all other components of the anholonomicity 2-form vanishing. Then χ is first class in Dirac's language, generating reparameterizations, while the u_α are not first class since $\Gamma^a_{\alpha\beta} u_a$ does not vanish. Nevertheless, $(\Gamma^a u_a)^2 = 2\chi \approx 0$, and hence there are first class constraints which are combinations of the u_β. Explicitly, these can be taken to be $\psi^\alpha = \Gamma^{a\alpha\beta} u_a u_\beta$ in flat superspace. ψ^α is the generator of the local fermionic symmetry discovered by Siegel.

In order to preserve the Siegel symmetry and parameter invariance in curved space, we need the curved analogs $\hat{\chi}$ and ψ^α. These should have a form which reduces in flat superspace to the above constraints, although in general they need not be in the flat space form. They must, however, be first class combinations of the constraints χ and u_α. To be first class they must bracket to 0 with all constraints. Thus

$$\left.\begin{aligned} \{\hat{\chi},\hat{\chi}\} \approx 0 \\ \{\hat{\chi},u_\alpha\} \approx 0 \\ \{\psi^\alpha,u_\beta\} \approx 0 \end{aligned}\right\} \quad \text{with} \quad \left\{\begin{aligned} \psi^\alpha &= F^{\alpha\beta}u_\beta \\ \hat{\chi} &= \chi + A^\alpha u_\alpha \ . \end{aligned}\right. \tag{17}$$

Then $H = \lambda\hat{\chi} + \psi^\beta\kappa_\beta$, for arbitrary λ and κ_β, is the most general Hamiltonian preserving the constraints under time translation, and we have the required set of gauge transformations.

First examine $\{\psi^\alpha, u_\beta\} \approx 0$ to find the constraints on F. Writing $F^{\alpha\gamma} = F_0^{\alpha\gamma} + F^{\alpha\gamma a}u_a$, where F_0 is independent of u_a, we need

$$F_0^{\alpha\gamma}\Omega_{\gamma\beta}{}^b u_b + F^{\alpha\gamma a}\Omega_{\gamma\beta}{}^b u_a u_b \approx 0 \ . \tag{18}$$

There is no constraint linear in u_b, but the trace of $u_a u_b$ is a constraint, so

$$\begin{aligned} \left(F_0\Omega^b\right)^\alpha{}_\beta &= 0 \\ \left(F^a\Omega^b + F^b\Omega^a\right)^\alpha{}_\beta &= C^\alpha{}_\beta\eta^{ab} \ . \end{aligned} \tag{19}$$

In flat space $\Omega^a = 2\Gamma^a$ and $C^\alpha{}_\beta = 4\delta^\alpha_\beta$, both of which are invertible. We assume they remain invertible in curved space, so $F_0 = 0$ and, if we renormalize $F \to C^{-1}F$, we have

$$F^a\Omega^b + F^b\Omega^a = 4\eta^{ab} \ . \tag{20}$$

This implies the $\tilde{\gamma}^a = \begin{pmatrix} 0 & \Omega^a/2 \\ F^a & 0 \end{pmatrix}$ form a representation of the 10-D Dirac algebra. Such a representation is unique so $\tilde{\gamma}^a$ differs from a standard set only by a similarity transformation. For the Ω^a piece, this implies there is a matrix U such that

$$\Omega_{\alpha\beta}{}^a = \pm 2(U^T\Gamma^a U)_{\alpha\beta} \ . \tag{21}$$

We see that $\Omega^a_{\alpha\beta}$ determines a spin structure, so it is time to think of equivalences to help put Ω in a prescribed form.

We have already noted that an arbitrary change in E^α is a background equivalence. In particular $\delta E^\alpha = E^\beta \Lambda_\beta{}^\alpha$ can be used to invoke an arbitrary U transformation on the matrices Ω^a. The sign can be adjusted if necessary by a Lorentz transformation $E^a \to -E^a$. Note that all these considerations are entirely local and we can require $\Omega_{\alpha\beta}{}^b = 2\Gamma^b_{\alpha\beta}$ at each point in superspace. We shall adopt this representative from now on.

Now consider

$$\{\hat{\chi}, u_\alpha\} \approx 0 \,. \tag{22}$$

Using the expression for $\hat{\chi}$, we have

$$-\Omega_{\beta a}{}^b u^a u_b + A^\alpha \Omega_{\alpha\beta}{}^b u_b \approx 0 \,. \tag{23}$$

If we decompose $\Omega_{\alpha ab}$ into pieces symmetric and antisymmetric in $a \leftrightarrow b$, it is clear that the antisymmetric piece is unconstrained, but there is a constraint on $\Omega_\alpha{}^{(ab)}$. As the constraints are homogeneous in u, we can take $A^\alpha = A^{\alpha a} u_a$. Then the symmetric piece is constrained up to a trace, so $\Omega_\alpha{}^{(ab)} = V_\alpha \eta^{ab} + A^{\beta(a}\Gamma^{b)}_{\beta\alpha}$, with $A^{\beta a}$ and V_α undetermined. The only real constraint here is that $\Omega_\alpha{}^{(ab)}$ is determined by its contraction $\Omega_\alpha{}^{(ab)}\Gamma_b^{\alpha\beta}$. This is precisely the kind of object which can be gauged away by a background equivalence $\delta E^\alpha = E^b \Lambda_b{}^\alpha$. We can therefore eliminate the A term, to get $\Omega_\alpha{}^{(ab)} = V_\alpha \eta^{ab}$, as Nilsson does, or we can go further and set $\Omega_\alpha{}^{(ab)} = 0$, as Witten does.

In our approach to the superparticle, constraints are naturally expressed in terms of the object of anholonomicity, while it is conventional to express constraints on the background fields in terms of the torsion tensor. These are related by introducing the $SO(9,1)$ connection 1-form $\omega_A{}^B = E^C \omega_{CA}{}^B$, defining

$$dE^A = \Omega^A = T^A - E^B \wedge \omega_B{}^A \,. \tag{24}$$

Because ω is a $SO(9,1)$ connection, we have

$$\omega_{ab} := \eta_{bc}\omega_a{}^c = -\omega_{ba}, \qquad \omega_b{}^\alpha = \omega_\beta{}^a = 0, \qquad \omega_\alpha{}^\beta = \frac{1}{4}\omega_{ab}\left(\Gamma^{ab}\right)_\alpha{}^\beta . \tag{25}$$

It follows from these definitions that $T_{\alpha\beta}{}^a = \Omega_{\alpha\beta}{}^a$, $T_{\alpha(ab)} = \Omega_{\alpha(ab)}$, and the constraints we have so far found can be written in terms of T:

$$T_{\alpha\beta}{}^a = 2\Gamma^a_{\alpha\beta} \,, \qquad T_\alpha{}^{(ab)} = B^{\beta(a}\Gamma^{b)}_{\beta\alpha}. \tag{26}$$

Both Witten and Nilsson use $\Omega_{\alpha a}{}^c = T_{\alpha a}{}^c - \omega_{\alpha a}{}^c$ to *define* the spinor component of the connection by choosing $T_{\alpha[ac]} = 0$. Nilsson also uses $\Omega_{ab}{}^c = T_{ab}{}^c - \omega_{ab}{}^c +$

$\omega_{ba}{}^{c}$ to define the vector component of the connection by choosing $T_{ab}{}^{c} = 0$. Hence, $\omega_{abc} = \frac{1}{2}(\Omega_{bca} + \Omega_{bac} - \Omega_{cab})$ in Nilsson's system of constraints. We have already mentioned that Nilsson's choice of $T_{\alpha}{}^{(ab)} = V_{\alpha}\eta^{ab}$ can be chosen by $\Lambda_{b}{}^{\alpha}$ transformation within the background equivalence class. The only remaining supergravity constraint required on T to get Nilsson's on-shell equations is $T_{\alpha\beta}{}^{\gamma} = \Gamma^{abcde}_{\alpha\beta}\Gamma^{\gamma\delta}{}_{abcde}T_{\delta}$.

We have imposed constraints on $\Omega_{AB}{}^{C}$ due to the requirement of first class constraints on the particle. As the Ω^{A}'s are exact forms, the coefficients are not independent but are further constrained by $d\Omega^{A} = 0$.

$$(-1)^{DE}\Omega_{[BC}{}^{E}\Omega_{D\}E}{}^{A} + V_{[B}{}^{M}\Omega_{CD\}}{}^{A}{}_{,M} = 0. \tag{27}$$

If we examine the 3-spinor piece of $d\Omega^{a}$, making use of the choice $\Omega_{\alpha\beta}{}^{a} = 2\Gamma^{a}_{\alpha\beta}$, we find

$$\Gamma^{b}_{(\alpha\beta}\Omega_{\gamma)b}{}^{a} = \Omega_{(\alpha\beta}{}^{\delta}\Gamma^{a}_{\gamma)\delta}. \tag{28}$$

The equation is slightly simpler in terms of the torsion. Plugging in

$$\begin{aligned} \Omega_{\gamma}{}^{ba} = S_{\gamma}{}^{ba} - \omega_{\gamma}{}^{ba} \qquad S_{\gamma}{}^{ba} := \Omega_{\gamma}{}^{(ba)} \\ \Omega_{\alpha\beta}{}^{\gamma} = T_{\alpha\beta}{}^{\gamma} - 2\omega_{(\alpha\beta)}{}^{\gamma} = T_{\alpha\beta}{}^{\gamma} - \frac{1}{2}\omega_{(\alpha}{}^{ab}\Gamma_{|ab|\beta)}{}^{\gamma} \end{aligned} \tag{29}$$

the ω terms cancel, and we have

$$\Gamma^{b}_{(\alpha\beta}S_{\gamma)b}{}^{a} = T_{(\alpha\beta}{}^{\delta}\Gamma^{a}_{\gamma)\delta}. \tag{30}$$

The analysis of this equation is complicated because tools for handling 4-spinor objects are not as convenient as for vectors. It is necessary to expand in terms of Γ's. We proceeded by expanding

$$T_{\alpha\beta}{}^{\gamma} = A_{\alpha}\delta^{\gamma}_{\beta} + B^{ab}_{\alpha}\Gamma_{ab}{}^{\gamma}{}_{\beta} + C^{abcd}_{\alpha}\Gamma_{abcd}{}^{\gamma}{}_{\beta} \tag{31}$$

and contracting with various Γ's. Any tensor with two lower and one upper spinor indices can be expanded in this form. The fact that $T_{\alpha\beta}{}^{\gamma}$ is symmetric under $\alpha \leftrightarrow \beta$ requires (and is implied by) a constraint that determines $C^{abcd}\Gamma_d$ in terms of A and B:

$$192C^{abcd}\Gamma_d = -2A\Gamma^{abc} + B^{de}\Gamma_{de}\Gamma^{abc} - 48B^{d[a}\Gamma_d\Gamma^{bc]} - 48B^{[ab}\Gamma^{c]}. \qquad (32)$$

Contractions with Γ's occur everywhere so we define the abbreviations

$$S^{a\alpha} = S^{ab}_{\beta}\Gamma^{\beta\alpha}_{b} \qquad S_{\alpha} = S^{a\beta}\Gamma_{a\beta\alpha}. \qquad (33)$$

The Bianchi equation can now be solved only when $S_{\alpha}{}^{ab} = F^{\beta(a}\Gamma^{b)}_{\beta\alpha}$ which we already required from the constraint algebra. In that case $T_{\alpha\beta}{}^{\gamma}$ is determined up to a spinor A_{δ},

$$\begin{aligned} T_{\alpha\beta}{}^{\gamma} &= \left(\frac{1}{12}S^{a\gamma} + \left(\frac{5}{48}S - 2A\right)_{\delta}\Gamma^{a\delta\gamma}\right)\Gamma_{a\alpha\beta} \\ &\quad + \left(-\frac{1}{26880}S + \frac{1}{840}A\right)_{\delta}\Gamma^{\delta\gamma}_{abcde}\Gamma^{abcde}_{\alpha\beta}. \end{aligned} \qquad (34)$$

Recall that the $\lambda_a{}^{\alpha}$ transformations permit us to adjust S^a to whatever we like. Thus for any value of A we may adjust S so that the coefficient of the $\Gamma^a\Gamma_a$ term vanishes. Then we have

$$T_{\alpha}{}^{ab} = V_{\alpha}\eta^{ab}, \qquad T_{\alpha\beta}{}^{\gamma} = \frac{1}{3360}\Gamma^{abcde}_{\alpha\beta}\Gamma^{\gamma\delta}_{abcde}V_{\delta}. \qquad (35)$$

Thus we have found that any field configuration which preserves the nature of the first class constraints for the particle lies in an equivalence class which includes a representative satisfying the supergravity torsion constraints of Nilsson.

In order to compare with Witten's equations, we need to work a bit harder. Plugging $S = 0$ into the equation for $\Omega_{\alpha\beta}{}^{\gamma}$ shows us that in general there is one undetermined spinor superfield. We also need to partially constrain $T_{a\alpha}{}^{\beta}$. Extracting the 2-spinor 1-vector piece of the $d\Omega^a$ Bianchi gives

$$-\Omega_{\alpha\beta}{}^{\gamma}\Omega_{\gamma b}{}^{a} - 2\Gamma^{e}_{\alpha\beta}\Omega_{e}{}^{ba} - 2\Omega_{(\alpha}{}^{be}\Omega_{\beta)e}{}^{a} - 4\Gamma^{a}_{\gamma(\alpha}\Omega^{b}{}_{\beta)}{}^{\gamma} + 2\Omega_{(\beta}{}^{ba}{}_{,\alpha)} = 0. \qquad (36)$$

This equation simplifies greatly upon symmetrization under $a \leftrightarrow b$, as we are taking $S^{ab} = 0$. We find

$$\Gamma^{e}_{\alpha\beta}\Omega_{e}{}^{(ab)} + 2\Gamma^{(a}_{\gamma(\alpha}\Omega^{b)}{}_{\beta)}{}^{\gamma} = 0. \qquad (37)$$

Writing $\Omega^{a}{}_{\beta}{}^{\gamma} = \left(\hat{\Omega}^a\right)_{\beta}{}^{\gamma}$, and contracting with $\Gamma^{\alpha\beta}_{c}$ and $\Gamma^{\alpha\beta}_{cdefg}$ gives $8\Omega_c{}^{(ab)} + \mathrm{Tr}\left(\Gamma^{(a}\Gamma_c\hat{\Omega}^{b)}\right) = 0$, $\quad \mathrm{Tr}\left(\Gamma^{(a}\Gamma_{cdefg}\hat{\Omega}^{b)}\right) = 0$. Ω^{cab} has no totally symmetric piece, which shows $\hat{\Omega}^a$ is traceless. The second equation constrains the 4-Γ piece in such a way that it matches $\Gamma^a\Phi$, leaving a 2-Γ term with a coefficient which can be *defined* to be ω_{abc}. Thus we have established Witten's constraint

$$T_{a\alpha}{}^{\beta} = \Gamma_{a\alpha\gamma}\Phi^{\gamma\beta}. \tag{38}$$

Now that we have the ω's, we do best working in terms of covariant derivatives

$$T^A = \mathcal{D}E^A = dE^A + E^B \wedge \omega_B{}^A. \tag{39}$$

The Bianchi identity is

$$\mathcal{D}T^A = -E^B \wedge R_B{}^A \tag{40}$$

where

$$R_B{}^A = d\omega_B{}^A + \omega_B{}^C \omega_C{}^A. \tag{41}$$

In terms of components

$$R_{[BCD\}}{}^A = 2T_{[BC}{}^E T_{|E|D\}}{}^A - \mathcal{D}_{[B}T_{CD\}}{}^A. \tag{42}$$

This equation has many pieces; by examining the ${}_{\beta\gamma d}{}^a$ and ${}_{\beta\gamma\delta}{}^\alpha$ components, recalling

$$R_\alpha{}^\beta = \frac{1}{4}R_a{}^b\Gamma^a{}_{b\alpha}{}^\beta, \tag{43}$$

and plugging in the constraints we already know, especially the fact that $T_{\alpha\beta}{}^\gamma$ is given in terms of a spinor superfield V_α, we can show

$$\mathcal{D}_\beta\Gamma^{\beta\gamma}_{rstuv}V_\gamma = 0. \tag{44}$$

This allows[11,12] one (at least locally) to write V_α as the spinor derivative of a scalar superfield

$$V_\alpha = \rho_{,\alpha}. \tag{45}$$

This we will eliminate via a ϕ translation, enabling us to get the last of Witten's constraints, $T_{\alpha\beta}{}^\gamma = 0$.

An obvious symmetry of the string action is a change of scale of the E^a with a compensating shift of ϕ. As changes of E^α are arbitrary anyway, we choose them in a convenient way, to leave $\Omega_{\alpha\beta}{}^a$ invariant. Then

$$\begin{aligned} \phi &\to \phi + u \\ E_M{}^a &\to e^{-u} E_M{}^a \\ E_M{}^\alpha &\to e^{-\frac{1}{2}u} E_M{}^\alpha , \end{aligned} \tag{46}$$

and

$$\begin{aligned} \Omega_{BC}{}^A \to & e^{(c_B + c_C - c_A)u} \Omega_{BC}{}^A - c_A \left(V_B{}^N u_{,N} \delta^A_C - (-1)^{BC} V_C{}^N u_{,N} \delta^A_B \right) \\ &= e^{(c_B + c_C - c_A)u} \Omega_{BC}{}^A - 2 c_A \, u_{,[B} \delta^A_{C]} , \end{aligned} \tag{47}$$

where $c_a = 1$, $c_\alpha = \frac{1}{2}$, and $u_{,B} := V_B{}^M u_{,M}$. The individual components transform by

$$\begin{aligned} \Omega_{\alpha\beta}{}^a &\to \Omega_{\alpha\beta}{}^a \\ \Omega_{ab}{}^\alpha &\to e^{\frac{3}{2}u} \Omega_{ab}{}^\alpha \\ \Omega_{\alpha a}{}^b &\to e^{\frac{1}{2}u} \Omega_{\alpha a}{}^b - u_{,\alpha} \delta^b_a \\ \Omega_{\alpha\beta}{}^\gamma &\to e^{\frac{1}{2}u} \Omega_{\alpha\beta}{}^\gamma - \delta^\gamma_{(\alpha} u_{,\beta)} . \end{aligned} \tag{48}$$

Notice that the ϕ translation messes up the $T_{\alpha a}{}^b = 0$ condition, which means we must accompany the u translation with a $\lambda_c{}^\alpha$ transformation, arranged to restore $T_{\alpha a}{}^b = 0$. This transformation changes the spinor connection as well, but the net effect on $T_{\alpha\beta}{}^\gamma$ is to scale V_α and also translate it by an amount proportional to $u_{,\alpha}$. Thus we can use the ϕ translation to impose the last Witten constraint $T_{\alpha\beta}{}^\gamma = 0$. Thus all the constraints on the torsion come from the superparticle.

4. The Superstring

I have spent so much time on the particle I don't have much time for the string. Again we use Dirac's procedure. The momenta conjugate to the supercoordinates and the zweibein are

$$\begin{aligned} P^M &= \frac{\partial \mathcal{L}}{\partial(\partial_0 Z^M)} = -\sqrt{-g} e^{2\phi} g^{0s} E_M{}^a \partial_s Z^N E_N{}^b \eta_{ab} + \epsilon^{01} \partial_1 Z^N B_{NM} , \\ \Pi_r{}^i &= \frac{\partial \mathcal{L}}{\partial \partial_0 v_i{}^r} = 0 . \end{aligned} \tag{49}$$

The momenta conjugate to the zweibein are primary constraints, as are the spinor components $u_\alpha \approx 0$ of

$$u_A = V_A{}^M \left(P_M - \partial_1 Z^M B_{NM} \right) = (u_a, u_\alpha). \tag{50}$$

The canonical Hamiltonian is

$$\mathcal{H}_C = \dot{Z}^M P_M - \mathcal{L} \approx \left(\frac{v_+{}^1}{v_+{}^0} \right) \chi_- - \left(\frac{v_-{}^1}{v_-{}^0} \right) \chi_+, \tag{51}$$

where we have introduced

$$\begin{aligned} \chi_\pm &= \frac{1}{4} \zeta^a_\pm \zeta_{\pm a} \approx 0, \\ \zeta^a_\pm &= e^{-\phi} u^a \pm e^\phi \left(\partial_1 Z^M \right) E_M{}^a. \end{aligned} \tag{52}$$

$\chi_\pm \approx 0$ follows from demanding that the constraints $\Pi_r{}^i$ are preserved by time translations $\{H, \Pi_r{}^i\} \approx 0$.

The constraints $\chi_\pm(\sigma)$, $u_\alpha(\sigma)$ obey the following algebra:

$$\begin{aligned} \{u_\alpha(\sigma), u_\beta(\sigma')\} =& \delta(\sigma - \sigma') \left(\Omega_{\alpha\beta}{}^A u_A + \partial_1 Z^A H_{A\alpha\beta} \right) \\ \{u_\alpha(\sigma), \chi_\pm(\sigma')\} \approx& \delta(\sigma - \sigma') \Big[\pm \frac{1}{2} \partial_1 Z^\beta \zeta_{\pm a} \left(e^\phi \Omega_{\alpha\beta}{}^a \pm e^{-\phi} H^a{}_{\alpha\beta} \right) \\ & + \frac{1}{2} \zeta_{\pm a} \zeta_{\mp b} \left(\partial_\alpha \phi \eta^{ab} + \Omega_\alpha{}^{(ab)} \mp \frac{1}{2} e^{-2\phi} H^b{}_\alpha{}^a \right) \Big] \\ \{\chi_\pm(\sigma), \chi_\pm(\sigma')\} =& \pm 2 \chi_\pm \delta'(\sigma - \sigma') \\ \{\chi_\pm(\sigma), \chi_\mp(\sigma')\} \approx& \frac{1}{4} \delta(\sigma - \sigma') \zeta_{\pm a} \zeta_{\mp b} \partial_1 Z^\gamma \left(\mp 2 \eta^{ab} \partial_\gamma \phi \mp 2 \Omega_\gamma{}^{(ab)} + H_\gamma{}^{ab} e^{-2\phi} \right). \end{aligned} \tag{53}$$

Even in flat superspace, χ_- is not first class, but one can add some of the u_α to make it so. In curved space we look for first class

$$\begin{aligned} \hat{\chi}_\pm &= \chi_\pm + A^\alpha_\pm u_\alpha \\ \psi^\alpha &= F^{\alpha\beta} u_\beta \,. \end{aligned} \tag{54}$$

The argument goes much as in the particle case. The role of $\Omega_{\alpha\beta}{}^a$ gets doubled, with $D^\pm{}_{\beta\gamma}{}^b = e^\phi \Omega_{\beta\gamma}{}^b \pm e^{-\phi} H^b{}_{\beta\gamma}$. The two D's obey anticommutation relations with two F's, but only one of these has a driving term in the flat space limit, and the opposite D's and F's have zero symmetric piece. The upshot of an involved argument[8,9] is

$$\begin{aligned} H_{\alpha\beta\gamma} &= 0, \\ \Omega_{\alpha\beta}{}^{a} &= 2\Gamma^{a}_{\alpha\beta}, \\ H_{\alpha\beta}{}^{a} &= 2e^{2\phi}\Gamma^{a}_{\alpha\beta}\,. \end{aligned} \tag{55}$$

From requiring $\hat{\chi}_{\pm}$ to be first class we again find the same constraint on $\Omega_{\alpha}{}^{(ab)}$, but we also get a corresponding constraint on H,

$$H_{\alpha}{}^{ab} = e^{2\phi}\left[(2\phi_{,\beta} + \frac{1}{6}\Omega_{\beta}{}^{c}{}_{c})\Gamma^{ab\beta}{}_{\alpha} + \frac{1}{6}\Omega_{\beta}{}^{(ca)}\Gamma_{c}{}^{b\beta}{}_{\alpha} - \frac{1}{6}\Omega_{\beta}{}^{(cb)}\Gamma_{c}{}^{a\beta}{}_{\alpha}\right]. \tag{56}$$

Imposing Nilsson's or Witten's constraint on $\Omega_{\alpha}{}^{(ab)}$ then gives the corresponding constraint on $H_{\alpha ab}$, where in Nilsson's case we also translate ϕ away. Thus we have satisfied all the required supergravity equations, starting from an arbitrary background consistent with the gauge invariance of the superstring.

5. Summary

We have seen that the existence of a local Siegel fermionic symmetry for the classical motion of strings in superspace imposes constraints on the background fields, requiring that they are equivalent to fields obeying supergravity equations of motion. The equivalence relation, defined by transformations of the background superspace fields which have no effect on the string action, is what is needed to relate different versions of superspace supergravity equations.

6. Acknowledgments

All of my work described here was done jointly with Cyrus C. Taylor. I would also like to thank Cyrus for his comments on this talk. I would like to thank the National Science Foundation for their support of this research under Grant Nos. NSF-PHY84-15534 and NSF-PHY88-18535.

REFERENCES

1. M. B. Green and J. H. Schwarz, *Phys. Lett.* **109B** (1982) 444.
2. L. Brink and J. H. Schwarz, *Phys. Lett.* **100B** (1981) 310.
3. W. Siegel, *Phys. Lett.* **128B** (1983) 397.
4. M. B. Green and J. H. Schwarz, *Phys. Lett.* **136B** (1984) 367.
5. E. Witten, *Nucl. Phys.* **B266** (1986) 245.
6. J. Harnad, J. A. Shapiro, S. Shnider, and C. Taylor, "Symplectic Reduction of the Minimally Coupled Massless Superparticle in D=10", to be published in L.-L Chau and W. Nahm, eds., *Differential Geometrical Methods in Theoretical Physics*, proceedings of conference held at Lake Tahoe, July 2-8, 1989.
7. J. A. Shapiro and C. C. Taylor, *Phys. Lett.* **181B** (1986) 67.
8. J. A. Shapiro and C. C. Taylor, *Phys. Lett.* **186B** (1987) 69.
9. J. A. Shapiro and C. C. Taylor, *Phys. Rep.*, to be published.
10. B. E. W. Nilsson, *Nucl. Phys.* **B188** (1981) 176.
11. S. J. Gates, private communication.
12. L.-L. Chau, and B. Milewski, "Light-Like Integrability and Supergravity Equations of Motion in D=10, N=1 Superspace", preprint UCD-87-05-R, unpublished.

METHOD OF EXACT SOLVING FOR A CLASS OF MODELS OF QUANTUM CONFORMAL THEORY IN D-DIMENSIONS

M.Ya. Palchik
Institute of Automation and Electrometry,
630090, Novosibirsk,USSR

ABSTRACT

Proposed is a method of solving a class of conformal quantum field theory models in D-dimensional Euclidean space. The method allows us to obtain closed differential equations for each Green function of any number of fundamental and composite fields. Each D> 2-model involves an analogue of the central charge, i.e. a special scalar field P of dimension d_p=D-2. When D=2 this field turns into a constant coinciding with a central charge. We also obtain a new class D=2 models with broken infinite parameter simmetry. As an illustration of how the method works we derive the Wess-Zumino-Witten model solution.

1. The basic statements assertions

An ample class of quantum field theory models in D dimensions exists which allows for an exact conformal invariant solution. We propose a method of deriving closed differential equations for each of the Green functions, and algebraic equations for scale dimensions of fundamental and composite fields in each of these models. When D=2, the class embraces some of the known two-dimensional models: the Thirring and Wess-Zumino-Witten models, and also the Ising model. We also obtain a new class of D = 2 models with broken infinite-parameter symmetry.

In each of the models in D > 2 there exists a special

composite scalar field $P(x)$; this field emerges as an operator Schwinger term when one calculates the commutator of different components of the energy-momentum tensor. This field is a D-dimensional analogue of the central charge in two dimensional theories.

Note that in application to the simplest of the D=2 models which we are going to consider, the closed equations coincide /1/ with the equations of ref. /2/ of the two-dimensional Ising model. The rest of the D=2 models of scalar fields are distinct from the minimal models of ref. /2/; the Virasore invariance is broked down in these models to the 6-parametric algebra of the conformal little group.

The main idea behind our method is based on four assertions, formulated as early as 1975-1978, see /3-5/:

1. Consider an operator product $\varphi_1(x_1)\varphi_2(x_2)$ of any two field φ_1 and φ_2. Suppose that the tensor field P_{σ_i} contributes in the operator expansion of the product

$$\varphi_1(x_1)\varphi_2(x_2) = [P_{\sigma_i}] + \ldots$$

where $\sigma_i = (l_i, s)$, l_i is the scale dimension, s_i is the rank of the tensor field P_{σ_i}. The assurtion is: all Green functions of this field are determined by equations /4,5/

$$\langle P_{\sigma_i}\varphi_3 \ldots \varphi_n\rangle = \operatorname*{res}_{\sigma=\sigma_i} \tilde{m}^{\sigma} \; (C^{\sigma}) \overset{\varphi_1}{\underset{\varphi_2}{=}} (G_n) \begin{matrix}\varphi_3\\ \vdots \\ \varphi_n\end{matrix} \tag{1.1}$$

where $G_n = \langle \varphi_1 \varphi_2 \varphi_3 \ldots \varphi_n \rangle$,

$C^{\sigma} \sim \langle P_{\sigma}(x_1)\varphi_1(x_2)\varphi_2(x_3)\rangle$ is a conformally invariant 3-point function, where P_{σ} is a tensor field with $\sigma = (l, s)$.

2. In any conformally invariant theory of a scalar field $\varphi(x)$, irrespectively of the type of interaction, there exists an infinite collection of symmetric traceless tensor fields /3,4/:

$$P_s = P^{d+s}_{\mu_1 \ldots \mu_s}(x)$$

induced by a conserved current or the energy-momentum tensor. These fields emerge when one expands the operator products

$$j_\mu(x_1)\,\varphi(x_2) \quad \text{or} \quad T_{\mu\nu}(x_1)\,\varphi(x_2)$$

in powers of the difference x_{12}. The scale dimension of these fields is equal to

$$d_s = d + s$$

When $s=0$, P_s coincides with the fundamental field $\varphi(x)$; the vector field P_μ^{d+1} can only appear in the product expansion of $j_\mu \varphi$, but not $T_{\mu\nu}\varphi$.

The P_s fields are singled out from all other conformal fields by the fact /3,4/ that their commutators with the energy-momentum tensor, $[T_{\mu\nu}, P_s]$ (or with the current, $[j_\mu, \varphi]$) contain the fields $P_{s'}$, with $s'<s$ and (derivatives of) the fundamental field $\varphi(x)$ in addition to the field P_s itself or its derivatives. Therefore these fields are analogous to certain combinations of the secondary fields of ref./2/.

The proof of the existence of these fields is based on the Ward identities.

3. The operator equality

$$P_s(x) = 0 \quad \text{for a fixed } s \tag{1.1a}$$

is equivalent to a set of closed differential equations for each of the Green functions involving the fundamental or composite fields. More precisely, each of the equations /5,6/

$$\langle P_s(x_1)\,\varphi_2(x_2)\ldots\varphi_n(x_n)\rangle = 0 \tag{1.2}$$

is equivalent to a number of differential equations of the form

$$L_n^{(s)}\left(x, \frac{\partial}{\partial x}\right)\langle\varphi(x_1)\,\varphi_2(x_2)\cdots\varphi_n(x_n)\rangle = 0 \tag{1.3}$$

where $L_n^{(s)}$ is a certain differential operator. In the general case od D > 2 this operator looks very complicated. Its derivation involves Ward identities and uses special representations for the Green functions of the P_s fields.

Consider, in particular, any one of the fields P_s which is generated by the energy-momentum tensor. This means that the field arises in the operator expansion of $T_{\mu\nu}(x_1)\,\varphi_1(x_2)$, and its Green functions can be expressed through the Green functions involving $T_{\mu\nu}(x)$

$$G^{d(n)}_{\mu\nu}(x\, x_1 \ldots x_n) = \langle T_{\mu\nu}(x)\,\varphi_1(x_1)\ldots \varphi_n(x_n)\rangle \tag{1.4}$$

according to the formula (1.1), *see (4.2) and [7,8] for details*

$$\langle P_s(x_1)\,\varphi_2(x_2)\ldots\varphi_n(x_n)\rangle = \underset{\ell=d+s}{\mathrm{res}} \; \overset{\ell,\,s}{\sim\!\sim}\!\left(\tilde{C}_{\mu\nu}\right)\!\overset{T_{\mu\nu}}{\underset{\varphi_1}{=}}\!\left(G^{d(n)}_{\mu\nu}\right)\!\begin{matrix}\varphi_2\\ \varphi_3\\ \varphi_n\end{matrix} \tag{1.5}$$

where $G^{d(n)}_{\mu\nu} = \langle P_s\, \tilde{\varphi}^{D-d}\, \tilde{T}_{\mu\nu}\rangle$. Using the Ward identity

$$-\partial^x_\nu G^{d(n)}_{\mu\nu}(x x_1 \ldots x_n) = \Big[\sum_{i=1}^{n} \delta(x-x_i)\,\partial^{x_i}_\mu - \partial^x_\mu \sum_{i=1}^{n} \frac{d_i}{D}\,\delta(x-x_i)\Big] \langle \varphi_1(x_1)\ldots\varphi_n(x_n)\rangle \tag{1.6}$$

where d_i is dimension of the field $\varphi_i(x)$, one can get from (1.5)

$$\langle P_s^{d_1+s}(x_1)\,\varphi_2(x_2)\ldots\varphi_n(x_n)\rangle = L^{(s)}_n\Big(x, \frac{\partial}{\partial x}\Big)\langle\varphi_1(x_1)\ldots\varphi_n(x_n)\rangle \tag{1.7}$$

whence (1.3) follows. The derivation of eq. (1.7) from eq. (1.5) is outlined briefly in the next sections. For simpler models (and the Thirring model in particular) the above results were obtained in /5/, while the general case was treated in /6/ and in more detail in /7/.

Special attention is deserved by the Green functions $\langle T_{\mu\nu} T_{\rho\sigma}\varphi\ldots\varphi\rangle$ whose Ward identities differ from eqs. (1.6) and contain anomalous terms, see Sec.III.

4. The equations (1.3) for the three point Green functions,

$$\langle \varphi\varphi T_{\mu\nu}\rangle, \quad \langle\varphi\varphi P_s^{\ell_\alpha}\rangle \tag{1.8}$$

in which $P_s^{\ell_\alpha}$, are composite fields of spin s and dimension ℓ_α, lead to algebraic equations for the scale dimensions and other parameters of the theory. This is a consequence of the known coordinate dependence of these

Green functions, which in turn follows from conformal invariance. These algebraic equations were derived and solved in refs./5/ and /8/ for the case of two-dimensional Thirring and Wess-Zumino-Witten models. In addition, several results pertaining to the general case $D > 2$ were also presented in ref./8/.

5. The models in D-dimensional space with $D > 2$ have a peculiar feature: nontrivial solutions are possible only if there is a scalar field $P(x)$ with a scale dimension

$$d_P = D - 2$$

This field is analogous to the central charge of the D=2 theories. It appears in the commutator of the energy momentum tensor components. Its three-point Green function

$$\langle \varphi(x_1)\varphi(x_2)P(x_3)\rangle = -\frac{1}{144\pi} C \left(\frac{x_{12}^2}{x_{13}^2 x_{23}^2}\right)^{\frac{D-2}{2}} \langle \varphi(x_1)\varphi(x_2)\rangle \quad (1.9)$$

where C is a constant, appears in the Ward identity for the Green function $\langle T_{\mu\nu} T_{\rho\sigma} \varphi\varphi\rangle$. The constant C is one of fundamental parameters of the theory, which are calculated in the course of solution. For D=2 the field $P(x)$ becomes a constant:

$$P(x)\Big|_{D=2} = -\frac{1}{144\pi} C$$

where C coincides with the central charge of D=2 theories.

2. Schwinger-Dyson equation system and operator equations

A conformal simmetry impose rigid restrictions on a field theory. In particular, it was found /4/, that Schwinger-Dyson renormed system of equations in a conformal field theory alows an exact solution.The system consists of an infinite set equations for different Green functions and its solution [4] contains an infinite set of arbitrary functions.

The analysis of the Ward identities fulfilled in /4,5/ have shown that they play an important role in a removal of this arbitrariness. The equations (1.2),(1.5) is one of the possible ways to additionally define the solution. Each operator equation (1.1a) is equivalent to an infinite set of restrictions for Green functions of the energy-momentum tensor and unambiguously fixes the field theory model taking into account the Ward identities. Firstly it has been demonstrated by an example of Thirring model in /5,13/ and by a trivial D-dimension model in /14/.

In the case of the Thirring model it has been shown in /5,13/, that the operator expansion of the product $j_\mu(x)\psi(x+\varepsilon)$ where ψ is a fundamental field with dimension d, has a form

$$\gamma_\mu j_\mu(x)\psi(x+\varepsilon) \sim \frac{\hat{\varepsilon}}{\varepsilon^2}\psi(x) + k\hat{\partial}\psi(x) + \psi_1(x) + \ldots \tag{2.1}$$

Here ψ_1 is a new spinor field with the dimension

$$d_1 = d+1$$

This results from the Ward identities which determines all Green functions $\langle j_\mu \psi \ldots \overline{\psi} \rangle$. The field ψ_1 belongs to the same class as the fields P_s, and a solution of the model is found from the operator equality /5,13/

$$\psi_1(x) = 0 \tag{2.2}$$

being an analogous to equations (1.1). To derive (2.2) define the right-hand-side of the model field equation

$$-\hat{\partial}\psi(x) = \lambda\gamma_\mu j_\mu(x)\psi(x) \tag{2.3}$$

as an $\varepsilon^2 \to 0$ limit of the product $\gamma_\mu j_\mu(x)\psi(x+\varepsilon)$, averaging over angles of the vector ε_μ and substitute here (2.1). Then one obtains (2.2), as a consequence of (2.3), see /5,13/ for detail.

To find the solution of the model from the operator

equation (2.2) one can write it for the Green functions

$$\langle \psi_1 \bar{\psi} j_\mu \rangle = 0\,, \quad \langle \psi_1 \psi \bar{\psi} \bar{\psi} \rangle = 0, \quad etc. \tag{2.4}$$

Now let us represent these equations in the form (1.1):

$$\langle \psi_1 \psi \bar{\psi} \bar{\psi} \rangle = \underset{l=d+1}{res}\ \overset{l}{\sim\!\sim} (\tilde{C}_\mu) \overset{j_\mu}{\underset{\psi}{=\!=}} (G_\mu) \begin{matrix} \to \bar{\psi} \\ \leftarrow \bar{\psi} \\ \to \bar{\psi} \end{matrix} = 0$$

The integrals over internal legs may be easiely calculated using Ward identities. The result is [5]

$$d = \frac{1}{2} + \frac{\lambda}{4\pi}(a - \bar{a})$$

$$-\hat{\partial}_{x_1} \langle \psi(x_1)\psi(x_2)\bar{\psi}(x_3)\bar{\psi}(x_4)\rangle = \lambda \{ a [-\hat{\partial}_{x_1} \ln x_{12}^2 + \hat{\partial}_{x_1} \ln x_{13}^2 +$$

$$+ \hat{\partial}_{x_1} \ln x_{14}^2] - \bar{a} \gamma_\mu \varepsilon_{\mu\rho} \partial_\rho^{x_1} [\gamma_5^{x_2} \ln x_{12}^2 + \gamma_5^{x_3} \ln x_{13}^2 +$$

$$+ \gamma_5^{x_4} \ln x_{14}^2] \} \langle \psi(x_1)\psi(x_2)\bar{\psi}(x_3)\bar{\psi}(x_4)\rangle = 0.$$

3. Wess-Zumino-Witten model

The model solution is derived by trivial modification /8/ of calculations of ref./5/. Consider Ward identities

$$\partial_\nu^x \langle j_\nu^a(x) g(x_1) \ldots g^{-1}(x_{2n}) \rangle = \\ = -\sum_{i=1}^{2n} \delta(x - x_i)\, t_{x_i}^a \langle g(x_1) \ldots g^{-1}(x_{2n}) \rangle \tag{3.1}$$

where $g(x)$ is the field, which takes values in a group G, t^a are the generators of G. Using the method of ref. /5/ one easily find that the operator product expansion of $j_\mu(x) g(x+\varepsilon)$ involves the additional vector field P_μ^{d+1} with the scale dimension $d+1$:

$$t^a j_\mu^a(x)\, g(x+\varepsilon) = \frac{\varepsilon_\mu}{\varepsilon^2} g(x) + k \partial_\mu g(x) + P_\mu^{d+1}(x) + \ldots \tag{3.2}$$

As above let us suppose that

$$P^{d+1}_{\mu}(x) = 0 \tag{3.3}$$

This operator equation unambiguously determines the model solution. Consider for example the next equations

$$\langle P^{d+1}_{\mu}(x_1)\, g^{-1}(x_2)\, j^a_\nu(x_3)\rangle = \langle P^{d+1}_{\mu}(x_1)\, g^{-1}(x_2)\, g(x_3)\, g^{-1}(x_4)\rangle = 0 \tag{3.4}$$

Represent their in the form

$$\langle P^{d+1}_{\mu}\, g^{-1}\, j_\nu\rangle = \underset{l=d+1}{\mathrm{res}}\; l \sim \left(C_{\mu\nu_1}\right) \overset{j_{\nu_1}}{\underset{g}{=}} \left(G_{\nu_1\nu}\right) \begin{matrix} j_\nu \\ g^{-1} \end{matrix} = 0 \tag{3.5}$$

where $G_{\nu_1\nu} = \langle g(x_1)\, g^{-1}(x_2)\, j_{\nu_1}(x_3)\, j_\nu(x_4)\rangle$

$$\langle P^{d+1}_{\mu}\, g^{-1} g\, g^{-1}\rangle = \underset{l=d+1}{\mathrm{res}}\; l \sim \left(C_{\mu\nu_1}\right) \underset{g}{=} \left(G_{\nu_1}\right) \begin{matrix} g^{-1} \\ g \\ g^{-1} \end{matrix} \tag{3.6}$$

where $G_{\nu_1} = \langle j_{\nu_1}\, g\, g^{-1}\, g\, g^{-1}\rangle$

Using the method of ref. /5/ (see /8/ for details) and Ward identities

$$\begin{aligned} \partial^{x_3}_{\nu_1}\langle g(x_1)\, g^{-1}(x_2)\, j^a_{\nu_1}(x_3)\, j^b_{\nu_2}(x_4)\rangle &= -\delta(x_{13})\, t^a \langle g(x_1)\, g^{-1}(x_2)\, j^b_\nu(x_4)\rangle \\ &+ \delta(x_{23})\langle g(x_1)\, g^{-1}(x_2)\, j^b_\nu(x_4)\rangle t^a + i\delta(x_{34})\, f^{abc}\langle g(x_1)\, g^{-1}(x_2)\, j^c_\nu(x_4)\rangle \\ &- \frac{k}{8\pi}\left(\delta_{\nu\rho} + i\beta\varepsilon_{\nu\rho}\right)\partial^{x_3}_\rho \delta(x_{34})\, \delta^{ab}\langle g(x_1)\, g^{-1}(x_2)\rangle \end{aligned} \tag{3.7}$$

where α, β are constants, one get from (3.6):

$$\alpha = \beta = 1, \quad d = \frac{2C_g}{C_v + K} \tag{3.8}$$

where $C_g = t^a t^a, \quad f^{abc} f^{a'bc} = C_v\, \delta^{aa'}$

Note, that the coupling constant λ of WZW-model also can be calculated from (3.6). The result is (see /8/ for detail):

$$\lambda^2 = \frac{4\pi}{K}$$

that coincides with conformal invariance condition in WZW-model.

Analogously, using the Ward identities (3.1) one find

$$\Big\{\tfrac{1}{2} C_g \partial_\tau^{x_1} + d\Big[\frac{(x_{12})_\tau}{x_{12}^2} t_1^b t_2^b - \frac{(x_{13})_\tau}{x_{13}^2} t_1^b t_3^b + \frac{(x_{14})_\tau}{x_{14}^2} t_1^b t_4^b\Big]\Big\}$$

$$\langle g(x_1) g^{-1}(x_2) g(x_3) g^{-1}(x_3)\rangle = 0 \tag{3.9}$$

Note, that (3.8) and (3.9) coincide with results of ref. /15/.

4. Derivation of closed equations for Green functions in D-dimension space

In general cas D> 2 the operator equation

$$P_s(x) = 0$$

can be represent taking in account (1.5) in the form

$$\operatorname*{res}_{\sigma=\sigma_s} \sim (\tilde{C}^\sigma_{\mu\nu}) \overset{T_{\mu\nu}}{=} (G_{\mu\nu}) \vdots = 0 \,, \qquad \sigma_s = (d+s, s) \tag{4.1}$$

As it is shown in /6,7/ the conformal invariant function $\tilde{C}^\sigma_{\mu\nu}$ has the next form:

$$\tilde{C}^\sigma_{\mu\nu}(x_1 x_2 x_3) \equiv \tilde{C}^l_{\mu\nu,\mu_1\ldots\mu_s}(x_1 x_2 x_3) = \langle P^l_{\mu_1\ldots\mu_s}(x_1) \tilde{\varphi}(x_2) \tilde{T}_{\mu\nu}(x_3)\rangle$$

$$= \partial_\mu^{x_3} B^l_{\gamma\mu_1\ldots\mu_s}(x_1 x_2 x_3) + \partial_\gamma^{x_3} B_{\mu_3\mu_1\ldots\mu_s}(x_1 x_2 x_3) - \frac{2}{D}\delta_{\mu\nu}\partial_\lambda^{x_3} B_{\lambda\mu_1\ldots\mu_s}(x_1 x_2 x_3) \tag{4.2}$$

where the function $B^\sigma_{\mu_3\mu_1\ldots\mu_s}$ is equal

$$B^\sigma_{\mu_3\mu_1\ldots\mu_s}(x_1 x_2 x_3) = B_{1\mu_3\mu_1\ldots\mu_s}(x_1 x_2 x_3) + \alpha B_{2\mu_3\mu_1\ldots\mu_s}(x_1 x_2 x_3) \tag{4.3}$$

where α is a constant that will be determine below,

$$B^\sigma_{1\mu_3\mu_1\ldots\mu_s} = \lambda_\mu^{x_3}(x_1 x_2) \lambda^{x_1}_{\mu_1\ldots\mu_s}(x_3 x_2)\, \Delta(x_1 x_2 x_3) \tag{4.4}$$

$$B^{\sigma}_{2\mu_3\mu_1\cdots\mu_s} = \left[\frac{1}{x_{13}^2}\sum_{K=1}^{s} g_{\mu\mu_K}(x_{13})\lambda^{x_1}_{\mu_1\cdots\hat{\mu}_K\cdots\mu_s}(x_3x_2) - traces\right]\Delta(x_1x_2x_3)$$

$$\Delta(x_1x_2x_3) = (x_{13}^2)^{-\frac{\ell+d-s-D-2}{2}}(x_{23}^2)^{\frac{\ell+d-s-D+2}{2}}(x_{12}^2)^{-\frac{\ell-d-s+D+2}{2}}$$

$$\lambda^{x_1}_{\mu_1\cdots\mu_s}(x_3x_2) = \lambda^{x_1}_{\mu_1}(x_3x_2)\cdots\lambda^{x_1}_{\mu_s}(x_3x_2) - traces, \quad \lambda^{x_1}_{\mu}(x_2x_3) = \frac{(x_{21})_\mu}{x_{12}^2} - \frac{(x_{31})_\mu}{x_{13}^2}, \quad g_{\mu\nu}(x) = \delta_{\mu\nu} - 2\frac{x_\mu x_\nu}{x^2}$$

It is essential that the invariant function $\tilde{C}^{\sigma}_{\mu\nu}$, eq. (4.2), can be represented as a derivative of another invariant function,

$$B^{\sigma}_{\mu_3\mu_1\cdots\mu_s}(x_1x_2x_3) \sim \langle P^{\ell}_{\mu_1\cdots\mu_s}(x_1)\tilde{\varphi}(x_2)\,T_\mu(x_3)\rangle$$

where $T_\mu(x)$ in a conformal vector of dimension -1. This allows us to rewrite the integral on the right-hand-side of eq. (4.1) as

$$-2\int dy_1\, dy_2\, B_{\mu_3\mu_1\cdots\mu_s}(x_1y_1y_2)\,\partial^{y_2}_{\nu}\langle T_{\mu\nu}(y_2)\varphi_1(y_1)\varphi_2(y_2)\cdots\varphi_n(y_n)\rangle \tag{4.5}$$

Now substitute $\partial_\nu\langle T_{\mu\nu}\cdots\rangle$ from the ward identity (1.6) into eq. (4.5). The integral over y_2 is easily done due to the δ-functions. We are left then with the integral over y_1 of a sum of terms proportional to derivatives of $\langle\varphi_1(y_1)\varphi_2(x_2)\cdots\varphi_n(x_n)\rangle$. This integration can be performed as well if one first calculates the residue $\underset{\ell=d+s}{res}$ and pulls it out of the integral. Expression (4.4) contains the singular factor at $\ell \to d+s$

$$(x_{13}^2)^{-\frac{\ell-d-s+D}{2}}\Big|_{\ell\to d+s} \simeq \frac{1}{(\ell-d-s)}\,\frac{4\pi^{D/2}}{\Gamma(\frac{D}{2})}\,\delta(x_{13})$$

Substituting this into (4.5), the integrand can be represented as a sum of derivatives of $\delta(x_1-y_1)$. Therefore, we get in the end the equation

$$\langle P_s(x_1)\varphi_2(x_2)\cdots\varphi_n(x_n)\rangle = L^{(s)}_n\left(x,\frac{\partial}{\partial x}\right)\langle\varphi_1(x_1)\cdots\varphi_n(x_n)\rangle = 0 \tag{4.6}$$

It should be clear from the above derivation that the operator $L^{(s)}_n$ depends on the number and the type of the fields $\varphi_i(x)$. In fact it depends on the dimensions and tensor structure of the fields, since the Ward identities (1.6) bear this dependence. It can be shown that in the general case of D> 2 the top degree of derivatives in $L_n^{(s)}(x, \frac{\partial}{\partial x})$ is $s+1$. It is essential that this operator depends also on an undetermined parameter α entering (4.3). This parameter is to be fixed along with the scale dimensions of fields, see Sec. 6.

All calculations become essentially more simple in D=2 case. As a result we get the next equations

$$\Big\{\frac{1}{2}(s-1)(s+1)(d_1+s-2)(\partial_\pm^{x_1})^s - \sum_{k=3}^{s+1} C_k^{s+1} \frac{\Gamma(d_1+s)}{\Gamma(d_1+s-k+1)} \times$$

$$\times \sum_{j=2}^{N} (x_{1j}^\pm)^{-(k-2)} \Big[\partial_\pm^{x_j} + \frac{1}{2}(k-2)d_j \frac{1}{x_{1j}^\pm}\Big](\partial_\pm^{x_1})^{s+1-k}\Big\} \times$$

$$\times \langle \varphi_1(x_1)\varphi_2(x_2)\ldots\varphi_N(x_N)\rangle$$

where $C_k^{s+1} = \frac{(s+1)!}{k!(s+1-k)!}$ and d_j is the dimension of the scalar field $\varphi_j(x)$, $x_\pm = x_1 \pm i x_2$ are lite-cone variables.

These equations are invariant under infinite-parameter conformal group at D=2 only for the cases $s=2,3$. When $s \geqslant 4$ the symmetry of the equation has broken up to six-parameter little conformal group.

5. The anomalous Ward identities

Consider the Green function

$$G_{1T} \equiv G_{\mu\nu\rho\sigma}^{1T}(x_1 x_2 x_3 x_4) = \langle T_{\mu\nu}(x_1)T_{\rho\sigma}(x_2)\varphi(x_3)\varphi(x_4)\rangle \quad (5.1)$$

The contribution of the terms coming from the commutator of the components of the energy-momentum tensor should be accounted for in the Ward identity satisfied by (5.1). Besides the usual terms, similar to those in eq. (1.6), there may also appear anomalous Schwinger terms, either c-number or operator ones. Thus, for instance two-dimensional field theory is notorious for giving rise to the central charge C. In general, when D> 2, one can show that an operator Schwinger term arises, which is proportional to the scalar field P of dimension d_P =D-2, see (1.9).

Therefore, the most general conformally invariant Ward identify satisfied by the Green function (5.1) is of the form

$$
\begin{aligned}
&\partial^{x_1}_{\mu}\langle T_{\mu\nu}(x_1)T_{\rho\sigma}(x_2)\varphi(x_3)\varphi(x_4)\rangle = -\Big\{\delta(x_{13})\partial^{x_3}_{\nu}+\delta(x_{14})\partial^{x_4}_{\nu}-\\
&-\frac{d}{D}\partial^{x_1}_{\nu}[\delta(x_{13})+\delta(x_{14})]+\delta(x_{12})\partial^{x_2}_{\nu}-a\,\partial^{x_1}_{\nu}\delta(x_{12})\Big\}\langle T_{\rho\sigma}(x_2)\varphi(x_3)\varphi(x_4)\rangle\\
&+\Big\{\Big[\frac{D}{4}(1-a)+\frac{1}{2}\Big]\partial^{x_1}_{\rho}\delta(x_{12})\langle T_{\nu\sigma}(x_2)\varphi(x_3)\varphi(x_4)\rangle+\\
&+\Big[\frac{D}{4}(1-a)-\frac{1}{2}\Big]\partial^{x_1}_{\tau}\delta(x_{12})\delta_{\nu\rho}\langle T_{\tau\sigma}(x_2)\varphi(x_3)\varphi(x_4)\rangle+(\rho\leftrightarrow\sigma)\text{-trace}\Big\}\\
&+\Big\{C_1\partial^{x_1}_{\nu}\delta(x_{12})\partial^{x_2}_{\rho}\partial^{x_2}_{\sigma}+2C_2\partial^{x_1}_{\rho}\delta(x_{12})\partial^{x_2}_{\nu}\partial^{x_2}_{\sigma}+2C_3\partial^{x_1}_{\tau}\delta(x_{12})\delta_{\nu\rho}\partial^{x_2}_{\sigma}\partial^{x_2}_{\tau}\\
&+e_1\partial^{x_1}_{\rho}\partial^{x_1}_{\sigma}\delta(x_{12})\partial^{x_2}_{\nu}+2e_2\partial^{x_1}_{\nu}\partial^{x_1}_{\sigma}\delta(x_{12})\partial^{x_2}_{\rho}+2e_3\delta_{\nu\rho}\partial^{x_1}_{\sigma}\partial^{x_1}_{\tau}\delta(x_{12})\partial^{x_2}_{\tau}\\
&+e_4\Box\delta(x_{12})\delta_{\nu\rho}\partial^{x_2}_{\sigma}+3f\partial^{x_1}_{\nu}\partial^{x_1}_{\rho}\partial^{x_1}_{\sigma}\delta(x_{12})+3\delta_{\nu\rho}\partial^{x_1}_{\sigma}\Box_{x_1}\delta(x_{12})\\
&+(\rho\leftrightarrow\sigma)-\text{trace}\Big\}\langle P(x_2)\varphi(x_3)\varphi(x_4)\rangle
\end{aligned}
\tag{5.2}
$$

where a is an aribtrary parameter, the Green function $\langle P\varphi\varphi\rangle$ is given in (1.9),

$$C_1=\frac{6}{(D-1)(D-2)},\quad C_2=-\frac{3D}{2(D-1)(D-2)},\quad C_3=C_2$$

$$e_1=\frac{3D}{(D-1)(D-2)},\quad e_2=\frac{3(D^2-4D+2)}{D(D-1)(D-2)},\quad e_3=-\frac{3D}{2(D-2)}$$

$$e_4=\frac{6}{(D-2)},\quad f=-\frac{3D-2}{D(D-1)}.$$

The above identity involves two independent parameters: a and C.

Therefore, the total number of independent parameters in the D > 2 theories under consideration is equal to four; these are

$$d,\ \alpha,\ a \ \text{ and } \ C \tag{5.3}$$

where d is the scale dimension of the fundamental field d and α is the parameter entering eq. (4.3).

Note that the above identity simplifies considerably at D=2. One can show that the sum of all terms proportional to a turns into zero identically. In addition to that, all the terms involving derivatives of $P(x)|_{D=2} \to$ const also drop out. One thus gets the following result (only at D=2):

$$\partial_\mu^{x_1}\langle T_{\mu\nu}(x_1)T_{\rho\sigma}(x_2)\varphi(x_3)\varphi(x_4)\rangle = -\{\delta(x_{13})\partial_\nu^{x_3}+\delta(x_{14})\partial_\nu^{x_4}-$$

$$-\frac{d}{D}\partial_\nu^{x_1}[\delta(x_{13})+\delta(x_{14})]+\delta(x_{12})\partial_\nu^{x_2}\}\langle T_{\rho\sigma}(x_2)\varphi(x_3)\varphi(x_4)\rangle$$

$$+[\partial_\rho^{x_1}\delta(x_{12})\langle T_{\nu\sigma}(x_2)\varphi(x_3)\varphi(x_4)\rangle+(\rho\leftrightarrow\sigma)-trace]$$

$$+C\left(\frac{1}{24}\partial_\nu^{x_1}\partial_\rho^{x_1}\partial_\sigma^{x_1}-\frac{1}{48}\delta_{\nu\rho}\partial_\sigma^{x_1}\Box_{x_1}\right)\delta(x_{12})\langle\varphi(x_3)\varphi(x_4)\rangle \tag{5.4}$$

with C being the central charge.

Thus, only three independent parameters,

$$d,\ \alpha,\ C \tag{5.5}$$

are present in the two dimensional case.

6. Equations for the scale dimensions

As mentioned above, the dimensions ℓ_α of the composite fields $P_s^{\ell_\alpha}$ can be determined from the equations

$$L_s\left(x,\frac{\partial}{\partial x}\right)\ \langle\varphi(x_1)\varphi(x_2)\,P_s^{\ell_\alpha}(x_3)\rangle = 0 \tag{6.1}$$

since the coordinate dependence of the Green functions $\langle\varphi\varphi P_s^{\ell_\alpha}\rangle$ is known. For technical reasons it is more convenient, however, to use eq. (1.5) (or eq. (2.1)) directly:

$$\underset{\ell=d+s}{\mathrm{res}}\ \overset{\ell,s}{\sim}\left(\tilde{C}^{\sigma}_{\mu\nu}\right)\overset{T_{\mu\nu}}{\underset{\varphi}{=}}\left(G^{T}_{\alpha}\right)\overset{\varphi}{-}\sim P_s^{\ell_\alpha} = 0 \tag{6.2}$$

and employ then the Ward identity for the Green function

$$G_{\alpha}^{\prime T} = \langle T_{\mu\nu}(x_1)\,\varphi(x_2)\,\varphi(x_3)\,P_s^{\ell_\alpha}(x_4)\rangle$$

The relation between dimensions d and Δ (of the field $\chi \sim \varphi^2$) is found in /8/.

The rest parameters (5.3) may be calculated from the equations

$$L_s^P(x,\partial_x)\langle\varphi\varphi P\rangle = 0, \quad L_s^T(x,\partial_x)\langle\varphi\varphi T_{\mu\nu}\rangle = 0 \qquad (6.3)$$

The last equation is equivalent to three algebraic equations. Really one has as it may be shown /3,8/

$$\sigma \sim\!\!\sim (\tilde{C}^{\sigma}_{\mu\nu})(G_{1T})^{T_{\mu\nu}} = A_1(\ell,s)\,C^{\sigma}_{1\mu\nu} + A_2(\ell,s)\,C^{\sigma}_{2\mu\nu} + A_3(\ell,s)\,C^{\sigma}_{3\mu\nu}$$

where $C^{\sigma}_{i\mu\nu}$ are three different conformally invariant functions of the type $\langle P^{\sigma}\varphi T_{\mu\nu}\rangle$. Hence the equation

$$\operatorname*{res}_{\ell=d+s} \sigma \sim\!\!\sim (\tilde{C}^{\sigma}_{\mu\nu})(G_{1T}) = 0$$

is equivalent to next three equations

$$\operatorname*{res}_{\ell=d+s} A_1(\ell,s) = \operatorname*{res}_{\ell=d+s} A_2(\ell,s) = \operatorname*{res}_{\ell=d+s} A_3(\ell,s) = 0 \qquad (6.4)$$

Thus the number of the algebraic equation coincides with the number of the independent parameters: four equation (following from (6.3)) for four parameters (5.3).

In conclusion we give the result of D=2 case calculations; for the central charge C we have :

$$\alpha = \frac{1}{2s}$$

$$C = 12\,\Gamma(s-1)\frac{\Gamma(d+1)}{\Gamma(d+s-1)}\Big\{(-1)^s\Big[1-\frac{1}{4}(s+1)d(d+s-2)\Big]$$

$$-\frac{1}{\Gamma(s+1)}(d-1)(d-2)\frac{\Gamma(d+s-1)}{\Gamma(d)}\Big(\frac{d+s-1}{d+s-2}-\frac{1}{s+1}\Big)\Big\}$$

It is not difficult to convince oneself that in the particular case of $s=2$ and $s=3$ the results of ref./2/ for minimal models are reproduced. At $s \geqslant 4$ the infinite-parametric conformal simmetry has broken up to six parametric little conformal group.

7. Does infinite-parametric simmetry exist in D dimensions?

To conclude, we point out a certain analogy between the field models in D dimensions, $D > 2$, and the minimal models in two dimensions.

1. As mentioned above, the fields P_s are similar to combinations of the secondary fields of two-dimensional theory (due to form of their commutator with energy-momentum tensor). The fundamental field $\varphi(x)$ inducing all P_s fields is analogue of the primary one.

2. The certain combinations of the states $P_s(x)|0\rangle$ (and their derivatives) are analogous to null-vectors of two-dimensional theory, and formally the solution of each of the models follows from equating to zero one of these states.

3. There exists at $D > 2$ an analogue of the central charge of D=2 theory. It is the field $P(x)$ wich appears in the commutator of the components of energy-momentum tensor.

It is not impossible that in the general case $D > 2$ there exists a symmetry with infinite number of parameters similar to the conformal group of two-dimension space. Then an analogy to the central extansion of the algebra of this group will be addition of the generators corresponding to the field $P(x)$.

The author is grateful to prof. E.S. Fradkin for numerous discussions.

REFERENCES

1. E.S. Fradkin, M.Ya. Palchik. Conformal Invariance in

D=2 field Theory. - In: Proc. of the 3-d Seminar "Group Theoretical Methods in Physics", Yurmala,1985, v.2, p.191, ed.M.A.Markov, V.J.Man'kov and V.V.Dodonov (VMU Science PRESS, The Netherlands, 1986).

2. A.A. Belavin, A.M. Polyakov, A.B.Zamolodchikov, Nucl. Phys. B241 (1984) 333.
3. E.S. Fradkin, M.Ya.Palchik. Nuovo Cim. A 34 (1976)438.
4. E.S. Fradkin, M.Ya.Palchik. Nucl. Phys. B 99 (1975) 317; 126 (1977) 477.
5. E.S.Fradkin, M.Ya.Palchik. Phys. Rep. 44 (1978) 249.
6. E.S.Fradkin, M.Ya.Palchik. New Results in Conformal Invariant Field Theory. - In: Proc. of 2-d Seminar "Group Theoretical Methods in Physica", Zvenigorod 1982,v.2,p.84, ed.M.A.Markov, V.J.Man'ko and A.E.Shabad (Harwood Academic Publisher; Chur, London.Paris, New York).
7. E.S.Fradkin, M.Ya.Palchik. Closed Set of Equations for Green Functions in Conformal Invariant Quantum Field Theory. In:"Wandering in the Fields".Festschrift for Prof. Kazuhiko Nishijima on the occasion of his sixtieth birthday; ed. by K.Karvarabayashi, A.Ukawa, 1987, World Scientific.
8. E.S.Fradkin, M.Ya.Palchik. J.Phys.and Geometry 2 (1989).
9. E.S.Fradkin, Zh.Eksp.Teor.Fiz. 26 (1954) 751; 29 (1955) 121.
10. A.A. Migdal. Landau Inst. for Theor. Phys.Preprint, 1972, Chernogolovka.
11. G.Mack, Renormalization and Invariance in Quantum Field Theory, ed. E.R. Caianiello (Plenum Press. New York, 1974) p. 123-157; J. de Physique 34, Suppl. N 10 (1973) 99.
12. V.K.Dobrev, G.Mack, V.B.Petkova, S.G.Petrova,J.T.Todorov.Lecture Notes in Physics,Springer, 1977.
13. M.Ya. Palchik. Proc. of the III Int. School in High-

-Energy Physics, Primorsko (Bulgaria), 1977; Sov. Phys. Lebedev Inst. Rep. 4 (1980) 19.

14. M.Ya. Palchik, V.N. Zaikin. Sov. Phys. Lebedev Inst. Rep., 10 (1981) 9; M.Ya. Palchik, M.C. Prati, V.N. Zaikin. Nuovo Cim. 72A (1982) 87.
15. V.G. Knizhnik, A.B. Zamalodchikov. Nucl. Phys.B 247 (1984) 83.
16. J.M. Gelfand, G.E. Shilov, V. 1. Generalized Functions. State Publ. House of Phys.-Math. Literature, Moscow, 1975.
17. E.S. Fradkin, M.Ya. Palchik. Class. Quantum Gravity 1 (1984) 131.

CONFORMAL FIELD THEORY, TRIALITY AND THE MONSTER GROUP*

L. Dolan

The Rockefeller University, New York, N.Y.10021†

ABSTRACT

Conformal field theories usually have a continuous group of automorphisms or symmetries, i.e. the operators and states of the field theory form representations of these symmetries, such as the gauge invariance in superstring theory. This Lie group is generated by the finite-dimensional subalgebra of the affine algebra, which is given by the operator product expansion of all weight one fields in the theory. In addition to such continuous symmetries, twisted conformal field theories often also admit a discrete group of automorphisms, of which one subgroup, called triality, is identified to be responsible for the equivalence of several of these theories, and in a particular $c = 24$ theory for the promotion of Conway's group to the Monster group, F_1, the largest sporadic finite simple group. First a brief outline of chiral bosonic conformal field theory is described; then in the Z_2-twisted theory, the vertex operators, i.e. the conformal fields for *all* the states (massive as well as massless) are given explicitly and their locality is proved; finally error correcting codes are used to organize the momentum lattices of these conformal field theories so as to exhibit their isomorphisms deduced from the triality.

† Supported in part by the U.S.Department of Energy under Contract No. DEAC 02-87-ER40325 TaskB1

1. INTRODUCTION

In various phenomenological models discrete symmetries are used to suppress certain scattering processes. A natural origin of such invariance is the automorphisms of the conformal field theory (CFT) of string models. These discrete automorphisms are endemic to twisted conformal field theory and thus to superstring theory.

In this talk, it is demonstrated that the Monster group, a unique mathematical structure, can be thought of as a particular ramification of triality, the discrete subgroup of symmetries which is common to many conformal field theories and whose analogue may thus have important physical implications when applied to superstrings, just as it has dramatic mathematical applications when applied to the bosonic theory.

First a brief review of chiral bosonic meromorphic CFT is described by outlining its required properties and subsequent features. Then, in the $\mathbf{Z_2}$-twisted CFT theory, the vertex operators for the states are given explicitly and their locality is proved[1]. The only complete proof of the existence of a twisted CFT is given by Frenkel, Lepowsky and Meurman[2], for a particular theory whose momentum lattice Λ is the Leech lattice (with dimension $d = 24$), and whose algebra of vertex operators posseses as its automorphism group the Monster group, F_1, the largest sporadic finite simple group.[3] They use an explicit construction of the vertex operators corresponding to states in the untwisted sector but deduce the vertex operators for states in the twisted sector using special features of the Leech lattice. Our approach is more general, because it applies whenever both Λ and $\sqrt{2}\Lambda^*$ (the dual of Λ) are even and d is a multiple of 8. We base our analysis on explicit expressions for the vertices for *all* of the states, including twisted ones, and this approach will be of value in gaining a deeper understanding of symmetry and fermion emission in string theory, where twisted sectors often play a crucial role in realistic models. Finally error correcting codes are used to organize the momentum lattices of these conformal field theories so as to exhibit their isomorphisms[4] deduced from the triality.

A chiral bosonic CFT, $\mathcal{H}$ is a set of basis states H together with a set of conformal fields, or vertex operators, $\mathcal{V}(\psi, z)$ defined for all $\psi \in \mathcal{F}(H)$ where $\mathcal{F}(H)$ is a dense subspace of the basis states, with the following two properties: 1) the defining property:

$$\mathcal{V}(\psi, z)|0\rangle = e^{zL_{-1}}\psi \tag{1}$$

and 2) locality:

$$\mathcal{V}(\psi, z)\mathcal{V}(\phi, \zeta) = \mathcal{V}(\phi, \zeta)\mathcal{V}(\psi, z) \tag{2}$$

in the sense of analytic continuation. Here $|0\rangle$ is the vacuum and L_{-1} one of the moments of the special vertex operator $V(\psi_L, z) = \sum_n L_n z^{-n-2}$, which satisfy

the Virasoro algebra. The vertex operators are in one-to-one correspondence with the basis states for the theory, i.e. with all the states on the Regge trajectories. From these two properties alone, all the features of a CFT follow, in particular 1) duality, 2) the operator product expansion (OPE) of the conformal fields, and 3) an affine Kac-Moody Lie algebra (KMA) associated with the weight one states:
1) duality:

$$V(\phi, z)V(\psi, \zeta) = V(V(\phi, z-\zeta)\psi, \zeta)\,. \tag{3}$$

2) OPE:

$$\mathcal{V}(\psi, z)\mathcal{V}(\phi, \zeta) = \sum_{n=0}^{\infty}(z-\zeta)^{n-h_\phi-h_\psi}\mathcal{V}(\chi_n, \zeta) \tag{4}$$

where

$$\chi_n = \mathcal{V}_{h_\phi - n}(\psi)\phi\,, \tag{5}$$

and the vertex operators are expanded in terms of moments:

$$\mathcal{V}(\psi, z) \equiv \sum_n \mathcal{V}_n(\psi) z^{-n-h}\,. \tag{6}$$

3) KMA:

$$[T^a_m, T^b_n] = i f^{abc} T^c_{m+n} + k\delta^{ab}\delta_{m,-n}\,, \tag{7}$$

where we let ψ^a, $1 \leq a \leq N$, form an orthogonal basis for the weight one states, $\langle \psi^a, \psi^b \rangle = k\delta^{ab}$, with $T^a_0 \psi^b = i f^{abc}\psi^c$, and define

$$\mathcal{V}(\psi^a, z) = T^a(z) \equiv \sum_n T^a_n z^{-n-1}\,. \tag{8}$$

Then T^a_0 are the elements of the Lie algebra $g(\mathcal{H})$, with structure constants f^{abc}, which generates a continuous group $G(\mathcal{H})$ of automorphisms of the conformal field theory:

$$U\mathcal{V}(\psi, z)U^{-1} = \mathcal{V}(U\psi, z) \tag{9}$$

where $U \equiv e^{i\sum_a \lambda^a T^a_0}$. Two CFT's $\mathcal{H}, \mathcal{H}'$ are isomorphic if there is a unitary mapping u such that

$$u\mathcal{V}(\psi, z)u^{-1} = \mathcal{V}'(u\psi, z) \tag{10}$$

for $\psi \in \mathcal{F}(\mathcal{H})$, where $V'(\psi', z)$, $\psi' \in \mathcal{F}(\mathcal{H}')$, denote the vertex operators for $\mathcal{H}'$, and $u\psi \in \mathcal{F}(\mathcal{H}')$. If $\mathcal{H} = \mathcal{H}'$ then u is an automorphism of $\mathcal{H}$. Thus in general, a CFT has the above continuous group of automorphisms and a finite group of discrete automorphisms.

2. UNTWISTED CFT and Z_2-TWISTED CFT

The untwisted conformal field theory $\mathcal{H}(\Lambda)$ associated with a lattice Λ of dimension d is defined to be that created by integrally-moded bosonic operators a_m^j, $1 \le j \le d$, $m \in \mathbf{Z}$ from momentum states $|\lambda\rangle$, $\lambda \in \Lambda$. The twisted conformal field theory $\widetilde{\mathcal{H}}(\Lambda)$ is defined for a $\mathbf{Z_2}$ reflection twist by keeping from $\mathcal{H}(\Lambda)$ only the $\theta = 1$ states, where $\theta|\lambda\rangle = |-\lambda\rangle$ and $\theta a_m^j \theta^{-1} = -a_m^j$ and adding in the $\theta = 1$ subspace of the space $\mathcal{H}_T(\Lambda)$ generated from an irreducible representation space $\mathcal{X}_0(\Lambda)$ for the gamma matrix algebra $\{\gamma_\lambda : \lambda \in \Lambda\}$ associated with Λ, by half-integrally moded oscillators c_r^j, $1 \le j \le d$, $r \in \mathbf{Z} + \frac{1}{2}$. In this case, the involution θ is defined by $\theta\chi_0 = (-1)^{\frac{d}{8}}\chi_0$, $\chi_0 \in \mathcal{X}_0(\Lambda)$, $\theta c_r^j \theta^{-1} = -c_r^j$. We define

$$\mathcal{H}^{\pm}(\Lambda) = \{\psi \in \mathcal{H}(\Lambda) : \theta\psi = \pm\psi\} \tag{11}$$

$$\mathcal{H}_T^{\pm}(\Lambda) = \{\chi \in \mathcal{H}_T(\Lambda) : \theta\chi = \pm\chi\}\,. \tag{12}$$

Then

$$\mathcal{H}(\Lambda) = \mathcal{H}^+(\Lambda) \oplus \mathcal{H}^-(\Lambda), \qquad \widetilde{\mathcal{H}}(\Lambda) = \mathcal{H}^+(\Lambda) \oplus \mathcal{H}_T^+(\Lambda)\,. \tag{13}$$

The weight, i.e. conformal spin of a momentum state $|\lambda\rangle$ is $\frac{1}{2}\lambda^2$, and of a twisted ground state is $\frac{d}{16}$. From the locality requirement of Eq. 2, we find that the untwisted CFT is a bosonic meromorphic theory provided that Λ is an even integral lattice. The twisted CFT is bosonic and meromorphic provided that $d = \dim\Lambda$ is a multiple of eight and both $\sqrt{2}\Lambda^*$ and Λ are even, a condition implied by self-duality of the lattice.[5,6] In particular if Λ is even and self-dual, then both $\mathcal{H}(\Lambda)$ and $\widetilde{\mathcal{H}}(\Lambda)$ are self-dual bosonic meromorphic CFT's. If Λ is the $d = 24$ Leech lattice, then $\widetilde{\mathcal{H}}(\Lambda)$ is the natural module for the Monster group. See the Table, and note that the number of states in each column add to give simple linear combinations of the irreducible representations of F_1.

In the untwisted CFT $\mathcal{H}(\Lambda)$, the vertex operators are given by

$$V(\psi, z) = :\left(\prod_{a=1}^{M} \frac{i}{(m_a - 1)!}\frac{d^{m_a}X^{j_a}(z)}{dz^{m_a}}\right)\exp\{i\lambda\cdot X(z)\}\sigma_\lambda : \tag{14}$$

for

$$\psi = \left(\prod_{a=1}^{M} a_{-m_a}^{j_a}\right)|\lambda\rangle\,, \tag{15}$$

where

$$X^j(z) = q^j - ip^j \log z + i\sum_{n\neq 0}\frac{a_n^j}{n}z^{-n}\,. \tag{16}$$

Leech Lattice Twisted Conformal Field Theory

L_0	0	1	$\frac{3}{2}$	2	$\frac{5}{2}$	3	...
I.	$\vert 0\rangle$	~~$a^i_{-1}\vert 0\rangle$~~		$a^i_{-1}a^j_{-1}\vert 0\rangle$ (300) ~~$a^i_{-2}\vert 0\rangle$~~ (24) $\vert p_2^+\rangle$ (98280) ~~$\vert p_2^-\rangle$~~ (98280)		~~$a^i_{-1}a^j_{-1}a^k_{-1}\vert 0\rangle$~~ (2600) ~~$a^i_{-3}\vert 0\rangle$~~ (24) $a^i_{-1}a^j_{-2}\vert 0\rangle$ (576) ~~$a^i_{-1}\vert p_2^+\rangle$~~ ($24x98284$) $a^i_{-1}\vert p_2^-\rangle$ ($24x98284$) $\vert p_3^+\rangle$ (8386560) ~~$\vert p_3^-\rangle$~~ (8386560)	
II.			~~$\vert 2^{12}\rangle$~~	$c^i_{-\frac{1}{2}}\vert 2^{12}\rangle$ (98304)	~~$c^i_{-\frac{1}{2}}c^j_{-\frac{1}{2}}\vert 2^{12}\rangle$~~	$c^i_{-\frac{1}{2}}c^j_{-\frac{1}{2}}c^k_{-\frac{1}{2}}\vert 2^{12}\rangle$ ($2600x2^{12}$) $c^i_{\frac{-3}{2}}\vert 2^{12}\rangle$ ($24x2^{12}$)	

In the twisted CFT, the untwisted and twisted sectors of $\widetilde{\mathcal{H}}(\Lambda)$ are the subspaces $\mathcal{H}^+(\Lambda)$ and $\mathcal{H}_T^+(\Lambda)$ on which $\theta = 1$. If $\psi \in \mathcal{H}^+(\Lambda)$, $\mathcal{V}(\psi, z)$ maps $\mathcal{H}^+(\Lambda) \to \mathcal{H}^+(\Lambda)$ and $\mathcal{H}_T^+(\Lambda) \to \mathcal{H}_T^+(\Lambda)$ whereas $\mathcal{V}(\chi, z)$ maps $\mathcal{H}^+(\Lambda) \to \mathcal{H}_T^+(\Lambda)$ and $\mathcal{H}_T^+(\Lambda) \to \mathcal{H}^+(\Lambda)$ if $\chi \in \mathcal{H}_T^+(\Lambda)$. Thus we can write these vertex operators in matrix form

$$\mathcal{V}(\psi, z) = \begin{pmatrix} V(\psi, z) & 0 \\ 0 & V_T(\psi, z) \end{pmatrix}, \qquad \mathcal{V}(\chi, z) = \begin{pmatrix} 0 & \overline{W}(\chi, z) \\ W(\chi, z) & 0 \end{pmatrix} \tag{17}$$

where $V(\psi, z) : \mathcal{H}^+(\Lambda) \to \mathcal{H}^+(\Lambda)$, $V_T(\psi, z) : \mathcal{H}_T^+(\Lambda) \to \mathcal{H}_T^+(\Lambda)$, $W(\chi, z) : \mathcal{H}^+(\Lambda) \to \mathcal{H}_T^+(\Lambda)$ and $\overline{W}(\chi, z) : \mathcal{H}_T^+(\Lambda) \to \mathcal{H}^+(\Lambda)$.

In this notation, the vertex operators of the twisted CFT $\widetilde{\mathcal{H}}(\Lambda)$ are given by, for the untwisted states:

$$\psi = \left(\prod_{a=1}^{M} a^{j_a}_{-m_a}\right)|\lambda\rangle\,, \tag{18}$$

$$V(\psi,z) = :\left(\prod_{a=1}^{M}\frac{i}{(m_a-1)!}\frac{d^{m_a}X^{j_a}(z)}{dz^{m_a}}\right)\exp\{i\lambda{\cdot}X(z)\}\sigma_\lambda : \tag{19}$$

$$= \sum_{\lambda'\in\Lambda}\langle\lambda'| : e^{F(-z)} : |\psi\rangle\sigma_{\lambda'} \tag{20}$$

and

$$V_T(\psi,z) = V_T^0(e^{\Delta(z)}\psi,z)\,, \tag{21}$$

$$= \sum_{\lambda'\in\Lambda}\gamma_{\lambda'}\langle\lambda'| : e^{B(-z)} : e^{A(-z)}|\psi\rangle\,, \tag{22}$$

where

$$V_T^0(\psi,z) = (4z)^{-\frac{1}{2}\lambda^2} : \left(\prod_{a=1}^{M}\frac{i}{(m_a-1)!}\frac{d^{m_a}R^{j_a}(z)}{dz^{m_a}}\right)\exp\{i\lambda{\cdot}R(z)\} : \gamma_\lambda \tag{23}$$

$$= \sum_{\lambda'\in\Lambda}(4z)^{-\frac{1}{2}\lambda'^2}\gamma_{\lambda'}\langle\lambda'| : e^{B(-z)} : |\psi\rangle\,, \tag{24}$$

and

$$R(z) = i\sum_{r=-\infty}^{\infty}\frac{c_r}{r}z^{-r}\,; \tag{25}$$

and for the twisted states:

$$\chi = \left(\prod_{a=1}^{M} c^{j_a}_{-m_a}\right)\mathcal{X}_0\,, \tag{26}$$

the analogue of the fermion emission operator is

$$W(\chi,z) = e^{zL^c_{-1}}\tilde{W}(\psi,z)\,, \tag{27}$$

where

$$\tilde{W}(\chi,z) = \sum_{\lambda\in\Lambda}\gamma_\lambda\langle\lambda| : e^{B(z)} : e^{A(z)}|\chi\rangle\,, \tag{28}$$

and

$$\overline{W}(\chi,z) = z^{-2h_\chi}W(e^{z^*L^c_1}\overline{\chi},1/z^*)^\dagger\,. \tag{29}$$

In the above expressions, define

$$A(z) = A_{00}(z) + \Delta(-z) \tag{30}$$

$$A_{00}(z) = -\frac{1}{2}p^2 \ln(-4z) \tag{31}$$

and

$$\Delta(z) = \frac{1}{2} \sum_{\substack{m,n\geq 0 \\ (m,n)\neq(0,0)}} \binom{-\frac{1}{2}}{m}\binom{-\frac{1}{2}}{n} \frac{z^{-m-n}}{m+n} a_m\cdot a_n \,. \tag{32}$$

Also, $B(z) = B_-(z) + B_+(z)$, with

$$B_\pm(z) = - \sum_{n\geq 0, s \gtrless 0} \binom{-s}{n} \frac{(-z)^{-n-s}}{s} a_n\cdot c_s \,, \tag{33}$$

and the function $F(z)$ is given by

$$F(z) = \tilde{F}_-(z) + \tilde{F}_+(z) \tag{34a}$$

$$\tilde{F}_-(z) = F_-(z) + ip_\alpha\cdot q \tag{34b}$$

$$\tilde{F}_+(z) = F_0(z) + F_+(z) \,, \tag{34c}$$

where

$$F_\pm(z) = - \sum_{n\geq 0, m \gtrless 0} \binom{-m}{n} \frac{z^{-m-n}}{m} \alpha_n\cdot a_m \tag{35a}$$

$$F_0(z) = p\cdot p_\alpha \ln z - p\cdot \sum_{n>0} \frac{(-z)^{-n}}{n} \alpha_n \,. \tag{35b}$$

Here the two sets of oscillators commute $[a_n, \alpha_m] = 0$; $\alpha_0 \equiv p_\alpha$. Also : $e^{F(z)} :\equiv e^{\tilde{F}_-(-z)} e^{\tilde{F}_+(-z)}$. Note that the special state ψ_L is given in these CFT's by $\frac{1}{2}a_{-1}\cdot a_{-1}|0\rangle$ and $L_n = \frac{1}{2}\sum_{m=-\infty}^{\infty} : a_m\cdot a_{n-m} :$, $L_n^c = \frac{1}{2}\sum_{m=-\infty}^{\infty} : c_m\cdot c_{n-m} : + \frac{d}{16}$. The cocycle operators σ_λ and γ_λ on the untwisted and twisted sectors respectively are defined and discussed comprehensively in Ref. [1]. This completes the list of the four components of the vertex operators which describe $\widetilde{\mathcal{H}}(\Lambda)$. Both the operator and matrix element forms are given for $V(\psi, z)$ and $V_T(\psi, z)$.

With respect to these components, defined via Eq. (17), the locality relation (2) which is to be proved becomes

$$V(\psi, z)V(\phi, \zeta) = V(\phi, \zeta)V(\psi, z) \,, \tag{36}$$

$$V_T(\psi, z)V_T(\phi, \zeta) = V_T(\phi, \zeta)V_T(\psi, z) \,, \tag{37}$$

$$V_T(\psi, z)W(\chi, \zeta) = W(\chi, \zeta)V(\psi, z) \,, \tag{38}$$

$$V(\psi, z)\overline{W}(\chi, \zeta) = \overline{W}(\chi, \zeta)V_T(\psi, z) \,, \tag{39}$$

$$\overline{W}(\chi, z)W(\varrho, \zeta) = \overline{W}(\varrho, \zeta)W(\chi, z) \,, \tag{40}$$

$$W(\chi, z)\overline{W}(\varrho, \zeta) = W(\varrho, \zeta)\overline{W}(\chi, z) \,, \tag{41}$$

for $\psi, \phi \in \mathcal{H}^+(\Lambda)$, $\chi, \varrho \in \mathcal{H}_T^+(\Lambda)$.

An essential feature of the work described in this talk is the proof of locality. This proof of Eq.'s 36-41 is rather detailed and the reader is deferred to Ref.[1], where it is shown that Eq.s 37,39,40 follow from the explicit check of Eq.'s 36,38,41. Eq. 41 is the bosonic analogue of the four-fermion scattering amplitude, which although at tree level nonetheless provides the constraint on the lattice Λ that both it and $\sqrt{2}\Lambda^*$ must be even, a condition strictly weaker than self-duality, yet implying a sort of modular invariance. The conformal field theories discussed here are defined on the the Riemann sphere, and are *chiral, i.e.* holomorphic, in that they involve only functions of z, and not its complex conjugate, z^*. One expects when the theories are extended to Riemann surfaces of higher genus, that locality then becomes synonymous with modular invariance.[7] These CFT's are also *meromorphic*, in that the correlation functions have only poles as functions of the arguments z_j of the fields. (For an account of meromorphic conformal field theory, see Ref. [8].)The more general subject of classification of rational conformal field theories[9-10] and fusion rules for higher genus Riemann surfaces[11] contain as a subset the trivial monodromy theories discussed here, theories which may themselves already contain important information about symmetries.

3. ERROR-CORRECTING CODES, LATTICES and ISOMORPHISMS in CFT's

A binary linear code $\mathcal{C}$ of length d is a set of d-dimensional vectors or codewords with coordinates 0 or 1, closed under coordinate-wise addition modulo 2, *i.e.* a subspace of the vector space $\mathbf{F}_2^d$ of dimension d over the field $\mathbf{F}_2$ of two elements. Such a code $\mathcal{C} \subset \mathbf{F}_2^d$ is self-dual if it coincides with its dual code $\mathcal{C}^* = \{c' \in \mathbf{F}_2^d : c \cdot c' \equiv 0 \bmod 2 \,\text{for all}\, c \in \mathcal{C}\}$. It is doubly-even if the weights $\mathbf{1}\cdot c \in 4\mathbf{Z}$ for all $c \in \mathcal{C}$ (not taken mod 2), where $\mathbf{1} = (1,1,\ldots,1) \in \mathbf{F}_2^d$. If $\mathcal{C}$ is doubly-even and self-dual, d must be a multiple of 8, and we can associate two even self-dual lattices, $\Lambda_\mathcal{C}$ and $\widetilde{\Lambda}_\mathcal{C}$, with $\mathcal{C}$ by an untwisted and a twisted construction respectively. If we set

$$\Lambda_0 = \left\{ \tfrac{1}{\sqrt{2}} c + \sqrt{2} x : c \in \mathcal{C}, x \in \mathbf{Z}^d, \mathbf{1}\cdot x \in 2\mathbf{Z} \right\}, \tag{42a}$$

$$\Lambda_1 = \left\{ \tfrac{1}{\sqrt{2}} c + \sqrt{2} x : c \in \mathcal{C}, x \in \mathbf{Z}^d, \mathbf{1}\cdot x \in 2\mathbf{Z}+1 \right\}, \tag{42b}$$

$$\Lambda_2 = \left\{ \tfrac{1}{2\sqrt{2}} \mathbf{1} + \tfrac{1}{\sqrt{2}} c + \sqrt{2} x : c \in \mathcal{C}, x \in \mathbf{Z}^d, \mathbf{1}\cdot x \in 2\mathbf{Z}+n+1 \right\}, \tag{42c}$$

$$\Lambda_3 = \left\{ \tfrac{1}{2\sqrt{2}} \mathbf{1} + \tfrac{1}{\sqrt{2}} c + \sqrt{2} x : c \in \mathcal{C}, x \in \mathbf{Z}^d, \mathbf{1}\cdot x \in 2\mathbf{Z}+n \right\}, \tag{42d}$$

where $d = 8n$. Then Λ_C and $\widetilde{\Lambda}_C$ are both even and self-dual:

$$\Lambda_\mathcal{C} = \Lambda_0 \cup \Lambda_1 ; \qquad \widetilde{\Lambda}_\mathcal{C} = \Lambda_0 \cup \Lambda_3 . \tag{43}$$

There is one doubly-even self-dual code of length 8, the Hamming code, e_8, and by uniqueness, the associated lattice is $\Lambda_{e_8} = \widetilde{\Lambda}_{e_8} = E_8$, the root lattice of E_8. In $d = 16$ there are two doubly-even self-dual codes, $e_8 \oplus e_8$ and d_{16}, with $\Lambda_{e_8 \oplus e_8} = \widetilde{\Lambda}_{d_{16}}$ and $\Lambda_{d_{16}} = \widetilde{\Lambda}_{e_8 \oplus e_8}$. For length 24, there are 9 doubly-even self-dual codes, including the Golay code, g_{24}, which is the unique shortest such code which has no codewords of weight 4. Its symmetry group is a sporadic simple group, the Mathieu group, M_{24}. The twisted lattice construction for this code is $\widetilde{\Lambda}_{g_{24}} = \Lambda_{24}$, the Leech lattice, which is the unique even, self-dual lattice with no points of length squared 2 with smallest dimension. Its automorphism group is $\mathrm{Aut}(\Lambda_{24})$. F_1 is the group of automorphisms of $\widetilde{\mathcal{H}}(\Lambda_{24}) = V^\natural$. This illustrates the non-triviality of the classes of theories we are considering. Frenkel, Lepowsky and Meurman have conjectured that $\widetilde{\mathcal{H}}(\Lambda_{24})$ is in some suitable sense unique, *i.e.* it is the unique bosonic self-dual (chiral) CFT with no weight one fields and the smallest value of c. This would give a characterization of F_1 comparable with that of Conway's sporadic simple group Co_1, which can be defined as $\mathrm{Aut}(\Lambda_{24})/\{\pm 1\}$.

We shall show how a certain subgroup of discrete symmetries, triality, which is the crucial ingredient in the construction of the Monster, can be understood in a more general context. It occurs as part of the automorphism groups of each of a class of conformal field theories (for c any multiple of 8) which are related to binary codes. The mechanism responsible for it also accounts for isomorphisms between CFT's, which one would otherwise expect to be distinct. Thus it plays an important role in the classification of CFT's, and perhaps this makes the Monster seem a little less mysterious.

Our main result is that $\mathcal{H}(\widetilde{\Lambda}_\mathcal{C}) \cong \widetilde{\mathcal{H}}(\Lambda_\mathcal{C})$, and that this isomorphism is induced by the triality group of automorphisms of $\mathcal{H}(\Lambda_\mathcal{C})$, which may be easily constructed, and in turn it induces a triality group of automorphisms of $\widetilde{\mathcal{H}}(\widetilde{\Lambda}_\mathcal{C})$. This may be shown by a sort of commutative diagram (Fig.1), in which untwisted constructions are represented by straight lines and twisted constructions by wavy lines.

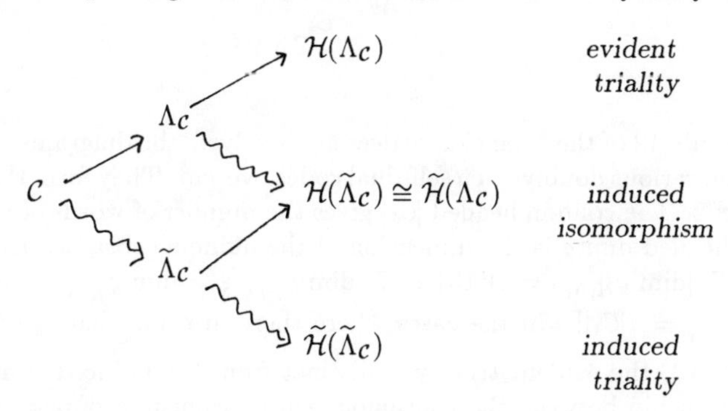

Fig. 1

We have defined from each doubly-even self-dual code, two even self-dual lattices, an untwisted construction Λ_C and a twisted construction $\widetilde{\Lambda}_C$, which are given in Eq. 43. Also, from each even self-dual lattice Λ we have defined two even self-dual CFT's, the untwisted $\mathcal{H}(\Lambda)$ and the twisted $\widetilde{\mathcal{H}}(\Lambda)$, which are given in Eq.13. In $d = 24$, only 12 of the 24 even self-dual (Niemeier) lattices are obtained from codes, suggesting by analogy that there are more than just the 39 *self-dual* bosonic CFT's with $c = 24$ constructed here (as one might have expected for a variety of other reasons). Two such additional theories have been conjectured by Schellekens and Yankielowicz[13], but they only produce a modular invariant partition function. To specify such a theory, one needs a local set of vertex operators, because for *eg.* the $E_8 \oplus E_8$ and D_{16} theories share the same partition function, demonstrating that there is not a one-to-one correspondence between partition functions and CFT's. Without more structure specified, the existence and nature of these theories remains in doubt.

$\lvert\mathcal{C}_4\rvert$	code	lattice	cft	$\dim g$
			E_8^3	744
		E_8^3	$D_{16}E_8$	744
42	e_8^3	$D_{16}E_8$		
42	$d_{16}e_8$		D_8^3	360
		D_8^3		
18	d_8^3		D_4^6	168
		D_4^6		
6	d_4^6		A_1^{24}	72
		A_1^{24}		
0	g_{24}		Λ_{24}	24
		Λ_{24}		
			$V^\natural$	0

Fig. 2a

Because only 12 of the Niemeier lattices are involved, the diagrams of the form of Fig.1 for the various doubly-even self-dual codes overlap. They form the patterns shown in Figs.2. The column headed $|\mathcal{C}_4|$ gives the number of words of weight 4 in $\mathcal{C}$, and that headed dim g is the dimension of the affine algebra described in Eq. 7 for the CFT [dim $g_{\mathcal{H}(\Lambda_C)} = 16|\mathcal{C}_4| + 72$, dim $g_{\widetilde{\mathcal{H}}(\Lambda_C)} =$ dim $g_{\mathcal{H}(\widetilde{\Lambda}_C)} = 8|\mathcal{C}_4| + 24$ and dim $g_{\widetilde{\mathcal{H}}(\widetilde{\Lambda}_C)} = 4|\mathcal{C}_4|$]. In the cases where there are concidences of the form $\widetilde{\mathcal{H}}(\widetilde{\Lambda}_C) \cong \mathcal{H}(\Lambda_{C'})$, the evident triality is distinct from the induced triality. Note also the relationship between the continuous automorphism group, generated by g, and the discrete group of automorphisms for a given CFT; for *eg.* when there is no continuous group, as in $V^\natural$, then the discrete group is maximal: F_1.

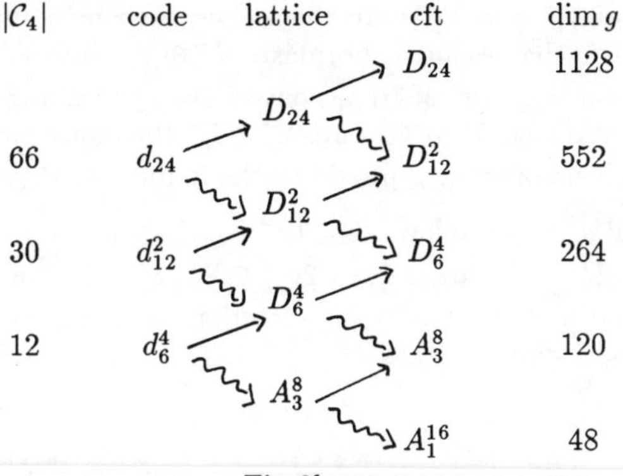

Fig 2b

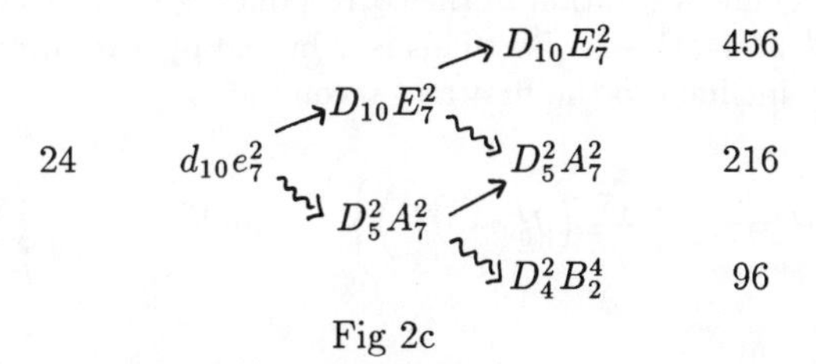

Fig 2c

In the case $\Lambda = \Lambda_{\mathcal{C}}$, $\mathcal{H}(\Lambda)$ has a triality structure. To study this and its implications, define $\mathcal{V}_j^{\pm} = \mathcal{H}^{\pm}(\Lambda_j)$, $0 \leq j \leq 3$, to be the 8 spaces generated by the a_m^j from the states $|\lambda\rangle$, $\lambda \in \Lambda_j$, with $\theta = \pm 1$. We can define 8 corresponding twisted spaces $\mathcal{T}_j^{\pm}$, $0 \leq j \leq 3$, starting from an irreducible representation $\mathcal{X}_0(\Lambda_0^*)$ of $\Gamma(\Lambda_0^*)$. [Since Λ_0^* is not integral, we need to amplify the definition of the twisted cocyles by specification of a symmetry factor $S(\lambda, \mu)$ as in [14]. We choose $S(\lambda, \mu) = e^{i\pi\lambda\cdot\mu}$ unless $\lambda \in \Lambda_1 \cup \Lambda_2$ and $\mu \in \Lambda_2 \cup \Lambda_3$, in which case $S(\lambda, \mu) = -e^{i\pi\lambda\cdot\mu}$.] $\mathcal{X}(\Lambda_0^*)$ splits into four representations $\mathcal{X}_j$, $0 \leq j \leq 3$, of $\Gamma(\Lambda_0)$ such that $\mathcal{X}_0 \oplus \mathcal{X}_1$ is a representation of $\Gamma(\Lambda_{\mathcal{C}})$ and $\mathcal{X}_0 \oplus \mathcal{X}_3$ is a representation of $\Gamma(\widetilde{\Lambda}_{\mathcal{C}})$. Then $\mathcal{T}_j^{\pm}$ is the $\theta = \pm 1$ space generated from $\mathcal{X}_j$ by the c_r^j. θ is defined to be $(-1)^{N/8}$ on $\mathcal{X}_a$, $a = 0, 1, 3$ and $-(-1)^{N/8}$ on $\mathcal{X}_2$. $\mathcal{T}_j^{\pm} = \mathcal{H}_T^{\pm}(\Lambda_j)$. With these definitions,

$$\begin{aligned}
\mathcal{H}(\Lambda_{\mathcal{C}}) &= \mathcal{V}_0^+ \oplus \mathcal{V}_0^- \oplus \mathcal{V}_1^+ \oplus \mathcal{V}_1^- , \\
\mathcal{H}(\widetilde{\Lambda}_{\mathcal{C}}) &= \mathcal{V}_0^+ \oplus \mathcal{V}_0^- \oplus \mathcal{V}_3^+ \oplus \mathcal{V}_3^- , \\
\widetilde{\mathcal{H}}(\Lambda_{\mathcal{C}}) &= \mathcal{V}_0^+ \oplus \mathcal{V}_1^+ \oplus \mathcal{T}_0^+ \oplus \mathcal{T}_1^+ , \\
\widetilde{\mathcal{H}}(\widetilde{\Lambda}_{\mathcal{C}}) &= \mathcal{V}_0^+ \oplus \mathcal{V}_3^+ \oplus \mathcal{T}_0^+ \oplus \mathcal{T}_3^+ .
\end{aligned} \tag{44}$$

Our strategy is to define a triality operator, σ_3, an automorphism of $\mathcal{H}(\Lambda_{\mathcal{C}})$

mapping each of $\mathcal{V}_0^+$ and $\mathcal{V}_1^-$ onto themselves, and mapping $\mathcal{V}_0^-$ onto $\mathcal{V}_1^+$, and *vice versa*. Thus it defines an isomorphism $\mathcal{V}_0^+ \oplus \mathcal{V}_0^-$ onto $\mathcal{V}_0^+ \oplus \mathcal{V}_1^+$. This can be extended to an isomorphism of $\mathcal{H}(\widetilde{\Lambda}_\mathcal{C})$ onto $\widetilde{\mathcal{H}}(\Lambda_\mathcal{C})$ preserving the columns of (44). Finally, the isomorphism $\mathcal{V}_0^+ \oplus \mathcal{V}_3^+$ onto $\mathcal{V}_0^+ \oplus \mathcal{T}_0^+$ thus constructed can be extended to an automorphism of $\widetilde{\mathcal{H}}(\widetilde{\Lambda}_\mathcal{C})$, providing the induced triality of this CFT.

We can define σ_3, as follows. For $1 \le j \le d$, there are weight one states $\xi_1^j = \frac{1}{\sqrt{2}} a_{-1}^j |0\rangle \in \mathcal{V}_0^-$, $\xi_2^j = \frac{i}{2}|\sqrt{2}\mathbf{e}_j\rangle + \frac{i}{2}|-\sqrt{2}\mathbf{e}_j\rangle \in \mathcal{V}_1^+$, $\xi_3^j = \frac{1}{2}|\sqrt{2}\mathbf{e}_j\rangle - \frac{1}{2}|-\sqrt{2}\mathbf{e}_j\rangle \in \mathcal{V}_1^-$, where $\mathbf{e}_j$ are the unit vectors in the directions of the axes of the $\mathbf{Z}^d$ lattice in (42), and corresponding currents $J^{ja}(z) = V(\xi_a^j, z)$, $1 \le a \le 3$, which generate an $su(2)^d$ affine algebra,

$$[J_m^{ja}, J_n^{kb}] = i\epsilon^{abc}\delta^{jk} J_{m+n}^{kc} + \tfrac{1}{2} m \delta_{m,-n}\delta^{ab}\delta^{jk} . \tag{45}$$

For each j, we define a rotation in the corresponding $su(2)$ which will send $J^{j1} \to J^{j2}$, $J^{j2} \to J^{j1}$ and $J^{j3} \to -J^{j3}$. This is achieved by a rotation through π about an axis equally inclined to the first and second axes:

$$\sigma^j = \exp\left\{ \frac{i\pi}{\sqrt{2}} \left(J_0^{j1} + J_0^{j2} \right) \right\} \qquad \text{and} \qquad \sigma_3 = \prod_{j=1}^{N} \sigma^j . \tag{46}$$

Then

$$\sigma_3 J_m^{j1} \sigma_3^{-1} = J_m^{j2} , \quad \sigma_3 J_m^{j2} \sigma_3^{-1} = J_m^{j1} , \quad \sigma_3 J_m^{j3} \sigma_3^{-1} = -J_m^{j3} . \tag{47}$$

σ_3 defines an automorphism of $\mathcal{H}(\Lambda)$: $\sigma_3 V(\psi, z)\sigma_3^{-1} = V(\sigma_3\psi, z)$, provided the cocyles are chosen appropriately.

Note that, although the individual $su(2)$'s, *i.e.* J_0^{ja} for fixed j, have half-integral spin representations on $\mathcal{H}(\Lambda)$, the diagonal group, with generators $\sum_j J_0^{ja}$, has only integral spins, because of the way the codewords correlate the occurrence of the half-integral spins. Thus $\sigma_3^2 = 1$.

It will turn out that we should modify the definition of σ_3 in the cases where d is an odd multiple of 8. In this case we change the definition of just one of the σ^j, $j = \ell$ say, by a further rotation through 2π (which is a central element of the corresponding $su(2)$),

$$\sigma^\ell = \exp\left\{ \frac{3i\pi}{\sqrt{2}} \left(J_0^{\ell 1} + J_0^{\ell 2} \right) \right\} , \tag{48}$$

and still define σ_3 as the product of the σ^j. This changes the action of σ_3 by a factor of -1 on the state with half-integral spin with repect to $J_0^{\ell a}$, leaving it unchanged on those with integral spin with respect to this $su(2)$. Then (47) still holds and we still have $\sigma_3^2 = 1$.

It can be shown that each of the sixteen spaces $\mathcal{V}_j^{\pm}$, $\mathcal{T}_j^{\pm}$, $0 \le j \le 3$, are irreducible as representation spaces for $\mathcal{V}_0^+ \equiv \mathcal{H}^+(\Lambda_0)$, *i.e.* $\{V(\psi, z) : \psi \in \mathcal{V}_0^+\}$, and that σ_3 maps $\mathcal{V}_0^+ \to \mathcal{V}_0^+$ and so maps this algebra of vertex operators onto itself.

Because $\sigma_3 \xi_1^j = \xi_2^j$, $\sigma_3 \xi_2^j = \xi_1^j$, $\sigma_3 \xi_3^j = -\xi_3^j$, we see that σ_3 maps $\mathcal{V}_0^+ \to \mathcal{V}_0^+$, $\mathcal{V}_0^- \to \mathcal{V}_1^+$, $\mathcal{V}_1^+ \to \mathcal{V}_0^-$ and $\mathcal{V}_1^- \to \mathcal{V}_1^-$. (Each of these spaces, is irreducible under $\mathcal{V}_0^+$, and this space is preserved by σ_3, it is only necessary to know what happens to one vector in each space to determine what happens to the whole of it.) σ_3 thus defines an isomorphism $\mathcal{H}(\Lambda_0) = \mathcal{V}_0^+ \oplus \mathcal{V}_0^- \to \mathcal{H}^+(\Lambda_C) = \mathcal{V}_0^+ \oplus \mathcal{V}_1^+$.

σ_3 defines an isomorphism $\mathcal{H}^+(\widetilde{\Lambda}_C) = \mathcal{V}_0^+ \oplus \mathcal{V}_3^+ \to \widetilde{\mathcal{H}}(\Lambda_0) = \mathcal{V}_0^+ \oplus \mathcal{T}_0^+$. It can be shown that $\mathcal{H}_T^+(\widetilde{\Lambda}_C) = \mathcal{T}_0^+ \oplus \mathcal{T}_3^+$ and $\widetilde{\mathcal{H}}(\Lambda_3) = \mathcal{V}_3^+ \oplus \mathcal{T}_3^+$ are equivalent as representations of $\mathcal{H}^+(\widetilde{\Lambda}_C)$ and $\widetilde{\mathcal{H}}(\Lambda_0)$, respectively, identified by σ_3. Since $\sigma_3(\mathcal{T}_0^+) = \mathcal{V}_3^+$, $\mathcal{T}_0^+$ corresponds to $\mathcal{V}_3^+$ and so the space $\mathcal{T}_3^+$ corresponds to $\mathcal{T}_3^+$. Thus $\sigma_3 : \mathcal{H}^+(\widetilde{\Lambda}_C) \to \widetilde{\mathcal{H}}(\Lambda_0)$ can be extended to an automorphism of $\widetilde{\mathcal{H}}(\widetilde{\Lambda}_C)$ and it can be established that this definition coincides with the definition of $\sigma_3 : \mathcal{T}_0^+ \to \mathcal{V}_3^+$ and $\sigma_3^2 = 1$. Thus we have the desired involution of $\widetilde{\mathcal{H}}(\widetilde{\Lambda}_C)$.

We can also define automorphisms σ_1 and σ_2 acting on $\mathcal{H}(\Lambda_C)$ by permuting ξ_1^j, ξ_2^j and ξ_3^j. σ_a, $a = 1,2,3$ generate a group isomorphic to the symmetry group of the cube, S_4, consisting of the 24 elements obtained by multiplying any of $1, \sigma_a, \sigma_1\sigma_2, \sigma_1\sigma_3$ by any of $1, \imath_b$, $a, b = 1,2,3$, where the automorphisms $\imath_b$ are defined by $\imath_1 = 1$ on $\mathcal{V}_0^+$ and $\mathcal{V}_0^-$, $\imath_1 = -1$ on $\mathcal{V}_1^+$ and $\mathcal{V}_1^-$, and $\imath_2, \imath_3$ are defined by cyclically permuting $\mathcal{V}_0^-$, $\mathcal{V}_1^+$ and $\mathcal{V}_1^-$. Then

$$\sigma_a \sigma_b \sigma_a = \imath_c \sigma_c \,, \qquad \imath_a \imath_b = \imath_c \,, \qquad \sigma_a \imath_a = \imath_a \sigma_a \,, \qquad \sigma_a \imath_b = \imath_c \sigma_a \,, \tag{49}$$

where a, b, c is any permutation of $1,2,3$. This group contains subgroups isomorphic to S_3, *e.g.* $\{1, \sigma_1, \sigma_2' = \sigma_2 \imath_2, \sigma_3, \sigma_1 \sigma_2', \sigma_1 \sigma_3\}$.

Note that, as an automorphism of $\widetilde{\mathcal{H}}(\widetilde{\Lambda})$, σ_a is defined up to an involution $\imath_a$, defined to be 1 on $\mathcal{V}_{02} \oplus \mathcal{V}_{a2}$ and -1 on $\mathcal{V}_{b2} \oplus \mathcal{V}_{c2}$ (where a, b, c is any permutation of $1,2,3$, as usual), so that $\imath_a \sigma_a = \sigma_a \imath_a$. It follows that, given a set of choices, either $\sigma_1 \sigma_3 \sigma_1$ equals either σ_2 or $\imath_2 \sigma_2$. In either case, $\sigma_1^2 = \sigma_3^2 = (\sigma_1 \sigma_3 \sigma_1)^2 = 1$, and the set $\{1, \sigma_1, \sigma_3, \sigma_1\sigma_3\sigma_1, \sigma_1\sigma_3, \sigma_3\sigma_1\}$ forms a group isomorphic to S_3, so that we have a triality group, S_3, of automorphisms of $\widetilde{\mathcal{H}}(\widetilde{\Lambda}_C)$. It is this triality group, or just the single element σ_3, which, together with Conway's group Co_1 (extended by a group associated with the cocycle operators), generates the Monster group, F_1.

Acknowledgement

The work described in this talk has been done in collaboration with Peter Goddard and Paul Montague.

4. REFERENCES

1. L. Dolan, P. Goddard, and P. Montague, Nucl. Phys. **B** *in press.*
2. I. Frenkel, J. Lepowsky and A. Meurman, *Vertex Operator Algebras and the Monster* (Academic Press, New York, 1988).
3. R. Griess, Invent. Math. **69** (1982) 1.
4. L. Dolan, P. Goddard, and P. Montague, Phys. Lett **B** *in press*; and *Conformal Field Theories, Representations and Lattice Constructions*, to appear.
5. E. Corrigan and T. Hollowood, Nucl. Phys. **B303** (1988) 135; Nucl. Phys. **B304** (1988) 77.
6. T. Hollowood, *Twisted Strings, Vertex Operators and Algebras* (Durham University Ph.D. Thesis, 1988).
7. D. Friedan, E. Martinec and S. Shenker, Nucl. Phys. **B271** (1986) 93.
8. P. Goddard, 'Meromorphic Conformal Field Theory' in *Infinite dimensional Lie algebras and Lie groups: Proceedings of the CIRM–Luminy Conference, 1988* (World Scientific, Singapore, 1989) 556.
9. A.A. Belavin, A.M. Polyakov and A.B. Zamolodchikov, Nucl. Phys. **B241** (1984) 333.
10. For a review, see G. Moore and N. Seiberg, *Lectures on RCFT* (1989) preprint RU-89-32, YCTP-P13-89.
11. E. Verlinde, Nucl. Phys. **B300** (1988) 360.
12. J.H. Conway and N.J.A. Sloane, *Sphere Packings, Lattices and Groups* (Springer-Verlag, New York, 1988).
13. A.N. Schellekens and S. Yankielowicz, Phys. Lett. **B226** (1989) 285.
14. P. Goddard, W. Nahm, D. Olive and A. Schwimmer, Comm. Math. Phys. **107** (1986) 179.

SPACE OF (2,2) STRING VACUA

Mirjam Cvetič*
Department of Physics
University of Pennsylvania
Philadelphia, PA 19104–6396

Abstract

Structure — connectedness, distance as specified by the Zamolodchikov metric, and boundaries — of $(2,2)$ superstring vacua is addressed. As an explicit example we show that the distance of the orbifold limit from the (blown up) orbifold is finite with a strictly positive definite metric. We also illustrate how the boundaries of the complex structure deformation effectively correspond to a "decompactified" string theory and are thus an infinite distance away in the space of $(2,2)$ superstring vacua.

CHAPTER 1: INTRODUCTION

When trying to connect superstring theories to the observable world, one inevitably stumbles on a problem; while starting with a fairly unique theory in ten dimensions, there are now many equally consistent, conformally invariant string vacua in four dimensions, ranging from the originally proposed Calabi-Yau manifolds to orbifolds, fermionic constructions, Gepner constructions, Kazama-Suzuki (coset) constructions, *etc.*.[1] Thus, one ends up with a vast variety of string vacua (models) with no principle to distinguish between them.

In these proceedings, A. Morozov[2] is proposing an approach to address this important question by studying the fluctuation away from the string models[2] (see Fig. 1). Here I am going to confine myself to the string models of two-dimensional theories in order to understand better how they are related to each other and, in particular, to shed light on the boundaries of string vacua. In that sense my approach is perpendicular to the one of Morozov's, since I am confining myself to a set of string vacua and studying their deformations in the z-direction of Fig. 1.

⋆ Supported by the Department of Energy contract # DOE–AC02–76–ERO–3071 and the University of Pennsylvania Research Foundation.

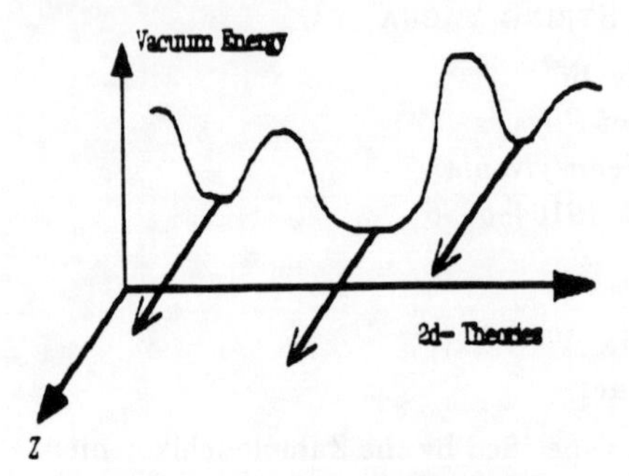

Fig. 1: Space of 2d Theories

Naively, one would think that string models form just a set of disconnected points (or maybe, at most, lines). However, there is a unique, generic feature of string vacua; namely, there is a large *degeneracy* in a certain class of string models. Such vacua are continuously connected, forming a multi-dimensional space. It is of vital importance to understand the structure of such vacua, including their boundaries in a stringy way, *i.e.* from the point of view of conformal field theory (CFT). This could eventually shed some light on the nature of the mechanism which chooses the preferred vacuum.

In my opinion, there are a few important problems that should be addressed within this context:

(*i*) **Effective Lagrangian of four-dimensional String Vacua.** One should gain insight into features of the effective Lagrangian in a stringy way. In particular, for supersymmetric theories, the structure of the Kähler potential (kinetic energy terms) and the superpotential for matter chiral supermultiplets should be well understood.

(*ii*) **Connectedness, Distance, and Boundaries of String Vacua.** This is an important problem which should be addressed within the context of CFT's. In particular I am going to concentrate on the question of distance (as specified by the Zamolodchikov metric) between different string vacua.

I am going to restrict myself to the string vacua with space-time supersymmetry. Such vacua possess at least $N = 2$ worldsheet supersymmetry[3] in the right-moving sector. Actually, when studying deformations, I will further restrict myself to $(2,2)$ vacua, *i.e.* those that along with $N = 2$ right-moving worldsheet supersymmetry possess as well $N = 2$ left-moving one. For any $(2,2)$ vacuum there are massless fields (scalar component of a chiral supermultipletwith conformal dimension $(1,1)$) – moduli M which have no potential, *i.e.* $V(M) \equiv 0$ to all orders in string loops.[4] Thus, any vacuum expectation value (VEV) of moduli still corresponds to the vacuum solution.

Moduli are classified according to the left-moving worldsheet $U(1)_l$ charge (associated with the left-moving $N = 2$ worldsheet supersymmetry), to have charge $+1$ and -1. For Calabi-Yau manifolds there is a one to one correspondence between scaling deformations of compactified spaces ($(1,1)$ forms) and complex structure deformations ($(2,2)$ forms) and moduli with $+1$ and -1 $U(1)_l$

charges, respectively. Also, moduli with $+1$ and -1 charge are in a one-to-one correspondence with **27** and $\overline{\mathbf{27}}$'s of the E_6 gauge group, repsectively.

From the point of view to CFT, there are points in moduli space where VEV's of most moduli are zero and CFT is solveable, *i.e.* correlation function of the corresponding operators can be calculated exactly at the string tree level (and potentially to all orders in string loops). There is now a stringy way to deform such vacua, *i.e.* by giving nonzero VEV's to moduli. Thus, the correlation function in the perturbed vacuum can be calculated exactly at the string tree level, but perturbatively in terms of VEV's of moduli. This is achieved by inserting an increasingly larger numbers of vertex operators of moduli.[5,6]

For example, Gepner's models[7] are points in the space of CFT's where most moduli have zero or special VEV's, namely not only the complex structure ((1,2) forms) and internal scaling deformations ((1,1) forms) are chosen symmetrically, but also the overall scale, *i.e.* radius R, is at a symmetric point with $R = \mathcal{O}(\sqrt{\alpha'})$, which is probably a dual point, $\frac{R}{\sqrt{\alpha'}} = \frac{\sqrt{\alpha'}}{\sqrt{2}R}$. Thus, it is hard to get the information about the boundaries of a particular Gepner model, since the study of moduli space within CFT can be done perturbatively even for radial deformations. On the other hand, for (2,2) orbifolds, CFT is defined exactly[8] on an arbitrarily deformed six-torus, T^6, on which the discrete point group of orbifolds acts as an automorphism while deformation of fixed points is treated perturbatively in terms of nonzero VEV's of the blowing-up modes.[5] Thus (blown-up) orbifolds are nontrivial examples where the boundaries of radial deformations can be studied explicitly by using the methods of CFT.

Another important question to address is connectedness of topologically distinct (2,2) CFT's. It turns out that the use of the Landau-Ginsburg potential is a useful method[9] to study topology change and the question of distance between different (topologically distinct) (2,2) compactifications. It was shown[10] by D. Kutasov and myself that such spaces are an infinite distance apart in the space of couplings of corresponding two-dimensional quantum field theories, where the metric is specified by the Zamolodchikov metric. This result seems to be different from the point-field limit approach when the distance in the space of metrics of Calabi-Yau manifolds was studied.[11]

In Chapter 2 we report results for the superpotential and the kinetic energy terms of the effective Lagrangian for the (2,2) string vacua. The study of kinetic energy terms for moduli naturally leads to the study of metric on moduli spaces which can be shown to be equivalent to the Zamolodchikov metric. The metric near the orbifold limit is calculated in Chapter 3 and a comparison with the point field limit is drawn. In Chapter 4 the boundaries of orbifold compactifcation are studied; in particula the complex structure deformation of the Z_4 orbifold is addressed.

CHAPTER 2: THE EFFECTIVE LAGRANGIAN

When addressing the effective potential, the natural thing is to study the effective superpotential of matter fields. While Yukawa couplings, *i.e.* three linear terms in the superpotential, have to be calculated explicitly, the nonrenormalizable sector, *i.e.* terms with more than three fields, possesses a special property. Namely, one can show[5,12] on general grounds that for the string vacua with overall scale $R >> \sqrt{\alpha'}$, such terms are exponentially suppressed by the scale of the compactified space R:

$$W_{\text{nonren}} \propto \frac{e^{-R^2/\alpha'}}{M_{\text{Pl}}^{k-3}} \phi_1 \ldots \phi_k \qquad k > 3 \ . \tag{1}$$

This is a clear signal of stringy effects in the study of the effective Lagrangian.

Other terms to be studied are the kinetic energy terms of matter fields.[13,14] In collaboration with J. Molera, J. Louis, and B. Ovrut, I addressed[13] this question by studying symmetries of the generating functional for string amplitudes. It turns out that such symmetries completely determine the structure of the kinetic energy terms for the untwisted sector of the Z_N orbifold.

However, when studying the twisted sector, one can do that only perturbatively in the VEV's – background values of the twisted fields. This naturally leads to the study of the Zamolodchikov metric[15] in the space of couplings of two-dimensional theories. Namely, kinetic energy of matter fields can be written as :

$$K = g_{ij}(\phi)\partial_\mu\phi_i\partial_\mu\phi_j \tag{2}$$

where $g_{ij}(\phi)$ defines the metric in the space of the fields, ϕ_i. On the other hand, the Zamolodchikov metric in a CFT is defined as :

$$g_{ij} = \lim_{w\to 0} \frac{1}{|w|^4} \left\langle V_{\phi_i(0)} V_{\phi_j(w)} e^{-S} \right\rangle \tag{3}$$

where V_{ϕ_i} are the vertex operators for the fields ϕ_i, which are physical fields with conformal dimension $(1,1)$, and S is the total string action with background fields for ϕ_i's turned on. It can be shown on general grounds[16] that the matrix in eqs. (2) and (3) are the *same*. This in turn implies that when studying the metric in the space of fields, one can as well use eq. (3) which is especially suitable when one is perturbing the total string action S around the free string action in terms of the background field-couplings, ϕ_i :[15]

$$S = S_0 + \phi_i \int d^2 z V_{\phi_i}. \tag{4}$$

We are now equipped to study the kinetic energy terms for all the fields in CFT. The results are exact in terms of CFT, however they are perturbative in terms of the background value of the fields. For example the connection between matter fields, *i.e.* **27** and $\overline{\mathbf{27}}$ of E_6, and moduli in the kinetic energy terms was calculated[17] for a class of orbifolds by using the above prescription. It explicitly confirms the equivalence of metrics (2) and (3).

Chapter 3 : The Metric in the Space of Moduli of (2,2) CFT's

The Zamolodchikov metric for moduli renders information about the structure of the space of moduli. This metric is on one side related to the kinetic energy of moduli (as physical fields in four dimensions) while on the other side it defines the distance in the space of the background values of moduli. Learning more about the metric of such a space could shed light on the boundaries of (2,2) vacua, *e.g.* this happens when corresponding background couplings of moduli $\to \infty$. On the other hand, this could perhaps illuminate the connectedness of different (2,2) string vacua.

Here I shall present the metric for the moduli of the sector associated with orbifold singularities (blowing-up modes) of Z_3 orbifold, done in part in collaboration with B. Ovrut. Calculation for all the Z_N orbifolds is completely analogous. As discussed in the introduction, this is done perturbatively in the background values of the blowing-up modes. For technical reasons one also takes the untwisted moduli (associated with the deformation of tori) to be at a special value $b_{ab} = \frac{\delta_{ab}}{2}$ for the three two-tori T^2.* Then the metric is evaluated as :

$$g_{i\bar{j}} = \lim_{w \to 0} \frac{1}{|w|^4} \left\langle V_{t_i}(0) V^*_{t_j}(w) e^{-S} \right\rangle, \tag{5}$$

where S is the total string action :

$$S = S_0 + \left[\delta t_i \int d^2 z V_{t_i}(z) + \delta b_{ab} \int d^2 z \left(\partial X^a \overline{\partial}\, \overline{X}^b + \overline{\psi}^a \overline{\partial} \psi^b + \overline{\lambda}^a \partial \lambda^b \right) + \text{c.c.} \right] \tag{6}$$

★ However, the metric for arbitrary background values of the untwisted moduli has been evaluated[13,14] by using the symmetry of the generating functional for the string amplitudes for the orbifolds.

and the free action is

$$S_0 = \frac{1}{2} \int d^2 z \left(\partial X^a \overline{\partial} \overline{X}^a + \overline{\psi}^a \overline{\partial} \psi^a + \overline{\lambda}^a \partial \lambda^a \right) + \text{c.c.}. \tag{7}$$

In equations (6–7), the sum over $i = 1, \ldots, 27$ (27 fixed points of the Z_3 orbifold) and $(a, b) = (1, 2, 3)$ (9 deformations of T^6 torus) are implied. Note that one necessarily has to include worldsheet supersymmetric partners of the bosonic coordinates ∂X^a and $\overline{\partial} \overline{X}^a$, which are the right-moving fermions ψ^a and the left-moving (gauge) fermions λ^a, respectively. Here the vertex operator for a blowing-up mode at a particular fixed point is :[8]

$$V_{t_i} = \sum_{I=1}^{3} \tau^I t^I \sum_{I'=1}^{3} \tilde{\tau}^{I'} \tilde{t}^{I'}, \tag{8}$$

where τ^I, $\tilde{\tau}^I$ are the excited right-moving and left-moving bosonic twist fields,[8] respectively defined through operator product expansion :

$$\begin{aligned} \partial X^I(z) \sigma_i(w) &= (z - w)^{k_I/N} \, \tau^I(w) \\ \overline{\partial} \overline{X}^I(\overline{z}) \tilde{\sigma}_i(\overline{w}) &= (\overline{z} - \overline{w})^{-k_I/N} \, \tilde{\tau}^I(\overline{w}) \end{aligned} \tag{9}$$

and t^I, $\tilde{t}^I$ are the corresponding right-moving and left-moving excited spin fields *e.g.* :

$$t^1 = e^{i[H_1(k_1/N - 1) + H_2 k_2/N + H_3 k_3/N]}. \tag{10}$$

Note that for the Z_3 orbifold, $\frac{k_i}{N} = \frac{1}{3}$, $(i = 1, 2, 3)$. Eq. (5) is evaluated perturbatively, in terms of background twisted fields δt_i and untwisted fields $b_{ab} = \left(\frac{1}{2}\delta_{ab} + \delta b_{ab}\right)$. In the lowest order one gets :

$$g^{(0)}_{i\bar{j}} = \lim_{w \to 0} \frac{1}{|w|^4} \left\langle V_{t_i}(0) V^*_{t_j}(w) e^{-S} \right\rangle = \delta_{ij}. \tag{11}$$

Though this result is to be expected since conformal fields are normalized so that their two-point function is normalized, actually the explicit evaluation through eq. (11) is more involved.

In the next order of the expansion in terms of δb_{ij}, one obtains :†

---

† The exact coefficient $-\delta_{ij}$ in front of $\mathrm{Tr}(\delta b + \delta b^\dagger)$ is consistent with the $N = 2$ four-dimensional supergravity space that moduli span.[18] It can be calculated from (12) only

$$\delta g^{(1)}_{i\bar{j}} = \delta b_{ab} \lim_{w\to 0} \frac{1}{|w|^4} \int d^2z \left\langle V_{t_i}(0) V^*_{t_j}(w) \left(\partial X^a \overline{\partial} \overline{X}^b + \overline{\psi}^a \overline{\partial} \psi^b \right) \right\rangle$$
$$+ \delta b^*_{ab} \lim_{w\to 0} \frac{1}{|w|^4} \int d^2z \left\langle V_{t_i}(0) V^*_{t_j}(w) \left(\partial \overline{X}^a \overline{\partial} X^b + \psi^a \overline{\partial}\, \overline{\psi}^b \right) \right\rangle$$
$$\delta g^{(1)}_{i\bar{j}} = -\, \delta_{ij}\, \mathbf{1}\, \mathrm{Tr} \left(\delta b + \delta b^\dagger \right). \tag{12}$$

Terms (11) and (12) add up to the metric :

$$g_{i\bar{j}} = \delta_{ij} \left(1 - \mathrm{Tr} \left(\delta b + \delta b^\dagger \right) \right)$$
$$\simeq \frac{\delta_{ij}}{\det \left(b + b^\dagger \right)}, \tag{13}$$

where we used $b_{ab} = \frac{1}{2}\delta_{ab} + \delta b_{ab}$. Eq. (13) can be written as :

$$g_{i\bar{j}} = \frac{\partial K}{\partial t_i \partial t^*_j},$$

where K is the Kähler potential :

$$K_{t\to 0} \simeq -\ln \left[\det \left(b + b^\dagger \right) - \sum_{i=1}^{27} t_i t^*_i \right] \tag{14}$$

to the lowest order in the t expansion.

A few comments are in order:

(*i*) The metric with Kähler potential (14) is strictly *positive* definite and *non-singular* at the orbifold limit ($t_i \to 0$).

---

by including *both* bosonic ∂X^a, $\overline{\partial} X^a$ *and* fermionic coordinates ψ^a, λ^a, respectively. In the evaluation of (12) one has to properly include the contact terms such as those from $\partial X^a \overline{\partial}\, \overline{X}^a$ correlators as well as properly taking care of the relevant O.P.E.'s in the presence of bosonic twist fields σ_i's. *Eg.*, the following expression is derived :

$$\frac{\left\langle \sigma(0) | \partial X^a(z_1) \overline{X}^a(z_2) | \sigma(w) \right\rangle}{\langle \sigma(0) | \sigma(w) \rangle} = \frac{1}{z_1 - z_2} \left(\frac{z_1 - w_1}{z_2 - w_2} \right)^{-(1-k_a/N)} \left(\frac{z_1}{z_2} \right)^{-k_a/N},$$

which, when it is integrated over $\int dz_2 \overline{\partial}_2$ properly, takes care of the contact terms and the cuts near the twist fields.

(*ii*) The distance from the points in moduli space corresponding to the nearby blown-up orbifolds to the point corresponding to the orbifold is *finite.*

(*iii*) Kähler potential (14) is certainly compatible with the $N = 2$ supergravity constraint,[19] however the expression is evaluated only up to first order in terms of background values of blowing-up modes t_i.

The most general Kähler potential compatible with $N = 2$ supergravity can be written[19] in terms of the holomorphic function $F(\phi)$ of moduli fields ϕ (those either associated with scaling or with complex structure deformations) as :

$$K = -\ln\left\{-\left[\frac{\partial F(\phi)}{\partial\phi_i} + \frac{\partial F^*(\phi)}{\partial\phi_i^*}\right](\phi_i + \phi_i^*) + 2\left(F(\phi) + F^*(\phi^*)\right)\right\} \tag{15}$$

while Yukawa couplings of the superpotential are related to the *same* $F(\phi)$ as :

$$W = \frac{\partial^3 F(\phi)}{\partial\phi_i \partial\phi_j \partial\phi_k}\tilde{\phi}_i\tilde{\phi}_j\tilde{\phi}_k. \tag{16}$$

Here $\tilde{\phi}_i$'s correspond to the matter fields, which are in one-to-one correspondence with the moduli; *i.e.*, $\tilde{\phi}_i$'s correspond to either **27**'s or $\overline{\mathbf{27}}$'s which are in a one-to-one correspondence to moduli with +1 and −1 $U(1)_l$ charges, respectively.

In our case Kähler potential (14) can be written as (15) where the holomorphic function F is of the form :

$$F_{t\to 0} = \epsilon_{abc}\epsilon_{a'b'c'}b_{aa'}b_{bb'}b_{cc'} + \frac{1}{4}\sum_{i=1}^{27} t_i^2. \tag{17}$$

At this stage we obtained $F_{t\to 0}$ by calculating the Zamolodchikov metric in the perturbative expansion of t_i background value of the blowing-up modes. However one can complete the expansion in terms of t for the case when $(b + b^*)/\sqrt{\alpha'} \gg 1$ and *for arbitrary values of* t without further calculation of the Zamolodchikov metric. Namely, in this case when $(b + b^*)/\sqrt{\alpha'} \gg 1$ one finds that also Yukawa couplings (16) for the twisted sector become b independent, *i.e.* $\frac{\partial^3 F}{\partial t_i \partial t_j \partial t_k} = \delta_{ij}\delta_{jk}$. Therefore in this case one can obtain the complete form for the holomorphic function F :

$$F_{b\to\infty} = F_{t\to 0} + \sum_{i=1}^{27} t_i^3, \tag{18}$$

which in turn determines the Kähler potential

$$K_{b\to\infty} = -\ln\left(\det\left(b+b^\dagger\right) - \sum_{i=1}^{27} t_i t_i^* - \sum_{i=1}^{27}(t_i+t_i^*)^3\right). \tag{19}$$

Let me emphasize again that this is the exact form of the Kähler potential for any value of the blowing-up modes as long as $(b+b^*)/\sqrt{\alpha'} \gg 1$.

I would now like to compare the forms (14) and (19) of the Kähler potential with the point-field limit result. Namely, the above Kähler potential is in apparent contradiction with the point-field limit result,[11] which is of the form :

$$K_{\left(\substack{b\to\infty \\ t\to\infty}\right)} = -\ln\left[\det(b+b^\dagger) - \sum_{i=1}^{27}(t_i+t_i^*)^3\right]. \tag{20}$$

Note that with $K_{\left(\substack{b\to\infty \\ t\to\infty}\right)}$ the metric in the twisted direction degenerates to $g_{ij}\equiv 0$ as the limit $t_i \to 0$ is taken. The resolution to the problem is the following.

In the point-field limit, one takes the background value of *all* the moduli associated with scaling deformations ($(1,1)$ forms) to be large compared to $\sqrt{\alpha'}$. Since the overall size as well as detailed structure of the Calabi-Yau manifold is much larger than $\sqrt{\alpha'}$. Thus, not only the size of the compactified space is taken to be very large; *i.e.* $(b+b^*)/\sqrt{\alpha'} \gg 1$, but also in this case $(t+t^*)/\sqrt{\alpha'} \gg 1$. In other words, one not only considers very large radii of tori, but also very large blowing-up. In this limit (20) is certainly compatible with (19), however it *is not valid when one takes* $t \to 0$ *!* In this case quadratic terms of t become important (see (14)). They are not detectable in the point-field limit approach, however they are calculable by using CFT perturbation.

Chapter 4: Complex Structure Deformations

Here I am going to report on preliminary work done in collaboration with J. Distler. One would like to study the boundaries of complex structure deformations. Motivation is two-fold. First one would like to see whether a generalization of duality picture emerges. Namely, the string may see complex structure deformation in the same way as scaling deformations. This would in turn imply that the boundaries of complex structure deformation correspond effectively to decompactification. Secondly, boundaries of complex structure deformation may be an infinite distance away in the space of couplings of two-dimensional field theories. This would in turn also be compatible with the notion of effective decompactification.

To study this phenomenon, it is most instructive to study a string vacuum that one can explicitly solve for arbitrary background values of the moduli corresponding to the complex structure deformation. The obvious example that sets in is an orbifold that allows for complex structure deformation of a torus, *e.g.*, the Z_4 orbifold, which is generated by a discrete rotation of $T^6 = T^2 \times T^2 \times T^2$ torus :

$$\Theta = \begin{bmatrix} i & & \\ & i & \\ & & -1 \end{bmatrix}. \tag{21}$$

Note that the third T^2 allows, along with the scale deformation, for complex structure deformations, whose contribution to the string action can be written as :

$$S^{(3)} = b_{(1,1)} \int d^2 z \partial X^3 \overline{\partial} \overline{X}^3 + b_{(1,2)} \int d^2 z \partial X^3 \overline{\partial} \overline{X}^3 + \text{c.c.}, \tag{22}$$

respectively. Note that the above parameters, $b_{(1,1)}$ and $b_{(1,2)}$ associated with corresponding moduli, are related to the parameters α and β of the torus (see Fig. 2).

$$\begin{aligned} X^3 &\rightarrow X^3 + \frac{1}{\alpha} = X^3 + \frac{1}{(\text{Im}\tau)^{1/2}} \\ X^3 &\rightarrow X^3 + \frac{1}{\alpha}(i+\beta) = X^3 + \frac{\tau}{(\text{Im}\tau)^{1/2}}. \end{aligned} \tag{23}$$

Note that the parametrization we chose is not standard. Here we restricted ourselves to complex structure deformations only, while keeping the overall "volume" constant ($\alpha\frac{1}{\alpha} = 1$). The boundary of the moduli space is reached when $\alpha \rightarrow 0$, and $\frac{\beta}{\alpha} \rightarrow 0$. This obviously leads to $b_{(1,2)} \rightarrow \infty$, while $\left|\frac{b_{(1,2)}}{b_{(1,1)}}\right| \leq 1$, thus letting a parameter $b_{(1,2)}$ in the string action go to infinity.

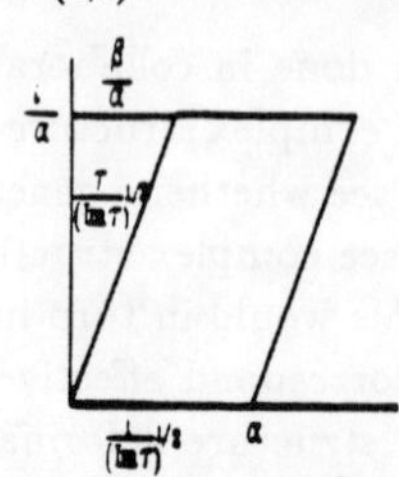

Fig 2: Parametrization of T^2 torus with fixed "volume".

It is known that for such a case,[10,13] distance in the space of the background couplings is infinitely far away (in other words, the Kähler potential $\rightarrow \infty$). On the other hand one can also see that there is an infinite number of states with internal momenta $p_L = p_R$ approaching zero, which clearly signals that the string states behave as if decompactification in the third complex direction of T^6 took place. Namely, the same phenomenon takes place as in the case when the overall volume of $T^2 \to \infty$, *i.e.* this direction is decompactified. This can be seen for the complex structure deformation by evaluating the mass levels of string states associated with the third complex direction of T^6 :

$$\frac{1}{2}p_L^2 = \frac{1}{4}\Bigg[m\alpha^2 m + \tilde{m}\alpha^2\tilde{m} + n\left(\frac{1}{\alpha^2} + \frac{\beta^2}{\alpha^2}\right)n + \\ 2\left(mn + \tilde{m}\tilde{n} + m\beta\tilde{n} + \tilde{m}\beta n\right)\Bigg] \tag{24a}$$

$$\frac{1}{2}p_R^2 = \frac{1}{4}\Bigg[m\alpha^2 m + \tilde{m}\alpha^2 m + n\left(\frac{1}{\alpha^2} + \frac{\beta^2}{\alpha^2}\right)n - \\ 2\left(mn + \tilde{m}\tilde{n} + m\beta\tilde{n} + \tilde{m}\beta n\right)\Bigg] \tag{24b}$$

$$p_L^2 - p_R^2 = 0. \tag{24c}$$

Here n and $\tilde{n}$ correspond to the eigenvalues of momenta while m and $\tilde{m}$ correspond to the winding numbers of $(1,2)$ and $(1,1)$ deformations, respectively. Thus as $\alpha \to \infty$, from (24a) one sees that one has to put winding number $n = 0$. Then (24c) implies $\tilde{n} = 0$ as well. In turn one concludes from (24a) and (24b) that now $p_L^2 = p_R^2 = 0$ for any m. Thus, we have seen that as $b_{(1,2)} \to \infty$ and the "volume" of the T^2 torus is kept constant, one obtains an infinite number of states whose mass levels $\to 0$.

What does the string vacuum at the boundary of this particular complex structure deformation correspond to? Preliminary study shows that the states associated with the twisted sector of the third complex direction of T^6 do not contribute to the physical spectrum of this string vacuum (essentially, they have *no* momentum in the third complex direction). We conjecture that the new vacuum corresponds to the singular K_3 space — four-dimensional Z_4 orbifold. This still has to be checked by comparing one-loop modular invariant partition functions.

I would like to thank my collaborators, J. Distler, D. Kutasov, and B. Ovrut, for many useful and fruitful discussions.

REFERENCES

1. For a review see, for example, L. Ibáñez, CERN–TH–5405/89 and references therein.
2. A. Morozov, these proceedings.
3. T. Banks, D. Friedan, L. Dixon, and E. Martinec, *Nucl. Phys.* **B299**, 613 (1988) .
4. See, for example, L. Dixon, *Proceedings of the Summer Workshop in High Energy Physics and Cosmology*, June 29–August 7, 1987, World Scientific, (J. C. Pati *et al.* eds.) and references therein.
5. M. Cvetič,*Proceedings of International Workshop on Superstrings, Cosmology, Composite Structure*, March 11 - 18, 1987, World Scientific (S. J. Gates, Jr. and R. N. Mohapatra eds.) and *Phys. Rev. Lett.* **59**, 1795 (1987).
6. A. Font, L. Ibáñez, P. Nilles, and F. Quevedo, *Phys. Lett.* **210B**, 101 (1988) .
7. D. Gepner, *Phys. Lett.* **199B**, 380 (1987) .
8. L. Dixon, D. Friedan, E. Martinec, and S. Shenker, *Nucl. Phys.* **B282**, 13 (1987) ; and S. Hamidi and C. Vafa, *Nucl. Phys.* **B279**, 465 (1987) .
9. C. Vafa and N. Warner, *Phys. Lett.* **218B**, 51 (1989) ; W. Lerche, C. Vafa and N. Warner, HUTP–89/A065; B. Greene, C. Vafa, N. Warner, *Nucl. Phys.* **B324**, 88 (1989) ; E. Martinec, *Phys. Lett.* **217B**, 431 (1989) ; V. G. Knizhnik Memorial volume, L. Brink editor.
10. M. Cvetič and D. Kutasov, PUPT–1157, UPR–410–T, (December 1989), to appear in *Phys. Lett. B*.
11. P. Candelas, P. Green, and T. Hübsch, *Proceedings of Strings 88 Meeting*, May 24 – 28, 1989, World Scientific, (S. J. Gates, Jr. *et al.* eds.) and Univ. of Texas Preprint, UTTG-10-89.
12. M. Cvetič, *Phys. Rev. Lett.* **59**, 2829 (1987) and *Phys. Rev. D* **37**, 2366 (1988) ; M. Dine and C. Lee, *Phys. Lett.* **203B**, 371 (1988) .
13. M. Cvetič, J. Louis, and B. Ovrut, *Phys. Lett.* **B206**, 229 (1988) and *Phys. Rev. D* **40**, 684 (1989) ; and M. Cvetič, J. Molera, and B. Ovrut, *Phys. Rev.* **D40**, 1140 (1989)
14. S. Cecotti, S. Ferrara, and L. Girardello, *Phys. Lett.* **213B**, 443 (1988) , *Nucl. Phys.* **B308**, 436 (1988) and *Int. J. Mod. Phys.* **A4**, 2475 (1989)
15. A. B. Zamolodchikov, *JETP Lett.* **43**, 730 (1986) .

16. N. Seiberg *Nucl. Phys.* **B303**, 286 (1988) ; M. Dine, P. Huet, and N. Seiberg *Nucl. Phys.* **B322**, 301 (1989) .

17. M. Cvetič, J. Louis, and B. Ovrut, *Phys. Rev. D* **40**, 684 (1989) .

18. L. Dixon, V. Kaplunovsky, and J. Louis, *Nucl. Phys.* **329B**, 27 (1990) .

19. E. Cremmer, C. Kounnas, A. van Proeyen, J. Deredinger, S. Ferrara, B. de Witt, and L. Girardello, *Nucl. Phys.* **B250**, 385 (1985) ; S. Ferrara and A. Strominger, *Proceedings of Strings '89 Workshop*, March 1989, World Scientific (R. Arnowitt *et al.* eds.).

WESS–ZUMINO–WITTEN FUSION RULES[1]

Mark A. Walton

Département de Physique, Université Laval
Québec, Canada G1K 7P4

Abstract

A simple method for calculating fusion rules in Wess-Zumino-Witten models is presented.

The importance of fusion rules in rational conformal field theories is by now well established. In general, however, they are difficult to calculate. To use the remarkable formula of Verlinde [1]:

$$N_{\lambda\mu}{}^{\nu} = \sum_{\varsigma} \frac{S_\lambda{}^{\varsigma} S_\mu{}^{\varsigma} S^{\dagger}{}_{\varsigma}{}^{\nu}}{S_0{}^{\varsigma}} \quad , \tag{1}$$

one must construct explicitly the matrix describing the transformation of the characters under the modular group generator $S : \tau \to -1/\tau$.

For a large class of rational conformal field theories, the coset conformal field theories [2], the S-transformation matrix may be constructed. It factors into a product of S-transformation matrices for Wess–Zumino–Witten (WZW) models, and Kač and Peterson have derived a formula for these [3]. The WZW primary fields are in one-to-one correspondence with the integrable highest weights $\lambda, \mu, \nu, \ldots$ of an affine Kač–Moody algebra $\hat{g}$. The Kač–Peterson formula involves a sum over the Weyl group $W(g)$ of the finite Lie algebra g that is the horizontal subalgebra of $\hat{g}$:

$$S_\lambda{}^{\mu} \propto \sum_{w \in W(g)} \epsilon(w) \, \exp\left\{ \frac{2\pi i}{k + h^\vee} (w(\lambda + \rho), \mu + \rho) \right\} . \tag{2}$$

(Here (λ, μ) is the usual Killing form, ρ is the sum of fundamental weights of $\hat{g}$, $h^\vee$ is the dual Coxeter nymber of g, and k is the level.) Since the Weyl group grows rapidly with rank, using this formula becomes difficult. Furthermore, the

[1] Work supported by NSERC of Canada and FCAR du Québec.

number of primary fields grows rapidly with increasing k, so that the dimension of the S-transformation matrix becomes very large.

The problem is alleviated by the existence of outer automorphisms of Kač–Moody algebras $\hat{g}$. Bernard [4] showed that the outer automorphisms, like the fusion rules, are diagonalised by the modular transformation S; if A is an element of the outer automorphism group $O(\hat{g})$ of $\hat{g}$, then

$$S_{A(\lambda)}{}^{\varsigma} = S_{\lambda}{}^{\varsigma} \exp\left[2\pi i \left(A\omega^0, \varsigma\right)\right] \quad . \tag{3}$$

Therefore, an outer automorphism A commutes with the fusion rules, $[A, N_\lambda] = 0$, i.e.

$$N_{\lambda A(\mu)}{}^{\nu} = N_{\lambda\mu}{}^{A(\nu)} \quad . \tag{4}$$

Using the symmetry of the fusion rules, we also have

$$N_{\lambda\mu}{}^{\nu} = N_{A(\lambda)A^{-1}(\mu)}{}^{\nu} = N_{A(\lambda)\mu}{}^{A(\nu)} \quad . \tag{5}$$

If we let $\tilde{\lambda}, \tilde{\mu}, \tilde{\nu}, \ldots$ be representatives of the orbits of $O(\hat{g})$, then all fusion rules are obtainable by equations (4,5) from the $N_{\tilde{\lambda}\tilde{\mu}}{}^{\tilde{\nu}}$. Moreover, using (3), we find that $N_{\tilde{\lambda}\tilde{\mu}}{}^{\tilde{\nu}}$ vanishes unless $\left(A\omega^0, \tilde{\lambda} + \tilde{\mu} - \tilde{\nu}\right) = 0 \bmod 1$; and then

$$N_{\tilde{\lambda}\tilde{\mu}}{}^{\tilde{\nu}} = \sum_{\tilde{\varsigma}} d(\tilde{\varsigma}) \frac{S_{\tilde{\lambda}}{}^{\tilde{\varsigma}} S_{\tilde{\mu}}{}^{\tilde{\varsigma}} S^{\dagger}{}_{\tilde{\varsigma}}{}^{\tilde{\nu}}}{S_0{}^{\tilde{\varsigma}}} \quad , \tag{6}$$

where $d(\tilde{\varsigma})$ is the dimension of the orbit of $\tilde{\varsigma}$ under the outer automorphism group $O(\hat{g})$. So it is only necesary to compute matrix elements of S for representatives of $O(\hat{g})$–orbits. But the calculations remain prohibitive, especially for large rank and level.

Here we show how to avoid these difficulties in the computation of WZW fusion rules [5]. There is no need to construct the S-transformation matrix. Furthermore, given two primary fields transforming as certain representations of the finite Lie algebra g, one calculation provides the fusion rules for all values of the level.

The method is motivated by the observation by Gepner and Witten [6] that the fusion rules in WZW models are in most cases exactly the coefficients $\bar{N}$ of tensor product decompositions:

$$\bar{\lambda} \otimes \bar{\mu} = \bar{N}_{\bar{\lambda}\bar{\mu}}{}^{\bar{\nu}}\, \bar{\nu} \quad (\text{sum over } \bar{\nu}) \; . \tag{7}$$

Here $\bar{\lambda}, \bar{\mu}, \bar{\nu}, \ldots$ are the highest weights of representations of g, and this equation says that the representation $\bar{\nu}$ appears $\bar{N}_{\bar{\lambda}\bar{\mu}}{}^{\bar{\nu}}$ times in the tensor product $\bar{\lambda} \otimes \bar{\mu}$. A simple procedure for calculating the tensor product coefficients $\bar{N}$ was described by Speiser [7]. We will first describe it with an example, illustrated in Figure 1. There the decomposition of the $su(3)$ tensor product $(20) \otimes (11)$ is computed.[2] First, the sum $\bar{\rho}$ of the fundamental weights of g is added to (20) to get (31). Then we add to this weight the weights $\{(11)\}$ of the representation (11). The resulting eight weights are shown. Denoted by dashed lines are the boundaries of the dominant Weyl sector. The weights (50) and (20) lie on a boundary, and so are ignored. The weight $(4,-1)$ lying outside the dominant sector is reflected back into the dominant sector by the Weyl reflection S_2 across the plane normal to the simple root α_2. It is reflected onto (31) and carries with it the sign of the reflection s_2, that is, -1. So it cancels one of the weights (31), and the remaining weights in the dominant sector yield the result. Subtracting $\bar{\rho} = (11)$ from them, we obtain $(20) \otimes (11) = (31) \oplus (12) \oplus (20) \oplus (01)$.

In general, the Speiser algorithm is as follows. One adds to $\bar{\lambda} + \bar{\rho}$ the set of weights $\{\bar{\mu}\}$ of the representation $\bar{\mu}$. Any weight $\bar{\lambda}+\bar{\rho}+\bar{\mu}'$, $\bar{\mu}' \in \{\bar{\mu}\}$, on the boundary of the dominant sector is ignored. Any in a sector other than the dominant one are reflected into the dominant sector, carrying along the signs of the Weyl group elements used. Then the highest weights of the tensor product decomposition are obtained by subtracting $\bar{\rho}$ from the resulting dominant weights.

Of course, it is not necessary to draw weight diagrams to use the Speiser method. So there is no obstruction to considering higher rank algebras, and the algorithm is simple and efficient.

Gepner and Witten also pointed out that the fusion rules differ from the $\bar{N}$ only because they must obey constraints imposed by the existence of null vectors for what they called "pseudospin". In fact, pseudospin is nothing but the $su(2)$ subalgebra in the direction of the simple root α_0 of $\hat{g}$. Now the $\bar{N}$ must obey similar constraints due to the existence of null vectors of the $su(2)$ subalgebras of g associated with the other simple roots of $\hat{g}$. But these must be taken care of by the Weyl reflections in the Speiser algorithm. So, one should try to incorporate the Weyl reflection s_0 across the hyperplane defined by α_0 into a Speiser-like algorithm. That is, the Weyl group $W(g)$ should be enlarged to the Weyl group $W(\hat{g})$ of the affine Kač–Moody algebra.

A natural modification of the example of Figure 1 is shown in Figure 2, where we choose the level $k = 2$. The finite projections of the $\hat{su}(3)$ weights are indicated, labelled by the full affine weights themselves. The boundaries of the finite projection of the affine dominant sector are drawn as dashed lines. Note that an extra boundary appears; it is the projection of the hyperplane with normal α_0. The weights are added in the same manner as above, and the zeroth Dynkin labels (coefficients of the fundamental weight ω^0) are adjusted so that the level of the resulting weights is $k + h^\vee = 5$. There are now three weights ($[320]$, $[050]$ and $[023]$) on boundaries, and an extra one $[-142]$ outside the dominant sector. It is reflected using s_0 into the

[2] If ω^1, ω^2 are the fundamental weights of $g = su(3)$, then $(ab) = a\omega^1 + b\omega^2$.

dominant sector and onto [131], so that both weights [131] are cancelled. Since [212] is the sole dominant weight remaining, the correct fusion rule is $[020] \bullet [011] = [101]$.

Note that essentially the same computation works for all values of the level k. In Figure 2, one only has to translate the boundary associated with s_0.

The general procedure is as follows. Denote by $\{\mu\}_0$ the set of level-zero affine weights having finite projection equal to $\{\bar{\mu}\}$. Then to find the fusion rules, one analyses $\lambda + \rho + \{\mu\}_0$. Again, weights on the boundary of the dominant (affine) Weyl sector are ignored, and those outside are reflected in with appropriate signs. In particular, any weight with zeroth Dynkin label less than zero may be reflected by s_0.

Unfortunately, we have no proof that the results are always the correct fusion rules. We have, however, the following evidence that this is so [5]. The algorithm reproduces the $\hat{su}(2)$ fusion rules [6] for all levels k. It yields the tensor product decompositions in the large k limit, as required. And we have verified that the results are correct for all simple untwisted Kač–Moody algebras at level one.

References

[1] E. Verlinde, Nucl. Phys. **B300** (1988) 389.

[2] P. Goddard, A. Kent and D. Olive, Phys. Lett. **B152** (1985) 88.

[3] V. Kač and D. Peterson, Adv. Math. **53** (1984) 125.

[4] D. Bernard, Nucl. Phys. **B288** (1987) 628.

[5] M. Walton, Laval University preprint (October, 1989).

[6] D. Gepner and E. Witten, Nucl. Phys. **B278** (1986) 493.

[7] D. Speiser, in *Lectures of the Istanbul Summer School of Theoretical Physics* (Gordon and Breach, New York, 1963), Vol. 1.

FIGURE 1. Speiser diagram for the $su(3)$ tensor product $(20) \otimes (11)$.

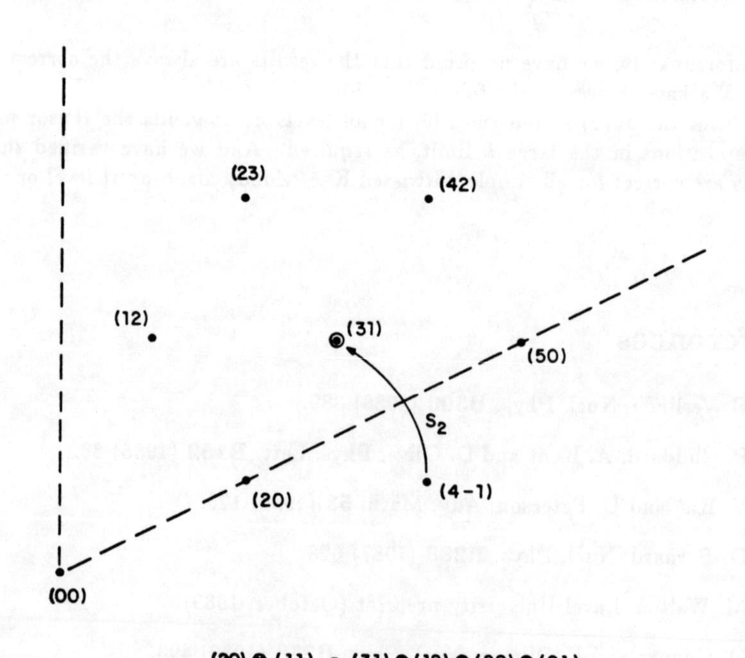

FIGURE 2. Diagram for computation of the $\hat{su}(3)$ fusion [020] • [011].

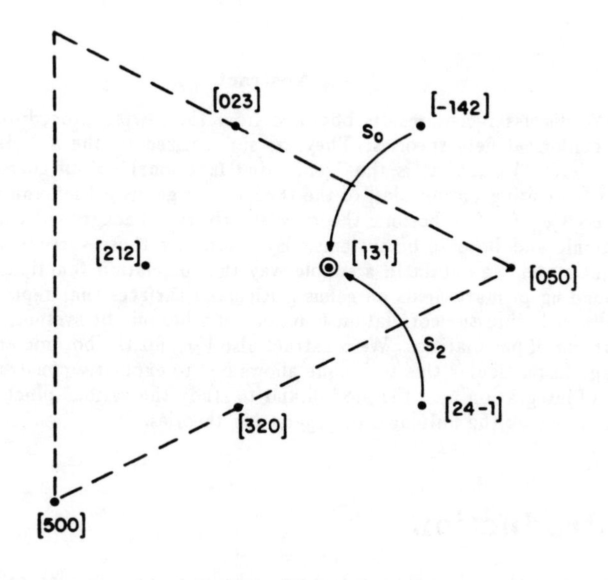

THE SEWING TECHNIQUE FOR STRINGS AND CONFORMAL FIELD THEORIES

P. Di Vecchia

NORDITA, Blegdamsvej 17, DK-2100 Copenhagen Ø, Denmark

Abstract

We discuss recent results obtained from the sewing procedure for strings and conformal field theories. They are summarized by the N Point [String] g loop Vertex $V_{N;g}$, that is the "generating functional" of all correlation functions [scattering amplitudes] of the theory on a genus g Riemann surface. We discuss $V_{N;g}$ for free bosonic theory with arbitrary background charge and for fermionic and bosonic bc systems. By saturating those vertices with highest weight states we obtain in a simple way the correlation functions of the corresponding primary fields on genus g Riemann surfaces that reproduce known results including the correlation functions of a bosonic bc system, that present a number of peculiarities. We construct also $V_{N;g}$ for the bosonic and fermionic string. In particular this technique allows one to explicitly construct the measure of integration over the moduli and to study the various pinching limits in order to check the finiteness of superstring theories.

1 Introduction

In this talk I want to discuss the sewing technique and summarize the results, that have been obtained in the last few years by means of this technique for string and conformal field theories.

In the case of a conformal field theory the starting point is the N Point Vertex $V_{N;0}$ that is the "generating functional" of all Green's functions of the theory on the sphere. Then applying the sewing procedure, that is explained in great detail in Refs. [1,2], where one can also find references to other approaches, one can compute the N Point g loop Vertex, that is the "generating functional" of all correlation functions of the theory on a genus g Riemann surface.

This vertex contains geometrical objects as Λ differentials, period matrix, prime form etc., that, however, are not obtained from geometrical considerations or from the knowledge of the theory of Riemann surfaces, but are instead explicitly derived in an algebraic way by applying the sewing procedure.

This technique has been successfully applied to free bosonic and fermionic systems [1,2] with arbitrary background charge.

It has recently been extended to minimal theories [3] and will presumably in the near future also be extended to Wess-Zumino-Witten models for which a free field representation has been recently found [4]. For those theory one can in fact use a free field representation being however careful not to allow all the states of the free theory to propagate in the internal lines. A truncation of the free theory must be performed as explained in detail for minimal theories by Felder [5].

There is some speculation that any rational conformal field theory can be obtained from a free theory after a suitable truncation.This means that the sewing technique could also be used in the future in order to compute correlation functions of arbitrary rational conformal field theories.

This technique has been also successfully applied for computing multiloop scattering amplitudes for strings. In this case $V_{N;0}[V_{N;g}]$ is the "generating functional" of all scattering amplitudes on the sphere[genus g Riemann surface].

It has the great advantage with respect to other methods to explicitly provide the integration measure on moduli space.

It does not give automatically, however, the domain of integration over the moduli, that must be fixed by hands as in the original calculation performed by Shapiro [6] of one loop in the closed bosonic string.

However the expressions for the scattering amplitudes are explicit enough that one could check if string theories are ultraviolet finite and compute for instance their Regge asymptotic behaviour.

The general structure of the Vertex for a theory containing a set of fields Φ is the following

$$V_N = \underbrace{\int dV(z_i \ldots)} < \Omega | \prod_i \delta(A_i) \exp\left\{ - \sum_{i,j=1}^{N} \oint_{z_i} du \oint_{z_j} dv \Phi_T^{(i)}(u) G(u,v) \Phi_T^{(j)} \right\} \quad (1.1)$$

where $G(u,v)$ is the Green's function of the theory, $< \Omega|$ is a suitable vacuum, the fields appearing in the exponential are the transported fields(defined in (2.12)) and the δ functions correspond to the zero modes of the theory in agreement with the Riemann-Roch theorem. In the case of a CFT one has to eliminate the integral appearing in front of the vertex, that instead must be there in the case of a string theory. The integral is meant over the punctures and moduli.

The talk is organized as follows.

In Sect. 2 we summarize the results obtained with the sewing technique for a free bosonic theory with a background charge and for fermionic and bosonic bc systems. We show in particular that the vertex $V_{N;g}$ reproduces the various correlators appearing in the literature for those theories.

In Sect. 3 we discuss the Vertex $V_{N;g}$ for the bosonic and fermionic string and the lacking steps toward a proof of finiteness of superstring theories at any order of perturbation theory.

Finally in Sect. 4 we briefly discuss a recently constructed Vertex $V_{N+2M;0}$ for the emission of N NS and $2M$ R states.

2 $V_{N;g}$ for conformal field theories

Let me start from a free bosonic theory with a background charge Q. It is described by the action

$$S[\varphi] = \frac{1}{2\pi}\int_{\Sigma} d^2z[-\bar{\partial}\varphi\partial\varphi - \frac{1}{4}Q\sqrt{g}R^{(2)}\varphi] \tag{2.1}$$

The N Point Vertex for this theory on the sphere has been written in Ref. [7] (See also Ref. [1]) and is given by:

$$V_{N;0} = \prod_{i=1}^{N}[\sum_{n_i} {}_i < n_i, O_a|]\delta[\sum_{i=1}^{N} N_i + Q]$$

$$\exp\left\{-1/2\sum_{i=1}^{N}\partial\varphi^{(i)}(z)[\alpha_0^{(i)} - Q]\log V_i'(z)\right\}$$

$$\exp\left\{\sum_{i<j}\oint_{z_i} du \oint_{z_j} [\partial\varphi^{(i)}]_T(u)\log(u-v)[\partial\varphi^{(j)}]_T(v)\right\} \tag{2.2}$$

where the transported fields are defined by

$$[\partial\varphi^{(i)}]_T(u) \equiv \partial_u\varphi^{(i)}[V_i^{-1}(u)] \tag{2.3}$$

and the projective transformation $V_i(z)$ corresponds to a choice of local coordinates around the puncture z_i and satisfies the property $V_i(0) = z_i$.

The δ function corresponds to the existence of a zero mode of the Laplace operator in agreement with the Riemann-Roch theorem. $N_i = -\alpha_0^{(i)}$ is the generator of $U(1)$ transformations for the i-th leg.

Finally the exponent contains the Green's function of the free scalar theory and the diagonal terms in the vertex are a consequence of the regularization procedure that preserves conformal invariance [8].

In the case of a bc system with background charge Q the Vertex is given by

$$V_{N;0} = \prod_{i=1}^{N}[{}_i\langle q = -Q|] \;:\; \exp\left\{-\sum_{i,j=1}^{N}\oint_{z_i} du \oint_{z_j} dv\; c_T^{(i)}(u)G(u,v)b_T^{(j)}(v)\right\} \;:$$

$$\prod_{r=1-\lambda}^{\lambda-1}\delta\left[\sum_{i=1}^{N}\oint_{z_I} dv\; b_T^{(i)}(v)\,v^{r+\lambda-1}\right] \tag{2.4}$$

The Green's function $G(u,v)$ is given by:

$$G(u,v)=\frac{<q=0|\prod_{i=1}^{2\lambda-1}e^{\epsilon\Phi(a_i)}b(u)c(v)|q=0>}{<q=0|\prod_{i=1}^{2\lambda-1}e^{\epsilon\Phi(a_i)}|q=0>}=\frac{\epsilon}{u-v}\prod_{i=1}^{2\lambda-1}\left(\frac{v-a_i}{u-a_i}\right) \qquad (2.5)$$

and $Q=\epsilon(1-2\lambda)$ with $\epsilon=1$ for a fermionic system and $\epsilon=-1$ for a bosonic system.

We have used the bosonization rules:

$$c(z)=e^{\Phi(z)} \qquad b(z)=e^{-\Phi(z)} \qquad (2.6)$$

for a fermionic system and

$$c(z)=e^{\Phi(z)}\eta(z) \qquad b(z)=e^{-\Phi(z)}\partial\xi(z) \qquad (2.7)$$

for a bosonic system. Remember that the scalar field $\Phi(z)$ has the contraction $<\Phi(z)\Phi(w)>=\epsilon\log(z-w)$.

The b and c fields appearing in (2.4) are the transported fields with conformal dimension equal to λ and $1-\lambda$ respectively:

$$c_T^{(i)}(u)\equiv[V_i^{-1'}(u)]^{1-\lambda}c^{(i)}[V_i^{-1}(u)] \qquad b_T^{(i)}(u)\equiv[V_i^{-1'}(u)]^{\lambda}b^{(i)}[V_i^{-1}(u)] \qquad (2.8)$$

The δ functions appearing in the Vertex correspond to the $(2\lambda-1)$ zero modes of the field c(z) required by the Riemann-Roch theorem on the sphere.

$V_{N;0}$ satisfies the following three very important properties:

1. Correlation functions

   The correlation functions of primary fields $\varphi_i(z_i)$ on the sphere are obtained by saturating $V_{N;0}$ with the corresponding highest weight states:

$$V_{N;0}\prod_{i=1}^{N}[V_i^{-1'}(0)]^{\Delta_i}|\varphi_i>_i=<0|\prod_{i=1}^{N}\varphi_i(z_i)|0> \qquad (2.9)$$

2. Overlap conditions

   The Vertex satisfies the following overlap conditions:

$$V_N[\phi_T^{(i)}(u)-\phi_T^{(j)}(u)]=0 \qquad (2.10)$$

   for any primary field $\phi(z)$ with conformal dimension Δ and arbitrary i and j and

$$V_N[\sum_{j=1}^{N}\oint_{z_i}duF(u)\phi_T^{(i)}(u)]=0 \qquad (2.11)$$

   where $F(u)$ is an arbitrary function and the transported field is defined by

$$\phi_T^{(i)}(u)\equiv[V_i^{-1'}(u)]^{\Delta}\phi^{(i)}[V_i^{-1}(u)] \qquad (2.12)$$

3. Factorization

If we sew together two vertices V_{N_1+1} and V_{N_2+1} after the insertion of a sewing operator we obtain the composite vertex $V_{N_1+N_2}$.

Once $V_{N;0}$ for a certain conformal field theory is known, the sewing procedure amounts to compute in a straightforward way $V_{N;g}$ that is the "generating functional" for all the correlation functions on a genus g Riemann surface.

$V_{N;g}$ is given by

$$V_{N;g} = \prod_{\mu=1}^{g} Tr_{(2\mu-1,2\mu)} \left[V^{\dagger}_{(N+2g);0} \prod_{\mu=1}^{g} P(x_\mu) \right] \qquad (2.13)$$

where for notational simplicity we label the first N legs of the vertex in the r.h.s. with an index i, running from 1 to N, and divide the remaining $2g$ legs into "odd" legs, labelled by $2\mu - 1$, and "even" legs labelled by 2μ with $\mu = 1, \ldots, g$. The expression (2.13) means that we sum over all states circulating in the loop by taking the trace in the Hilbert spaces $2\mu - 1$ and 2μ, that are identified after the insertion of a sewing operator $P(x_\mu)$.

Since the original vertex $V_{N+2g;0}$ depends on the $(N + 2g - 3)$ variables of the punctures, after the insertion of g sewing operators depending on g additional variables x_μ we get a vertex $V_{N;g}$, that depends on $[N + 3(g - 1)]$ variables corresponding to the N punctures of the external legs and to the $3g - 3$ moduli of a Riemann surface of genus g.

The computation of the traces in (2.13) is in general rather laborious, although straightforward. The technical details can be found in Refs. [1,2]. Here we give just the results for various theories.

In the case of a free scalar theory described by action (2.1) one gets the following N-point g-loop vertex :

$$\begin{aligned} V_{N;g} = & [\det \bar{\partial}_0]^{-1/2} \prod_{i=1}^{N} [\sum_{n_i} {}_i\langle n_i, O_a|]\ \delta(\sum_{i=1}^{N} N_i - (g-1)Q) \\ & \exp\left\{ -\frac{1}{2} \sum_{i}^{N} \oint_0 dz \partial\varphi^{(i)}(z) [\alpha_0^{(i)} - Q] \log[V_i'(z)] \right\} \\ & \exp\left\{ \frac{1}{2} \sum_{i,j=1}^{N} \oint_0 dz \oint_0 dy \partial\varphi^{(i)}(z) \log \frac{E(V_i(z), V_j(y))}{V_i(z) - V_j(y)} \partial\varphi^{(j)}(y) \right\} \\ & \exp\left\{ \frac{1}{2} \sum_{\substack{i,j=1 \\ i \neq j}}^{N} \oint_0 dz \oint_0 dy \partial\varphi^{(i)}(z) \log(V_i(z) - V_j(y)) \partial\varphi^{(j)}(y) \right\} \\ & \left[\Theta \begin{pmatrix} \alpha \\ \beta \end{pmatrix} \left(\left[\frac{1}{2\pi i} \sum_{i=1}^{N} \oint_0 dz \partial\varphi^{(i)}(z) [\int_{z_0}^{V_i(z)} \omega_\mu] - Q\Delta_\mu^{z_0} \right] |\tau \right) \right] \end{aligned}$$

$$\exp\left\{Q\sum_{i=1}^{N}\oint_0 dz\partial\varphi^{(i)}(z)\log\sigma\left[V_i(z)\right]\right\} \tag{2.14}$$

We have introduced the standard Riemann Θ-function with characteristics α and β, given by

$$\Theta\begin{pmatrix}\alpha\\ \beta\end{pmatrix}(z|\tau)=\sum_{\{n_\mu\}}\exp 2\pi i\left\{\sum_{\mu,\nu=1}^{g}\frac{1}{2}(n_\mu+\alpha_\mu)\tau_{\mu\nu}(n_\nu+\alpha_\nu)+\sum_{\mu=1}^{g}(n_\mu+\alpha_\mu)(z_\mu+\beta_\mu)\right\} \tag{2.15}$$

with n_μ integer.

The vertex $V_{N;g}$ can be just expressed in terms of geometrical objects as the period matrix $\tau_{\mu\nu}$, the holomorphic differentials ω^μ, the vector of Riemann constants $\Delta_\mu^{z_0}$, the function σ and the prime form $E(u,v)$. Their explicit expression in terms of the Schottky parametrization of the Riemann surface can be found in Ref. [1]. Finally the determinant is given in this parametrization by

$$[\det\bar{\partial}_0]^{-1/2}=\prod_\alpha{}'\prod_{n=1}^{\infty}\left(\frac{1}{1-k_\alpha^n}\right) \tag{2.16}$$

where the product is over the primary classes of the Schottky group.

By saturating the Vertex (2.14) with the states

$$|q_i>_i\equiv\lim_{z\to z_i}[V_i^{-1\prime}(z)]^{\frac{1}{2}q_i(q_i+Q)}:\exp\{q_i\varphi[V_i^{-1}(z)]\}:|0>_i \tag{2.17}$$

for $i=1,\ldots,N$ with $q_i=\pm1$ one gets the following correlation functions

$$<q=0|\prod_{i=1}^{N_1}:e^{-\varphi(z_i)}:\prod_{h=1}^{N_2}:e^{\varphi(w_h)}:|q=0>_g=$$

$$\prod_\alpha{}'\prod_{n=1}^{\infty}\left(\frac{1}{1-k_\alpha^n}\right)\frac{\prod_{i,j=1,\,i<j}^{N_1}E(z_i,z_j)\prod_{h,k=1,\,h<k}^{N_2}E(w_h,w_k)}{\prod_{i=1}^{N_1}\prod_{h=1}^{N_2}E(z_i,w_h)}\frac{\prod_{h=1}^{N_2}\sigma(w_h)^Q}{\prod_{i=1}^{N_1}\sigma(z_i)^Q}\delta(N_1-N_2+(g-1)Q)$$

$$\left[\Theta\begin{pmatrix}\alpha\\ \beta\end{pmatrix}\left(\left[-\frac{1}{2\pi i}\sum_{i=1}^{N_1}\int_{z_0}^{z_i}\omega_\mu+\frac{1}{2\pi i}\sum_{h=1}^{N_2}\int_{z_0}^{w_h}\omega_\mu-Q\Delta_\mu^{z_0}\right]|\tau\right)\right] \tag{2.18}$$

reproducing the result obtained with other methods [9]. Notice that both the Vertex (2.14) and the correlation functions (2.18) involve arbitrary spin structures.

As discussed in Ref. [9] (2.18) transforms as a differential of rank λ $[1-\lambda]$ in the variables z_i $[w_h]$. This is a consequence of the fact that the prime form contains the 1/2 differentials and that σ transforms as a $\frac{g}{2}$ differential. (2.18) has poles when $z_i \to w_h$ for any i and h and zeroes when $z_i \to z_j$ or $w_h \to w_k$. It has therefore the right structure for representing the following fermionic correlation functions:

$$< q = 0| \prod_{i=1}^{N_1} b(z_i) \prod_{h=1}^{N_2} c(w_h) |q = 0 > \tag{2.19}$$

The equality between (2.18) and (2.19) is an obvious consequence of the bosonization rules (2.6).

In the case of a bosonic and fermionic bc system one can also compute $V_{N;g}$ with the sewing procedure. The detailed calculations are explained in detail in Ref. [2] and, apart the exceptional cases of $\lambda = 1$ and $g = 1$ for integer λ that must be treated separately[See Ref [2] for details], the result is

$$V_{N;g} = (\det \bar{\partial}_{1-\lambda})^{\epsilon} \prod_{i=1}^{N} [{}_i\!< q = -Q|] \; : \; \exp\left\{ -\sum_{i,j=1}^{N} \oint_{z_i} du \oint_{z_j} dv \; c_T^{(i)}(u) G(u,v) b_T^{(j)}(v) \right\} \; : \\ \prod_{a=1}^{(2\lambda-1)(g-1)} \delta\left[\sum_{i=1}^{N} \oint_{z_i} du \; c_T^{(i)}(u) \Lambda_a(u) \right] \tag{2.20}$$

where $G(u,v)$ is the Green's function of the (b,c)-system on a Riemann surface of genus g and $(\det \bar{\partial}_{1-\lambda})$ gives the correct normalization. The subindex of $\bar{\partial}$ means that $\bar{\partial}$ acts on conformal fields with dimension equal to $1-\lambda$.

In this case the calculation of $V_{N;g}$ has been explicitly performed only for those spin structures that do not involve spin field exchanges in the loops. However the structure of the Vertex is always given by (2.20) except exceptional cases as $\lambda = 1$, $g = 1$ for integer λ and $\lambda = 1/2$ for odd spin structures.

The determinant in the Schottky parametrization is given by:

$$\det \bar{\partial}_{1-\lambda} = \prod_{\mu=1}^{g} \prod_{n=1-\lambda}^{\lambda-1} (1 - k_\mu^n) \prod_{\alpha}{}' \prod_{n=\lambda}^{\infty} [1 - (-1)^{N_\alpha} k_\alpha^n][1 - (-1)^{-N_\alpha} k_\alpha^n] \tag{2.21}$$

for those spin structures that do not involve spin field exchanges in the loops. $N_\alpha = \sum_{\mu=1}^{g} N_\alpha^\mu (1/2 - \beta_\mu)$ where N_α^μ denotes the number of factors S_μ in the generic element of the Schottky group and β_μ labels the spin structures around the B cycles.

Moreover, we have the product of $(2\lambda - 1)(g-1)$ linear terms in the c-fields which saturate precisely the $(2\lambda-1)(g-1)$ zero modes of the b-fields, parametrized by the Λ-differentials $\Lambda_a(u)$. This particular structure is clearly in agreement with the Riemann-Roch theorem for genus g.

The explicit expressions of the Λ differentials and of the Green's function in the Schottky parametrization of the Riemann surface can be found in Ref. [2] for those spin structures that do not involve spin field exchanges in the loops.

The correlation functions (2.19) can also be computed in the fermionic theory described by the Vertex (2.20). They are obtained by saturating (2.20) with the following states:

$$\lim_{z\to z_i} b_T^{(i)}(z)|q=0>_i \qquad i=1,2,\dots,N_1$$

$$\lim_{z\to z_j} c_T^{(j)}(z)|q=0>_j \qquad j=1,2,\dots,N_2 \tag{2.22}$$

The result is

$$<q=0|\prod_{i=1}^{N_1} b(z_i)\prod_{h=1}^{N_2} c(w_h)|q=0>=\det\bar{\partial}_{1-\lambda}\det B \tag{2.23}$$

where B is the following matrix

$$B=\begin{pmatrix} \Lambda_1(z_1) & \Lambda_1(z_2) & \dots & \Lambda_1(z_{N_1}) \\ \Lambda_2(z_1) & \Lambda_2(z_2) & \dots & \Lambda_2(z_{N_1}) \\ . & . & \dots & . \\ . & . & \dots & . \\ \Lambda_N(z_1) & \Lambda_N(z_2) & \dots & \Lambda_N(z_{N_1}) \\ G(z_1,w_1) & G(z_2,w_1) & \dots & G(z_{N_1},w_1) \\ . & . & \dots & . \\ . & . & \dots & . \\ G(z_1,w_{N_2}) & G(z_2,w_{N_2}) & \dots & G(z_{N_1},w_{N_2}) \end{pmatrix} \tag{2.24}$$

and $N\equiv N_1-N_2=(2\lambda-1)(g-1)$.

Since the Λ differentials transform as conformal tensors of rank λ and the Green's function transforms as a conformal tensor of rank $\lambda[1-\lambda]$ in the variables $z_i[w_h]$ it is easy to check that the correlation functions (2.23) transform as conformal tensors of rank $\lambda[1-\lambda]$ in the variables $z_i[w_h]$. In addition they are vanishing when $z_i\to z_j$ or $w_h\to w_k$ for any choice of the indices and, because of the presence of the Green's function, they have simple poles when $z_i\to w_h$ (for any i and h).

Bosonization implies that (2.18), (2.19) and (2.23) must be equal. They are certainly equal for $\lambda=1/2$ as a consequence of eq. (4.33) at p. 33 of Fay's book [10]. For arbitrary λ we have not been able to check directly their identity.

Finally the Vertex for a bosonic bc system can also be used for computing correlation functions of this theory. For instance the correlation functions:

$$<q=0|\prod_{i=1}^{N_1}\delta[\beta(z_i)]\prod_{j=1}^{N_2}\delta[\gamma(w_j)]|q=0> \tag{2.25}$$

can be computed by saturating the Vertex with the following highest weight states:

$$[V_i^{-1\prime}(0)]^{-\lambda} \int \frac{dp_i}{2\pi} e^{ip_i\beta^{(i)}_{-\lambda}} |q=0>_i \qquad i=1...N_1$$

$$[V_i^{-1\prime}(0)]^{\lambda-1} \int \frac{dq_j}{2\pi} e^{iq_j\gamma^{(j)}_{1-\lambda}} |q=0>_j \qquad j=1...N_2 \tag{2.26}$$

corresponding to the primary fields in (2.25) with $N_1 - N_2 = (2\lambda-1)(g-1)$.

The calculation is straightforward and one obtains the inverse of the correlation function (2.23) for the fermionic bc system. Then, by using the identity between (2.18) and (2.23) one gets the same expression as in the free bosonic theory with background charge where now, however, the bosonic field has "wrong" metric.

More complicated correlation functions can also be computed [11] obtaining agreement with known results [12,13].

3 $V_{N;g}$ for string theories

The sewing procedure can also be applied for computing multiloop amplitudes in string theories.

In the bosonic string the result has been known for some time and has been obtained by letting both the orbital and the ghost degrees of freedom to propagate in the loops. The ghost degrees of freedom of the external legs are then eliminated by saturating the resulting vertex with the ghost state $c_1^{(i)}|q=0>_i$ for $i=1,2..N$, that is the ghost part of a representative of the homology classes of physical states annihilated by the BRST charge Q.

In this way one arrives at the following $V_{N;g}$, that summarizes all scattering amplitudes of the closed bosonic string with an arbitrary number of loops [14,15,16,17]

$$V_{N;g} = \int dV \prod_{i=1}^{N} [\int d^{26}p_i \; {}_i\!<p_i, O_a, O_{\bar{a}}|] \; \delta(\sum_{i=1}^{N} p_i)$$

$$\exp\left\{-\frac{i}{2}\sum_i^N \oint_0 dz\partial x^{(i)}(z)\alpha_0^{(i)} \log[V_i'(z)]\right\} \exp\left\{-\frac{i}{2}\sum_i^N \oint_0 d\bar{z}\partial\bar{x}^{(i)}(\bar{z})\bar{\alpha}_0^{(i)} \log[V_i'(\bar{z})]\right\}$$

$$\exp\left\{-\frac{1}{2}\sum_{i,j=1}^{N}{}' \oint_{z_i} du \oint_{z_j} dv [\partial x^{(i)}]_T(u) \log E(u,v) [\partial x^{(j)}]_T(v)\right\}$$

$$\exp\left\{-\frac{1}{2}\sum_{i,j=1}^{N}{}' \oint_{\bar{z}_i} d\bar{u} \oint_{\bar{z}_j} d\bar{v} [\partial \bar{x}^{(i)}]_T(\bar{u}) \log E(\bar{u},\bar{v}) [\partial \bar{x}^{(j)}]_T(\bar{v})\right\}$$

$$\exp\left\{\sum_{i,j=1}^{N} \left[\oint_{z_i} du [\partial x^{(i)}]_T(u) \int^u \omega^\mu + \oint_{\bar{z}_i} d\bar{u} [\partial\bar{x}^{(i)}]_T(\bar{u}) \int^{\bar{u}} \bar{\omega}^\mu\right] [\pi Im\tau]^{-1}_{\mu\nu}\right.$$

$$\left[\oint_{z_j} dv[\partial x^{(j)}]_T(v) \int^v \omega^\nu + \oint_{\bar{z}_j} d\bar{v}[\partial \bar{x}^{(j)}]_T(\bar{v}) \int^{\bar{v}} \bar{\omega}^\nu\right]\Bigg\} \tag{3.1}$$

$\sum'$ means that we have only the terms $i \neq j$ in the case of the contribution of the identity in the prime form and the transported fields are given by

$$[\partial x^{(i)}]_T(u) \equiv \partial_u x^{(i)}[V^{-1}(u)] \tag{3.2}$$

The measure of integration

$$dV = \prod_{\mu=1}^{g}\left[\frac{d^2k_\mu d^2\eta_\mu d^2\xi_\mu}{|k_\mu|^2|\xi_\mu - \eta_\mu|^4}\right]\prod_{i=1}^{N}\left[\frac{d^2z_i}{|V_i'(0)|^2}\right]\frac{1}{dV_{abc}}$$

$$\prod_{\mu=1}^{g}\left[\frac{|1-k_\mu|^4}{|k_\mu|^2}\right]\prod_\alpha{}'\left[\prod_{n=1}^{\infty}|1-k_\alpha^n|^{-52}\prod_{n=2}^{\infty}|1-k_\alpha^n|^4\right](\det(Im\tau))^{-13} \tag{3.3}$$

consists of a piece coming from the integration measure of the propagators and of the Koba-Nielsen variables of the original vertex and of another piece that corresponds to the contribution of the determinant coming from the orbital modes (2.16)

$$|\det\bar{\partial}_0|^{-26} = \prod_\alpha{}'\prod_{n=1}^{\infty}|1-k_\alpha^n|^{-52} \tag{3.4}$$

and from the ghost modes:

$$|\det\bar{\partial}_{-1}|^2 = \prod_{\mu=1}^{g}\left[\frac{|1-k_\mu|^4}{|k_\mu|^2}\right]\prod_\alpha{}'\prod_{n=2}^{\infty}|1-k_\alpha^n|^4 \tag{3.5}$$

The divergences of the bosonic string can be read immediately in the measure since they correspond to the singular behaviour in the pinching limits $k_\mu \to 0$ and $\xi_\mu \to \eta_\mu$ for $\mu = 1, 2...g$.

The integration measure (3.3) has also been explicitly computed by using the path integral technique in the case of two and three loops.

In fact in this case the moduli space can be parametrized with the independent elements of the period matrix and the volume of integration is given by [18]:

$$dV = \prod_{i=1}^{N}\left[\frac{d^2z_i}{|V_i'(0)|^2}\right]\frac{1}{dV_{abc}}\frac{d^2\tau_{11}d^2\tau_{12}d^2\tau_{22}}{[\det(Im\tau)]^{13}}\prod_{\alpha,\beta}\left|\Theta\begin{pmatrix}\alpha\\\beta\end{pmatrix}(0|\tau)\right|^{-4} \tag{3.6}$$

for $g = 2$. An analogous expression can also be written for $g = 3$.

By expanding for some small moduli it has been shown [19] that the two expressions (3.3) and (3.6) are coincident.

$V_{N;g}$ for the bosonic string given by (3.1) has been extended to the case of the superstring only for a subset of 2^g out of the 2^{2g} spin structures, that correspond to the exchanges of NS states in the loops by using a superfield formalism.

It is given by [20,21]:

$$V_{N;g} = \int dV \prod_{i=1}^{N} [\int d^{10}p_i \; {}_i\langle p_i, O_a, O_{\bar{a}}, O_{\psi}, O_{\bar{\psi}} |] \; \delta(\sum_{i=1}^{N} p_i)$$

$$\exp\left\{ -\frac{i}{2} \sum_{i}^{N} \oint_0 dz D X^{(i)}(Z) \alpha_0^{(i)} \log[D V_i^{\theta}(Z)] \right\}$$

$$\exp\left\{ -\frac{i}{2} \sum_{i}^{N} \oint_0 d\bar{z} \bar{D} \bar{X}^{(i)}(\bar{Z}) \bar{\alpha}_0^{(i)} \log[\bar{D} \bar{V}_i^{\theta}(\bar{Z})] \right\}$$

$$\exp\left\{ -\frac{1}{2} \sum_{i,j=1}^{N}{}' \oint_{z_i} dZ \oint_{z_j} dW [DX^{(i)}]_T(Z) \log \mathcal{E}(Z,W) [DX^{(j)}]_T(W) \right\}$$

$$\exp\left\{ -\frac{1}{2} \sum_{i,j=1}^{N}{}' \oint_{\bar{z}_i} d\bar{Z} \oint_{\bar{z}_j} d\bar{W} [\bar{D}\bar{X}^{(i)}]_T(Z) \log \mathcal{E}(\bar{Z},\bar{W}) [\bar{D}\bar{X}^{(j)}]_T(\bar{W}) \right\}$$

$$\exp\left\{ \sum_{i,j=1}^{N} \left[\oint_{z_i} dZ [DX^{(i)}]_T(Z) \int^{Z} \Omega^{\mu} + \oint_{\bar{z}_i} d\bar{Z} [\bar{D}\bar{X}^{(i)}]_T(\bar{Z}) \int^{\bar{Z}} \bar{\Omega}^{\mu} \right] [\pi \hat{T}]^{-1}_{\mu\nu} \right.$$

$$\left. \left[\oint_{z_j} dW [DX^{(j)}]_T(W) \int^{W} \Omega^{\nu} + \oint_{\bar{z}_j} d\bar{W} [\bar{D}\bar{X}^{(j)}]_T(\bar{W}) \int^{\bar{W}} \bar{\Omega}^{\nu} \right] \right\} \tag{3.7}$$

where $X(Z)$ and $\bar{X}(\bar{Z})$ are superfields depending from the supercoordinates $Z \equiv (z, \theta)$ and $\bar{Z} \equiv (\bar{z}, \bar{\theta})$ respectively and the supersymmetric covariant derivative is defined by

$$D = \frac{\partial}{\partial \theta} + \theta \frac{\partial}{\partial z} \tag{3.8}$$

$\mathcal{E}$ is the super prime form, Ω^{μ} are the superholomorphic differentials and $T_{\mu\nu}$ is the superperiod matrix defined in Ref. [20]. Finally [21]

$$\hat{T} = \frac{1}{2i} [T_{\mu\nu} - \bar{T}_{\mu\nu}] \tag{3.9}$$

and the volume of integration is given by [21,20]

$$dV = \prod_{\mu=1}^{g} \left[\frac{d^2 K_{\mu} d^2 U_{\mu} d^2 V_{\mu}}{|K_{\mu}|^2 |U_{\mu} - V_{\mu}|^2} \right] \prod_{i=1}^{N} \left[\frac{d^2 Z_i}{|D V_i^{\theta}(0)|^2} \right] \frac{1}{dV_{abc}}$$

$$(\det \hat{T})^{-5} \prod_{\mu=1}^{g} \left[\frac{|1 - K_{\mu}|^4}{|K_{\mu}| (1 - K_{\mu}^{1/2})^2 (1 - \bar{K}_{\mu}^{1/2})^2} \right]$$

$$\prod_{\alpha}{}' \left[\prod_{n=1}^{\infty} \frac{(1 - K_{\alpha}^{n-1/2})^{10} (1 - \bar{K}_{\alpha}^{n-1/2})^{10}}{|1 - K_{\alpha}^{n}|^{20}} \prod_{n=2}^{\infty} \frac{|1 - K_{\alpha}^{n}|^4}{(1 - K_{\alpha}^{n-1/2})^2 (1 - \bar{K}_{\alpha}^{n-1/2})^2} \right] \tag{3.10}$$

The previous expression is explicit enough that one can compute the Regge limit [$s \to \infty$ and t fixed] of the four graviton scattering with g loops. In this limit one needs to consider only the 2^g spin structures corresponding to the exchanges of NS states in the loops. The other spin structures are negligible at high energy because they correspond to the exchanges of the gravitino Regge pole, that has an intercept smaller by a factor 1/2 with respect to the graviton Regge pole.

The previous calculation has been recently performed by Cristofano, Fabbrichesi and Roland [22] who obtained the following result:

$$A(s,t) \sim (g_D)^{2g+2}(\epsilon_b \cdot \epsilon_c)(\epsilon_1 \cdot \epsilon_2)\beta(t)[e^{-i\pi/2}s]^{\alpha_g(t)} \left(\frac{2\pi}{\log|s|}\right)^{g\frac{D-2}{2}} \tag{3.11}$$

where

$$\alpha_g(t) = g + 2 + \alpha_c{}' \frac{t}{g+1} \qquad \alpha_c{}' = \frac{\alpha'}{2} \tag{3.12}$$

and α_c' is the slope of the closed string Regge trajectory.

The contribution of spin structures where we have space time fermions propagating in the loops has not yet been computed in complete detail by using the Schottky parametrization of the Riemann surface. It is only known for special amplitudes in the special cases of genus 2 by using other techniques [23,24].

There exist only partial results [25,26] by using the bosonized Vertex $V_{N;g}$ that is known for arbitrary spin structures as discussed in the previous section.

In this way it has been shown [25] that the divergences occurring in the pinching limits $k_\mu \to 0$ are eliminated when one sums over the NS spin structures. It remains to sum also over the Ramond spin structures and see if the remaining divergences occurring for $\xi_\mu \to \eta_\mu$ also cancel. This is the missing step toward a proof of finiteness in superstring theories at any order of perturbation theory.

4 Vertex with the emission of spin fields

In this section we discuss the Vertex $V_{N+2M,0}$ constructed in Ref. [27] that corresponds to the emission of N NS and $2M$ R states.

It is given by

$$V_{N+2M} = \prod_{i=1}^{M} [{}_{2i-1}\langle 1/2| \; {}_{2i}\langle -1/2| + {}_{2i-1}\langle -1/2| \; {}_{2i}\langle 1/2|] \prod_{2M+1}^{N+2M} [{}_{i}\langle 0|]$$

$$: \exp\left\{ -\sum_{i,j=1}^{N} \oint_{z_i} du \oint_{z_j} dv \; \psi_T^{(i)}(u) G(u,v) \bar{\psi}_T^{(j)}(v) \right\} : \tag{4.1}$$

where $G(u,v)$ is the Green's function of the fermion field in the presence of $2M$ spin field given by:

$$G(u,v) = \frac{<0|\psi(u)\bar{\psi}(v)\prod_{i=1}^{M}\frac{1}{2}[S(z_{2i-1})\bar{S}(z_{2i}) + \bar{S}(z_{2i-1})S(z_{2i})]|0>}{<0|\prod_{i=1}^{M}\frac{1}{2}[S(z_{2i-1})\bar{S}(z_{2i}) + \bar{S}(z_{2i-1})S(z_{2i})]|0>} \tag{4.2}$$

It can be easily computed by using the bosonized expressions for the fermion and spin fields.

The normal ordering in (4.1) is defined in the usual way for the non zero modes, while for the zero modes it implies antisymmetrization.

The vacuum in eq. (4.1) for the spin field implies that the zero modes for the states $(2i-1)$ and $(2i)$ have to be identified. The exact relation can be found in Ref. [27].

The Vertex (4.1) satisfies the important property, when saturated with $N+2M$ highest weight states, of giving the correlation function of the corresponding primary fields with a proportionality factor that is equal to the inverse of the correlation function of the $2M$ spin fields. In other words we get:

$$V_{N+2M} \prod_{i=1}^{N+2M} [\Phi_i(0)] \prod_{i=1}^{N+2M} [|0>_i] = \tag{4.3}$$

$$\frac{<0|\prod_{i=1}^{N+2M}[V_i'(0)]^{\Delta_i}\phi_i(z_i)|0>}{<0|\prod_{i=1}^{M}\frac{1}{2}\left[S(z_{2i-1})\bar{S}(z_{2i}) + \bar{S}(z_{2i-1})S(z_{2i})[V_{2i-1}'(0)V_{2i}'(0)]^{1/8}\right]|0>}$$

Therefore by saturating the Vertex containing the emission of spin field with highest weight states we get a ratio of correlation functions. This is a new feature brought in by the presence of spin fields, that makes the factorization properties of the Vertex to be somewhat more complicated than in the case with no spin field (as explained in Ref. [27]).

References

[1] P. Di Vecchia, F. Pezzella, M. Frau, K. Hornfeck, A. Lerda and S. Sciuto, Nucl. Phys. **B322** (1989) 317.

[2] P. Di Vecchia, F. Pezzella, M. Frau, K. Hornfeck, A. Lerda and S. Sciuto, Nordita preprint 89-29/P. To be published in Nuclear Physics.

[3] M. Frau, A. Lerda, J. McCarthy and S. Sciuto, Phys. Lett. **228B** (1989) 205 and MIT preprint CTP 1803 (1989).

[4] D. Bernard and G. Felder, Zürich preprint ETH-TH/89-26(1989).
P. Bouwknegt, J. McCarthy and K. Pilch, MIT preprints CTP-1796 and CTP-1797.
J. Distler and Z. Qiu, Cornell preprint CLNS 89/911(1989).

K. Ito, Tokyo-Komaba preprint UT-Komaba 89-13 (1989).
A. Gerasimov, A. Marshakov, A. Morozov, M. Olshanetsky and S. Shatashvili, ITEP preprint(1989).

[5] G. Felder, Nucl. Phys. **B317** (1989) 215.

[6] J. Shapiro, Phys. Rev. **D5** (1972) 1945.

[7] A. D'Adda, M. Rego Monteiro and S. Sciuto, Nucl. Phys. **B294** (1987) 573.
U. Carow-Watamura and S. Watamura, Nucl. Phys. **B301** (1988) 132.

[8] P. Di Vecchia, R. Nakayama, J.L. Petersen, J. Sidenius and S. Sciuto, Nucl. Phys. **B287** (1987) 621.

[9] E. Verlinde and H. Verlinde, Nucl. Phys. **B288**(1987)357.

[10] J.D. Fay, Theta functions on Riemann surfaces, Springer Verlag.

[11] P. Di Vecchia, to appear.

[12] E. Verlinde and H. Verlinde, Phys. Lett. **B192** (1987) 95.

[13] J. Atick and A. Sen, Nucl. Phys. **B309** (1988) 361.
A. Morozov, Nucl. Phys. **B303** (1988) 343.
A. Losev, Phys. Lett. **B226** (1989) 62.
U. Carow-Watamura, Z.F. Ezawa, K. Harada, A. Tezuka and S. Watamura, Phys. Lett. **B227** (1989) 73.

[14] P. Di Vecchia, M. Frau, A. Lerda and S. Sciuto, Phys. Lett. **B199** (1987) 49.
P. Di Vecchia, K. Hornfeck, M. Frau, A. Lerda and S. Sciuto, Phys. Lett. **B206** (1988) 643.
P. Di Vecchia, K. Hornfeck, M. Frau, A. Lerda and S. Sciuto, in *Perspectives in String Theory*, P. Di Vecchia and J.L. Petersen editors, World Scientific (1988) pag. 422.

[15] J.L. Petersen and J. Sidenius, Nucl. Phys. **B301** (1988) 247.

[16] G. Cristofano, F. Nicodemi and R. Pettorino, Phys. Lett. **B200** (1988) 292 and Int. Jour. of Mod. Phys. **A4** (1989) 857.
G. Cristofano, R. Musto, F. Nicodemi and R. Pettorino, Phys. Lett. **B211** (1988) 417.

[17] H. Konno, Phys. Lett. **B212** (1988) 165.

[18] A.A Belavin, V. G. Knizhnik, A. Morozov and A. Perelomov, Phys. Lett. **177B** (1986) 324.
A. Kato, Y. Matsuo and S. Odake, Phys. Lett. **179B** (1986) 241.

[19] J.L. Petersen, K.O. Roland and J. Sidenius, Phys. Lett. **B205** (1988) 262.
K. O. Roland, Nucl. Phys. **B313** (1989) 432.

[20] P. Di Vecchia, M. Frau, K. Hornfeck, A. Lerda and S. Sciuto, Phys. Lett. **B211** (1988) 301.

[21] J.L. Petersen, J. Sidenius and A. Tollstén, Nucl. Phys. **B317** (1989) 109, Phys. Lett. **213B** (1988) 30 and Phys. Lett. **B214** (1988) 533.
See also: J.L. Petersen, CERN preprint CERN-TH 5309/89 (1989) to appear in the memorial volume for V.G. Knizhnik.

[22] G. Cristofano, M. Fabbrichesi and K.O. Roland, Nordita preprint 89-43 P. To be published in Phys. Letters B.

[23] R. Iengo and C. J. Zhu, Phys. Lett. **212** (1988) 309,313.
C. J. Zhu, Int. Jour. of Mod. Phys. **A4** (1989) 3977.

[24] O. Lechtenfeld, Nucl. Phys. **B309** (1988) 361.
O. Lechtenfeld and A. Parkes, City College preprint CCNY-HEP-89/04.

[25] G. Cristofano, R. Musto, F. Nicodemi and R. Pettorino, Phys. Lett. **B217** (1989) 59.

[26] U. Carow-Watamura, Z. F. Ezawa and S. Watamura, Tohoku University TU-88-328.

[27] P. Di Vecchia, R. Madsen, K. Hornfeck and K. O. Roland, Nordita preprint 89/41 P. To be published in Physics Letters B.

RENORMALIZATION GROUP AND STRING LOOPS

J. RUSSO
International School for Advanced Studies,
Strada Costiera 11, 34014 Trieste, Italy.

and

A. A. TSEYTLIN*/
Department of Mathematics, King's College,
Strand, London WC 2R2LS, UK.

ABSTRACT

The program of renormalization of modular infinities in string theory is studied. Renormalizability becomes non trivial at the order $\log^2\epsilon$. Explicit calculations are made to prove the renormalizability of the generating functional to this order. We show that the $\log^2\epsilon$ counterterms are universal (e.g., the same counterterms provide finiteness both of two-loop scattering amplitudes and of the three-loop partition function) and are related to the $\log\epsilon$-counterterms (beta-functions) in the standard way dictated by the renormalization group. This is a strong indication that the 2d renormalization group acts consistently within the first-quantized theory, and thus relates different loop orders; this is highly non-trivial and may provide a way to resum string perturbation theory. An ansatz for the resummation of the partition function (whereby the sum over all topologies is reduced to a path integral on the sphere) is investigated.

1. Introduction

A remarkable property of string theory is that the tree level string dynamics is dictated by the fixed points of the 2d renormalization group (RG)[1-3]. The extension of this equivalence to the string loop level would be important since it would imply the existence of "non-perturbative conformal theories", which move out of the tree level fixed point to determine the true string vacuum structure. The mechanism to implement this idea was suggested by Fischler and Susskind [4], who observed that in general renormalization of string loop tadpole infinities can be reinterpreted as renormalization of string σ-model couplings (for 'old' suggestions,

---

*/ Permanent address: Department of Theoretical Physics, Lebedev Physical Institute, Moscow 117924, USSR.

see ref.5; for recent reviews and more references see, e.g., refs. 6,7). However, it is by no means trivial that string combinatorics is such that tadpole loop infinities can be consistently cancelled out by renormalizing σ-model couplings. This is a conjecture; a complete proof of tadpole renormalizability is presently unknown. In order for the renormalization group (RG) to be realized in a consistent way, the coefficients of $\log^n \epsilon$-infinities should satisfy certain relations, namely they should be expressed in terms of the $\log \epsilon$-coefficients (β-function coefficients). In sect.2 we present a general formulation of the problem. The renormalization of 1-tadpole factorization ($O(\log \epsilon)$) is considered in sect. 3. In sect. 4 we will perform a non-trivial check of the realization of RG in string loops taking as an example the $\log^2 \epsilon$ massless tadpole infinities in two loop tachyonic amplitudes in the closed bosonic string theory [8]. We shall compute the coefficient of $\log^2 \epsilon$ -infinities and compare it with the one which is predicted by the RG. As in field theory, the non-trivial structure of renormalizability first shows up at 2-loop level ($O(\log^2 \epsilon)$). The analysis to order $\log^2 \epsilon$ is rather technical, but a crucial step.

First-quantized string theory is incomplete: the relative weights in the sum over topologies are not fixed a priori. RG might act consistently on string loops only for special values of these weights. Indeed, RG "glues" together different orders of the string coupling constant. This suggests a reformulation of 1st quantized string theory, in which the RG is obvious (as in field theory), perturbation theory is improved (i.e.; we have automatic "exponentiation") and, probably, everything can be reduced to a path integral on S^2, with a complicated non-local structure. This is discussed in sect.5.

Renormalization procedure in string theory is very different from that of field theory. Let us consider for example the "φ^3" theory (in D=6):

$$\mathcal{L} = \frac{1}{g^2}\left(\frac{1}{2}\varphi\Delta\varphi + \frac{1}{6}\varphi^3\right). \tag{1.1}$$

Expanding around a classical solution: $\varphi = \bar{\varphi} + \psi$, one obtains

$$\mathcal{L}_{quant} = \frac{1}{g^2}\left(\frac{1}{2}\psi(\Delta + \bar{\varphi})\psi + \frac{1}{6}\psi^3\right). \tag{1.2}$$

Thus

$$\Gamma(\bar{\varphi}) = S(\bar{\varphi}) + \frac{1}{2}\log det(\Delta + \bar{\varphi}) + g^2 \text{``}\ominus\text{''} + O(g^4) \tag{1.3}$$

$$= S - \frac{1}{2}\int_\epsilon \frac{dT}{T}\int_{\bigcirc} [dx] e^{-I(x,\bar{\varphi})} + g^2 \int_\epsilon dT_1 dT_2 dT_3 \int_{\ominus} [dx] e^{-I(x,\bar{\varphi})} + \ldots \tag{1.4}$$

where

$$I = \frac{1}{2}\int_0^1 d\tau (e^{-1}\dot{x}^2 + e\bar{\varphi}(x)). \tag{1.5}$$

Infinities are eliminated by a renormalization of g. The action I is *not* renormalized. For a general field theory one has

$$g = g_R + T_1(g_R)\log\epsilon + T_2(g_R)\log^2\epsilon + \dots \ . \tag{1.6}$$

If the theory is renormalizable, one can introduce the β-function which satisfies

$$\beta \equiv \frac{dg}{d\log\epsilon} = \beta(g)\,. \tag{1.7}$$

This leads to the "pole equations"

$$\beta = T_1(g)\ ; \quad T_2 = \frac{1}{2}T_1\frac{dT_1}{dg}\ , \quad \text{etc.} \tag{1.8}$$

In the present example, from

$$\Gamma = \int \frac{1}{6} g_R^{-2}\bar\varphi^3 + \dots = \int \frac{1}{6} g^{-2}\bar\varphi^3 + loops \tag{1.9}$$

it follows

$$g_R = g + (c_1 g^3 + c_2 g^5 + \dots)\log\epsilon + (b_1 g^5 + \dots)\log^2\epsilon + \dots \ . \tag{1.10}$$

Hence $\beta = -c_1 g^3 - c_2 g^5 + \dots$, $b_1 = c_1^2$, etc.

The moral of this example is that renormalization and RG exist only for properly chosen coefficients in the 1st quantized expansion (1.4) (dictated by Wick expansion and unitarity). Renormalization in string theory is very different: one renormalizes the string action I (g goes into string action; also α' renormalises), "counting" rules are different, etc. But again renormalization holds only for particular weights associated to the topologies.

2. Renormalization group in the σ-model approach

The generating functional for string amplitudes is

$$\hat Z = \sum_{n=0}^{\infty} c_n \int_{M_n} d\mu_n \int [dX] exp(-I)\ . \tag{2.1}$$

We have explicitly indicated the weight factors c_n corresponding to each topology. In eq.(2.1), c_0 is understood to account somehow for the Möbius group volume factor. I can be written as $I = I_0 + \varphi^i V_i$, where I_0 is the string action in a particular vacuum and V_i are the corresponding vertex operators for massless particles. Expanding $\hat Z$ in powers of the bare couplings φ^i and putting them on shell we get

$$\hat{Z}\Big|_{\Box\varphi^i=0} = \sum_N A_N \varphi^N \,, \tag{2.2}$$

with the on-shell amplitudes A_N as coefficients.

Including a graviton and dilaton background we get the σ-model action

$$I(G,\Phi) = \frac{1}{4\pi\alpha'}\int d^2z\sqrt{h}h^{ab}\partial_a X^\mu \partial_b X^\nu G_{\mu\nu}(X) + \frac{1}{4\pi}\int d^2z\sqrt{h}R^{(2)}\Phi(X). \tag{2.3}$$

The generating functional $\hat{Z}$ completely defines the theory; thus it should be renormalizable:

$$\hat{Z}(G(\epsilon),\Phi(\epsilon);\epsilon) = \hat{Z}_R(G^R,\Phi^R). \tag{2.4}$$

Then one can introduce β-functions, RG and study the correspondence with the string equations of motion. Eq.(2.4) implies

$$\frac{\partial \hat{Z}}{\partial \log\epsilon} - \beta^G_{\mu\nu}\frac{\partial \hat{Z}}{\partial G_{\mu\nu}} - \beta^\Phi \frac{\partial \hat{Z}}{\partial \Phi} = 0 \ . \tag{2.5}$$

The effective action (EA) S should be nothing but the renormalized part $\hat{Z}_R$ of the generating functional, since renormalization corresponds to a substraction of massless exchanges [1,9,10]. The β-functions are related to the effective action by[11]

$$\beta^i = \kappa^{ij}\frac{\partial S}{\partial \varphi^j}\,, \quad (\varphi^i = \{G,\Phi\}) \tag{2.6}$$

where the leading term in the matrix κ^{ij} accounts for the mixing between the metric and the dilaton in the kinetic term of the tree level EA S_0.

Modular divergences corresponding to shrinking of a trivial cycle can be classified as momentum dependent singularities, which appear when the number of particles on both sides of the degenerating Riemann surface is greater than two (these are the standard physical poles present only for particular values of the external momenta), and momentum independent. The latter are present for arbitrary values of external (on-shell) momenta. They appear when one part of the degenerating surface has either only one vertex (external leg divergence) or none (tadpole divergence). External leg divergences are absorbed into a multiplicative renormalization of the vertex operators or, equivalently, a multiplicative renormalization of the couplings φ^i [12]. The bare couplings will be related to the renormalized ones by[5-7]

$$\varphi^i = \varphi^i_R + \delta_{tad}\varphi^i + \delta_{e.l.}\varphi^i + \delta_{local}\varphi^i, \tag{2.7}$$

where the "external leg" piece is

$$\delta_{e.l.}\varphi^i = e^i_j(\epsilon)\varphi^j_R \, .$$

The same cutoff is used to regularize "local" and "modular" infinities. The parametrization and regularization chosen for the integral over the moduli must be so as to not spoil renormalizability of $\hat{Z}$. It is convenient to *not* fix $SL(2,C)$ Möbius symmetry but regularize the resulting infinity so as to get "$1/Vol\ SL(2,C)$" as a common factor for all genera. This can be accomplished by using an 'extended' Schottky parametrization (see sect. 5)

The β-functions contain the "local" and "modular" contributions [5–10,13,14]

$$\beta^i = \beta^i_0 + \beta^i_{mod}, \quad (\beta^G_{mod})_{\mu\nu} = F_1(\Phi)G_{\mu\nu} + ..., \quad \beta^\Phi_{mod} = F_2(\Phi) + ... \tag{2.8}$$

$$F_1 = \sum_{n=1}^{\infty} \lambda_n e^{2n\Phi} \, , \quad F_2 = \sum_{n=1}^{\infty} \nu_n e^{2n\Phi} \, .$$

Using eq.(2.6), one can find the field-theory prediction for the "modular" corrections to the β-functions

$$\begin{aligned} (\beta^G_{mod})_{\mu\nu} &= -d_0^{-1}(\frac{d_1}{2}g_0^2 + d_2 g_0^4)G_{\mu\nu} + ... \\ \beta^\Phi_{mod} &= -\frac{Dd_1}{8d_0}g_0^2 - \frac{D-1}{4d_0}g_0^4 d_2 + ... \, . \end{aligned} \tag{2.9}$$

The RG requires that the β-functions, $\beta^i \equiv -\frac{\partial\varphi^i}{\partial \log\epsilon}$, depend only on φ, but not explicitely on ϵ (cf.eqs.(1.7,8)). Thus assuming the validity of the RG and using eqs. (2.8) we expect to find the following $\log^2 \epsilon$-terms in the bare couplings

$$G_{\mu\nu} = G^R_{\mu\nu}(1 - F_1(\Phi^R)\log\epsilon + \frac{1}{2}(F_1^2 + F_2 F_1')\log^2\epsilon + O(\log^3\epsilon)), \quad F' \equiv \frac{dF}{d\Phi} \, , \tag{2.10a}$$

$$\Phi = \Phi^R - F_2(\Phi^R)\log\epsilon + \frac{1}{2}F_2 F_2' \log^2\epsilon + O(\log^3\epsilon) \, . \tag{2.10b}$$

Computing the amplitudes with the action (2.3) and regularizing the massless scalar tadpole divergences with the help of a cutoff ϵ we should be able to check that all the ϵ dependence cancels out. In terms of the renormalized fields, the action (2.3) reads

$$I(G(\epsilon), \Phi(\epsilon)) = I(G^R, \Phi^R) + \delta I(G^R, \Phi^R; \epsilon). \tag{2.11}$$

The counterterm action δI will be given as an expansion in the renormalized string coupling $g = e^{\Phi^R_{(const)}}$ ($\Phi^R_{(const)}$ is the constant part of the renormalized dilaton field) and α'. The "tadpole" part of the counterterm will have the form

$$\delta I = \frac{1}{4\pi\alpha'}\int d^2z\sqrt{h}\partial_a X^\mu \partial^a X^\nu \delta G^R_{\mu\nu} + \frac{1}{4\pi}\int d^2z\sqrt{h}R^{(2)}\delta\Phi^R, \tag{2.12}$$

where

$$\delta G^R_{\mu\nu} = G^R_{\mu\nu}(-(\lambda_1 g^2 + \lambda_2 g^4 + ...)\log\epsilon + \frac{1}{2}((\lambda_1^2 + 2\nu_1\lambda_1)g^4 + ...)\log^2\epsilon + O(\log^3\epsilon)), \tag{2.13a}$$

$$\delta\Phi^R = -(\nu_1 g^2 + \nu_2 g^4 + ...)\log\epsilon + \frac{1}{2}(2\nu_1^2 g^4 + ...)\log^2\epsilon + O(\log^3\epsilon)\ . \tag{2.13b}$$

Let us introduce the following notation for the genus-n contribution to the vacuum amplitude ($\chi_n = 2 - 2n$ is the Euler number)

$$Z_n = \hat{d}_n g^{-\chi_n}\ , \quad \hat{d}_n = d_n + d_n^{(1)}\log\epsilon + d_n^{(2)}\log^2\epsilon + ... + d_n^{(n-1)}\log^{n-1}\epsilon, \tag{2.14}$$

where the $d_n^{(k)}$ are finite (complex) numbers*. The analysis of 1-tadpole ($\log\epsilon$) factorization of string amplitudes [14] implies that the counterterm should have the form [7]

$$\delta I = \sum_{n=1}^{\infty} g^{2n} b_n \mathcal{O}_n \log\epsilon + O(\log^2\epsilon)\ ; \tag{2.15}$$

$$\mathcal{O}_n = \frac{1}{4\pi\alpha'}\int d^2z\sqrt{h}[2n : \partial_a X^\mu \partial^a X^\mu : + \alpha'(1-n)R^{(2)}], \tag{2.16}$$

where normal ordering "::" denotes substraction of the corresponding correlator at genus zero,

$$: \partial_a X^\mu \partial^a X^\mu := \partial_a X^\mu \partial^a X^\mu + \frac{D}{4}\alpha' R^{(2)}\ . \tag{2.17}$$

Comparing eqs. (2.15), (2.16), with (2.12), (2.8), (2.10) we find that λ_n and ν_n are related by

$$\lambda_n = -2nb_n\ , \quad \nu_n = -(1 + \frac{n}{2}(D-2))b_n\ . \tag{2.18}$$

Comparing eq.(2.18) with the field theory prediction, eq. (2.9), we see that there is a complete agreement with the EA approach if the constants b_n are equal to bd_n, where

$$b = \frac{1}{4d_0}\ . \tag{2.19}$$

---

* We assume the analytic continuation prescription of refs.15 and 16 to eliminate the divergences associated with the tachyon mode.

Thus we can rewrite the lowest order counterterms in the form

$$\delta I = J_1 \log \epsilon + J_2 \log^2 \epsilon + \dots , \tag{2.20}$$

with

$$\begin{aligned} J_1 &= J_1^{(1)} + J_1^{(2)} + \dots = b_1 g^2 \mathcal{O}_1 + b_2 g^4 \mathcal{O}_2 + \dots \\ J_2 &= b_1^2 g^4 [\frac{D+2}{4\pi\alpha'} \int d^2z \sqrt{h} h^{ab} \partial_a X^\mu \partial_b X^\mu + \frac{D^2}{16\pi} \int d^2z \sqrt{h} R^{(2)}] + \dots . \end{aligned} \tag{2.21}$$

We have found the $\log^2 \epsilon$-counterterm just by requiring consistency with the RG and correspondence with the effective field theory. In the next two sections we will show that this is precisely the counterterm needed to render finite the scattering amplitudes, including the contributions from genus 0, 1 and 2, and at the same time to render finite the three-loop partition function.

3. Finiteness of scattering amplitudes

3.1.*Finiteness conditions.* A scattering amplitude of M external particles is defined as

$$A_M = \sum_{n=0}^{\infty} c_n g^{2n-2} \int_{F_n} d\mu_n A_M[\Sigma_n]; \tag{3.1}$$

$$A_M[\Sigma_n] \equiv g^M < V_1 \dots V_M >_n . \tag{3.2}$$

(Henceforth the factor g^M will be absorbed into a redefinition of the vertex operators). Both the measure $d\mu_n$ and the integrand $A_M[\Sigma_n]$ are assumed to be invariant under the modular transformations. The weights c_n will be fixed so that to have correspondence with the effective action.

It is convenient to use the "period matrix" representation for the two loop measure [17] (see Appendix for notation and formulas).

Let us adopt the following notation

$$< X_M >_n \equiv c_n g^{2n-2} \int_{F_n} d\mu_n A_M[\Sigma_n] ; \quad X_M \equiv V_1 \dots V_M . \tag{3.3}$$

Taking into account the (tadpole) counterterm in eq. (2.7) , the scattering amplitude can be represented as

$$A_M = < X_M e^{-\delta I} >_0 + < X_M e^{-\delta I} >_1 + < X_M e^{-\delta I} >_2 + \dots . \tag{3.4}$$

and should be finite. Expanding in powers of g we have

$$A_M = < X_M >_0 - < X_M \delta I >_0 + < X_M \frac{(\delta I)^2}{2} >_0 + < X_M >_1$$

$$- < X_M \delta I >_1 + < X_M >_2 \ . \tag{3.5}$$

A_M can be rewritten as (we shall restrict consideration to the g^2-order. Note that the tree level contribution is $\sim g^{-2}$)

$$A_M = (A_M)_{\text{finite}} + (A_M)_{\log\epsilon} + (A_M)_{g^2\log\epsilon} + (A_M)_{g^2\log^2\epsilon} \ . \tag{3.6}$$

Inserting δI, given by eq. (2.20), (2.21) we find for the singular parts in eq.(3.6)

$$(A_M)_{\log\epsilon} = -\log\epsilon < X_M J_1^{(1)} >_0 + < X_M >_1 \Big|_{\log\epsilon} \ , \tag{3.7a}$$

$$(A_M)_{g^2\log\epsilon} = -\log\epsilon < X_M J_1^{(2)} >_0 - \log\epsilon < X_M J_1^{(1)} >_1^F + < X_M >_2 \Big|_{\log\epsilon} \ , \tag{3.7b}$$

$$(A_M)_{g^2\log^2\epsilon} = \log^2\epsilon < X_M Q >_0 - \log\epsilon < X_M J_1^{(1)} >_1 \Big|_{\log\epsilon} + < X_M >_2 \Big|_{\log^2\epsilon} \ , \tag{3.7c}$$

$$Q \equiv \frac{1}{2}(J_1^{(1)})^2 - J_2 \ ,$$

where "$<>_n^F$" denotes finite part. In the remainder of this section we will show that

$$< X_M >_1 \Big|_{\log\epsilon} = \log\epsilon < X_M J_1^{(1)} >_0 \ , \tag{3.8a}$$

$$< X_M >_2 \Big|_{\log\epsilon} = \log\epsilon < X_M J_1^{(2)} >_0 + \log\epsilon < X_M J_1^{(1)} >_1^F , \tag{3.8b}$$

and hence $(A_M)_{\log\epsilon} = (A_M)_{g^2\log\epsilon} = 0$. In the next section we shall prove that all three $\log^2\epsilon$ contributions in eq.(3.7c) are proportional with the relative coefficients 1, 2 and 1, respectively, i.e.,

$$< X_M J_1^{(1)} >_1 \Big|_{\log\epsilon} = 2\log\epsilon < X_M Q >_0 \ ,$$

$$< X_M >_2 \Big|_{\log^2\epsilon} = \log^2\epsilon < X_M Q >_0 , \tag{3.8c}$$

thus establishing $(A_M)_{g^2\log^2\epsilon} = 0$ as well. Thus we shall explicitly check the renormalizability assumption to the given order.

3.2. *One-tadpole analysis.* Consider first the simplest case, eq.(3.7a). The vanishing of the r.h.s. of eq. (3.7a) is actually automatic, since the counterterm (2.15) has been chosen precisely for this purpose. In fact, for single tadpole factorization, we may use the formula from ref.14 (see also ref.7)

$$< X_M >_n \Big|_{tadpole} = g^2 \log\epsilon \, (-\frac{1}{d_0 D} < X_M V_g >_{n_1} < V_g >_{n_2} + \frac{1}{8d_0} \chi_{n_1} \chi_{n_2} < X_M >_{n_1} Z_{n_2}) \, , \tag{3.9}$$

where n_1, n_2 are the genera of the separating Riemann surfaces ($n_1 + n_2 = n$). Using eqs. (3.9) and (A.12) one finds

$$< X_M >_1 \Big|_{\log\epsilon} = \frac{d_1}{2d_0} g^2 \log\epsilon < X_M V_g >_0 \tag{3.10}$$

Using (2.21), (2.16) we find that $(A_M)_{\log\epsilon} = 0$.

Eq. (3.10) can be used to fix the constant c_0 in (3.1), which shall appear frequently in the computations. We may consider, for example, the case of scattering amplitudes for M tachyons which is given by

$$< X_M >_1 = c_1 \int_{F_1} d\mu_1 \, \tau_2 \int \prod_{i=1}^{M-1} d^2\nu_i \prod_{i<j}^{M} \chi(\nu_{ij}|\tau)^{\alpha' k_i . k_j} \; ; \quad \nu_M \equiv 0. \tag{3.11}$$

The notation is as follows

$$\nu_{ij} \equiv \nu_i - \nu_j \quad ; \quad \chi(\nu_{ij}|\tau) \equiv exp[-\pi\tau_2^{-1}(Im\nu_{ij})^2] \left| \frac{\theta_1(\nu_{ij}|\tau)}{\theta_1'(0|\tau)} \right| \; ;$$

$$\theta_1'(u|\tau) \equiv \frac{\partial}{\partial u}\theta_1(u|\tau) \; ; \quad \tau_2 \equiv Im\tau \, . \tag{3.12}$$

In order to extract the $\log\epsilon$-divergence corresponding to the limit in which all external particles approach each other, it is convenient to define the variables η_r

$$\epsilon\eta_r = \nu_r \; ; \quad \eta_{M-1} \equiv \epsilon e^{i\phi} = \nu_{M-1} \, . \tag{3.13}$$

Expanding the integrand of eq.(3.11) in powers of ϵ one gets for the massless-exchange contribution

$$< X_M >_1 \Big|_{\log\epsilon} =$$

$$-2\pi^2 c_1 \int_{F_1} d\mu_1 \log\epsilon \int \prod_{r=1}^{M-2} d^2\eta_r \prod_{r<s}^{M} |\eta_r - \eta_s|^{\alpha' k_r . k_s} \sum_{r,s=1}^{M} \alpha' k_r . k_s \eta_r \bar{\eta}_s . \tag{3.14}$$

[Here $\eta_{M-1} \equiv 1$, $\eta_M \equiv 0$]. Eq. (3.14) can be expressed as a scattering amplitude on the sphere:

$$< X_M >_1 \Big|_{\log\epsilon} = \left(\frac{-4\pi^3}{c_0}\right) 2d_1 g^2 \log\epsilon < X_M V_g >_0 .$$

Comparing this with eq. (3.10) one finds that $c_0 = -16\pi^3 d_0$.

Consider now the more subtle case of the $O(g^2 \log\epsilon)$-terms. There are two inequivalent limits for single tadpole factorization of the amplitude $< X_M >_2$: (i) All external particles go close to each other; (ii) $\Omega_{12} \to 0$. From eq. (3.9) we get

$$\lim_{(i)} < X_M >_2 = \frac{d_2}{d_0} g^4 \log\epsilon (< X_M V_g >_0 - \frac{1}{2} < X_M >_0) ; \tag{3.15}$$

$$\lim_{(ii)} < X_M >_2 = \frac{d_1}{2d_0} g^2 \log\epsilon < X_M V_g >_1^F . \tag{3.16}$$

The total $\log\epsilon$ part in $< X_M >_2$, given by the sum of (3.15) and (3.16), is thus

$$< X_M >_2 \Big|_{\log\epsilon} = \log\epsilon < X_M J_1^{(2)} >_0 + \log\epsilon < X_M J_1^{(1)} >_1^F , \tag{3.17}$$

i.e. is in agreement with the statement made above.

4. Order $\log^2 \epsilon$ contributions to scattering amplitudes

In order to extract the divergent parts of $< X_M J_1 >_1$ and $< X_M >_2$ we shall perform an explicit calculation of scattering amplitudes in the factorization limit (for the definition of J_1 see eqs.(2.20), (2.21)).

4.1. *One loop analysis of factorization.* The scattering amplitude for M-tachyons and 1 zero-momentum massless scalar vertex is given by

$$< X_M \hat{V} >_1 = \frac{c_1}{4\pi} \int_{F_1} d\mu_1 \; \tau_2 \int (\prod_{i=1}^{M-1} d^2\nu_i) d^2\nu_{M+1} \prod_{i<j}^{M} \chi(\nu_{ij}|\tau)^{\alpha' k_i . k_j}$$

$$\times (\sum_{i,j=1}^{M} \alpha' k_i . k_j K(\nu_{i,M+1}|\tau) \bar{K}(\nu_{j,M+1}|\tau)) \tag{4.1}$$

where $\hat{V} \equiv V_g + \frac{1}{2} nD$ (this definition amounts to a substraction of the self- contraction; cf. eq.(A.12), (A.14)) and

$$K(\nu_{i,j}|\tau) \equiv \frac{\theta_1'(\nu_{i,j}|\tau)}{\theta_1(\nu_{i,j}|\tau)} - \pi\tau_2^{-1}(\nu_{i,j} - \bar{\nu}_{i,j}) \ . \tag{4.2}$$

We can distinguish two degeneration limits:
A) All ν_i, $i = 1, ..., M+1$ approach each other.
B) The ν_i, $i = 1, ..., M$, corresponding to the tachyon vertices, approach each other.

By a suitable change of variables, similar to the one made in the previous section, we find [8]

$$\lim_A < X_M \hat{V} >_1 = \frac{d_1}{2d_0} g^2 \log\epsilon < X_M V_g V_g >_0 + \frac{d_1 D}{4d_0} g^2 \log\epsilon < X_M >_0 \ . \tag{4.3}$$

$$\lim_B < X_M \hat{V} >_1 = -\frac{d_1}{2d_0} g^2 \log\epsilon < X_M V_g >_0 \ . \tag{4.4}$$

Using eqs. (3.10), (4.3), (A.13) and (4.4) we obtain for the $\log\epsilon$ part of $< X_M V_g >_1$

$$\begin{aligned} < X_M V_g >_1 \Big|_{\log\epsilon} &= \frac{d_1}{2d_0} g^2 \log\epsilon < X_M V_g V_g >_0 + \frac{d_1 D}{4d_0} g^2 \log\epsilon < X_M >_0 \\ &- \frac{d_1}{2d_0} g^2 \log\epsilon < X_M V_g >_0 - \frac{d_1 D}{4d_0} g^2 \log\epsilon < X_M V_g >_0 \ . \end{aligned} \tag{4.5}$$

Therefore

$$< X_M J_1^{(1)} >_1 \Big|_{\log\epsilon} = 2\log\epsilon < X_M Q >_0 \tag{4.6}$$

where Q was defined in (3.7). This confirms what was asserted in sect. 3.

4.2. *Two loop calculation of factorization.* The two loop contribution to the M-tachyon scattering amplitude can be written as

$$< X_M >_2 = c_2 g^2 \int_{F_2} d\mu_2 \int \prod_{i=1}^{M} d^2 z_i \prod_{i<j} F(z_i, z_j)^{\alpha' k_i . k_j} \tag{4.7}$$

where

$$F(z_i, z_j)^2 = |E(z_i, z_j)|^2 exp[\frac{\pi}{2} \int_{z_i}^{z_j} (\omega_a - \bar{\omega}_a) \int_{z_i}^{z_j} (\omega_b - \bar{\omega}_b)(\Omega_2^{-1})_{ab}] \ ; \ \ \Omega_2 \equiv Im\Omega \ . \tag{4.8}$$

Consider the limit $\Omega_{12} \to 0$. The genus-2 Riemann surface degenerates to two tori Σ_1, Σ_2. The behaviour of the measure is displayed in eq.(A.6). As concerns the integrand, we are going to use the following formulas, derived in ref.8:

$$E(z,w) \to E_1(z,w) + O(\Omega_{12}^2), \qquad \text{if } \ z,w \in \Sigma_1, \tag{4.9}$$

$$\omega_1(z) \to \nu(z) + O(\Omega_{12}^2), \qquad \text{if } \ z \in \Sigma_1, \tag{4.10}$$

$$\omega_2(z) \to \frac{\Omega_{12}}{2\pi i}\frac{\partial}{\partial z}\left(\frac{\theta_1'(z-p_1|\tau)}{\theta_1(z-p_1|\tau)}\right) dz + O(\Omega_{12}^2), \quad \text{if } \ z \in \Sigma_1, \tag{4.11}$$

where $E_1(z,w) = \frac{\theta_1(z-w|\tau)}{\theta_1'(0|\tau)}(dz)^{-1/2}(dw)^{-1/2}$, $\nu(z) = dz$ are respectively the prime-form and abelian differential on the torus Σ_1 and $\tau = \Omega_{11} + O(\Omega_{12}^2)$.

In particular, using eq. (4.9), we obtain

$$\prod_{i<j}^{M} |E(z_i,z_j)|^{\alpha' k_i . k_j} \to \prod_{i<j}^{M} |E_1(z_i,z_j)|^{\alpha' k_i . k_j} + O(\Omega_{12}^2) \ ; \quad z_i, z_j \in \Sigma_1 \tag{4.12}$$

where $E_1(z_i,z_j) \equiv \frac{\theta_1(z_i-z_j|\tau)}{\theta_1'(0|\tau)}$ is the prime-form on the torus Σ_1 ($\tau \equiv \Omega_{11}$). Using eqs.(4.10) and (4.11), it is a straightforward exercise to obtain

$$\sum_{i,j=1}^{M} \pi\alpha' k_i . k_j \int_{z_i}^{z_j} (\omega_a - \bar{\omega}_a) \int_{z_i}^{z_j} (\omega_b - \bar{\omega}_b)(\Omega_2^{-1})_{ab} =$$

$$\sum_{i,j=1}^{M} \pi\alpha' k_i . k_j \left[\frac{1}{Im\Omega_{11}}(z_{ji} - \bar{z}_{ji})^2 - \frac{|\Omega_{12}|^2}{2\pi^2 Im\Omega_{22}} K(z_j|\tau)\bar{K}(z_i|\tau) + ...\right] \tag{4.13}$$

where we have also used energy-momentum conservation, $\sum_{i=1}^{M} k_i = 0$.

From eqs.(4.12) and (4.13), one can see that the amplitude (4.7) has the form

$$< X_M >_2 \sim \int_\epsilon \frac{d^2\Omega_{12}}{|\Omega_{12}|^4} + \int_\epsilon \frac{d^2\Omega_{12}}{|\Omega_{12}|^2} + \text{finite}\,.$$

The first term corresponding to a tachyon exchange, the second term to a massless scalar exchange, and the remaining to higher-level contributions. Explicitly, the massless exchange contribution is

$$< X_M >_2 \Big|_{\log \epsilon} = \frac{c_2}{2} g^2 \frac{4\pi}{(2\pi^2)^D} (\frac{2}{\pi})^4 \int_{F_1} \frac{d^2\Omega_{22}}{(Im\Omega_{22})^{14}|\Delta_{(1)}|^{16}} \int_\epsilon \frac{d^2\Omega_{12}}{|\Omega_{12}|^2}$$

$$\times \left(\int_{F_1} \frac{d^2\Omega_{11}}{(Im\Omega_{11})^{14}|\Delta_{(1)}|^{16}} \int \prod_{i=1}^{M} d^2 z_i \prod_{i<j}^{M} \chi(z_{i,j}|\tau)^{\alpha' k_i . k_j} \right.$$

$$\left. \left[\frac{D}{4} - \sum_{i,j} \frac{\alpha' k_i . k_j}{16\pi} K(z_j|\tau)\bar{K}(z_i|\tau)\right] \right). \tag{4.14}$$

Thus

$$< X_M >_2 \Big|_{\log\epsilon} =$$

$$-\frac{d_1}{2d_0} g^2 \log\epsilon (\frac{D}{4} < X_M >_1 - \frac{1}{2} < X_M \hat{V} >_1) = \frac{d_1}{4d_0} g^2 \log\epsilon < X_M V_g >_1 \ . \quad (4.15)$$

Thus

$$< X_M >_2 \Big|_{\log^2\epsilon} = \frac{1}{2} \log\epsilon < X_M J_1^{(1)} >_1 \Big|_{\log\epsilon} \ . \quad (4.16)$$

This completes the proof of eq.(3.8c).

4.3.*Finiteness of the three-loop partition function.* The singularities arising in the partition function at genus 3 should also be cancelled by the same counterterm δI discussed in sect.2. This time the finiteness condition is

$$< e^{-\delta I} >_0 + < e^{-\delta I} >_1 + < e^{-\delta I} >_2 + < e^{-\delta I} >_3 + ... = \text{finite}. \quad (4.17)$$

The first term on the l.h.s. of eq. (4.17) vanishes, since it involves insertions of zero-momentum massless vertices on the sphere. Expanding $e^{-\delta I}$ and retaining terms up to order g^4 we arrive at the following three conditions:
$g^2 \log\epsilon$:

$$-\log\epsilon < J_1^{(1)} >_1 + < 1 >_2 \Big|_{\log\epsilon} = 0, \quad (4.18a)$$

$g^4 \log\epsilon$:

$$< 1 >_3 \Big|_{\log\epsilon} - \log\epsilon < J_1^{(2)} >_1 - \log\epsilon < J_1^{(1)} >_2^F = 0 \ , \quad (4.18b)$$

$g^4 \log^2\epsilon$:

$$< 1 >_3 \Big|_{\log^2\epsilon} + \log^2\epsilon < Q >_1 - \log\epsilon < J_1^{(1)} >_2 \Big|_{\log\epsilon} = 0 \quad (4.18c)$$

(cf. (2.21), (3.7)). Using formulas (A.12), (A.15), we get for the coefficients of the divergent terms in the partition function (2.14)
$g^2 \log\epsilon$:

$$d_2' = -bDd_1^2 = -\frac{Dd_1^2}{4d_0} \ , \quad (4.19a)$$

$g^4 \log\epsilon$:

$$d_3' = -4bDd_2d_1 = -\frac{Dd_2d_1}{d_0} \ , \quad (4.19b)$$

$g^4 \log^2 \epsilon$:

$$d_3'' = b^2 D^2 d_1^3 = \frac{D^2 d_1^3}{16 d_0^2} , \qquad (4.19c)$$

which is in perfect agreement with the results obtained from the effective action [8].

5. Resummation of topologies: Ansatz for partition function based on "topological fixture insertion" operators*

It is quite reasonable to expect that the leading $O((g^2 \log \epsilon)^n)$-divergence in $A_M^{(n)}$ will always be proportional to the correlator on the sphere $< V_1 ... V_M Q >_0$, $Q \sim V_g^n + ...$, where Q may be computed by simply using the 1-loop β-functions and the RG. (We have seen in previous sections that this is true at least to the order $O(\log^2 \epsilon)$). Thus the RG, if it exists, would determine the leading $\log \epsilon$ approximation to all loop orders, in the same way as it does in field theory. This suggests that there should exist a reformulation or resummation of the 1st quantized string topological expansion in which the RG is incorporated in a manifest way, e.g. as in field theory, whereby renormalizability is evident, everything is reduced to the complex plane, and tadpole divergences are automatically "exponentiated".

For this purpose, it is convenient to represent the loop part of the generating functional by insertions of operators corresponding to holes, handles, etc. [18,19,7]. So let $I_0 = \frac{1}{2} \int d^2 z \sqrt{g} X \Delta X, \Delta = -\nabla^2$ be the free string action on an arbitrary curved 2-surface (for clarity in the notation we shall often set $2\pi\alpha' = 1$ and do not explicitly indicate the index of X^μ). Consider the expectation value

$$< F[X] > \equiv \int [dX] e^{-I_0} F[X] \, , \quad < 1 >= 1 \, . \qquad (5.1)$$

Let us split the Green function $G = \Delta^{-1}$ in two parts $G = G_0 + \bar{G}$, where $G_0 = -\log |z-w|^2$,and $\bar{G}$, depends on moduli. Substituting into eq.(5.1) we get

$$< F[X] > = < F[X] Q_n[X] >_0 \, , \qquad (5.2)$$

where $Q_n[X]$ depends on the equivalence class of Riemann surface $[\Sigma_n]$ (n=genus). Explicitly

$$Q_n[X] = N \, exp(-\frac{1}{2} \int X \bar{\Delta} X) \, , \quad \Delta = \Delta_0 + \bar{\Delta} \, , \quad \bar{\Delta} = - \frac{\Delta_0 \cdot \bar{G} \cdot \Delta_0}{1 + \Delta_0 \cdot \bar{G}} \, ; \qquad (5.3)$$

$$N = \left(\frac{det \Delta}{det \Delta_0} \right)^{-D/2} \, .$$

* A more detailed presentation of the contents of this section is given in ref.7.

Defining $\hat{h} = -\frac{1}{2}\int X\Delta_0 \cdot \bar{G} \cdot \Delta_0 X$, $Q_n[X]$ can be written as

$$Q_n[X] =: e^{\hat{h}} : . \tag{5.4}$$

Let us consider as an example the torus in the 'extended' Schottky parametrization (in which $SL(2, C)$ symmetry is not fixed and the Möbius volume Ω is a common factor for all genera) The torus is represented as the complex plane with two holes; let us denote ξ, η the distance to the origin of the respective centers. Then [18]

$$\hat{h} = \frac{\pi}{\log|k|}[X(\xi) - X(\eta)]^2 - 4\pi \sum_{n=1}^{\infty} \left(\frac{k^n}{(1-k^n)n!(n-1)!}[\partial^{n-1}(\xi-\eta)^n \partial X](\xi) \right.$$

$$\left. \times [\partial^{n-1}(\xi-\eta)^n \partial X](\eta) + c.c. \right) \tag{5.5}$$

where $k = e^{2\pi i \tau}$ (related to the radii of the boundary of the holes), and τ is the standard modular parameter in the "parallelogram" representation. In the limit $\xi \to \eta$ we have the expansion

$$: e^{\hat{h}} := 1 + \frac{: \partial X \partial X : (\xi)}{4 \log|k|} |\xi - \eta|^2 + \dots \ . \tag{5.6}$$

Using eq.(5.6) and eq. (5.2) one obtains

$$(\hat{Z}_1)_{\log\epsilon} \sim \frac{1}{\Omega} \int [d\tau] \int \frac{d^2\eta}{|\xi-\eta|^2} < \int d^2\xi : \partial X \bar{\partial} X : e^{-I} >_0 . \tag{5.7}$$

Then a counterterm $\delta I = d_1 \mathcal{O}_1 \log\epsilon$ is added in order to renormalize. Note that

$$\hat{Q}_1 \equiv \int [d^2k d^2\xi d^2\eta] Q_1(k, \xi, \eta) \sim d_1 \log\epsilon \mathcal{O}_1 + \text{finite} . \tag{5.8}$$

RG suggests exponentiation of the "handle" operator. In order for the RG to operate at string loop level tadpole divergences should in fact *exponentiate* to become counterterms for the string action. It is thus natural to expect that in some regions of moduli space (far separated handles) the generating functional can be written as ($I_{int} = \varphi^i V_i$)

$$\hat{Z} \sim \frac{1}{\Omega} \sum_n \int [d(\xi, \eta, k)]_n < Q_n e^{-I_{int}} >_0 \sim \frac{1}{\Omega} < exp(\hat{Q}_1 - I_{int} >_0 + \dots,$$

$$\sim \frac{1}{\Omega} < e^{\sum_n \hat{Q}_n} e^{-I_{int}} >_0 ; \tag{5.9}$$

i.e., everything is reduced to a path integral on the sphere with a complicated (non-local) action $\tilde{I} = I + \mathcal{F}$, with

$$\mathcal{F} \equiv -\sum_n \hat{Q}_n \simeq \sum_n b_n g^{2n} \mathcal{O}_n \log\epsilon + O(\log^2\epsilon) + (\textit{nonlocal terms}) \qquad (5.10)$$

(cf. eq.(2.15)). In terms of $\tilde{I}$, eq. (5.9) takes the simple form:

$$\hat{Z} = c_0 g^{-2}\Omega^{-1} < e^{-(\mathcal{F}+I_{int})} >_0 = \Omega^{-1}\int [dX] e^{-(\tilde{I}+I_{int})_0} \,. \qquad (5.11)$$

This works only in parametrizations where Ω^{-1} is a common factor.

If one assumes renormalizability, all $\log^n \epsilon$ terms are determined from the 1-loop β-functions (cf. eq. (1.8)) and all local terms are absorbed into $I(G(\epsilon), \Phi(\epsilon))$, as in eq. (2.11), where $\beta^i \equiv -\frac{\partial \varphi^i}{\partial \log \epsilon}$ depends on φ, but not explicitly on ϵ.

Remarks

(i) This ansatz fixes automatically all relative weights of topologies.

(ii) It explains *exponentiation* of "tadpole", etc. divergences.

(iii) It helps to prove renormalizability to all orders: the question about renormalizability of $\hat{Z} = \sum_n \hat{Z}_n$ is reduced to that about renormalizability of $\tilde{I}$.

(iv) Renormalizability of $\tilde{I}$ on the sphere should follow from the operator product expansion. Corrections to β-functions contain all powers of φ^i ($I_{int} = \varphi^i V_i$).

6. Concluding remarks

Renormalizability of *tadpole* modular infinities is a matter of "combinatorics". The present results give strong evidence -which was absent earlier- that the combinatorics of string diagrams, with the particular choice of weights for string loop contributions implied by the correspondence with the effective action, is precisely what we need in order to have RG. We have checked this by a computation to $O(\log^2 \epsilon)$, which reveals the nontrivial structure of the renormalization procedure (which is different from what we have in field theory).

A proof that RG works at string loop level would strongly suggest that string theory might be formulated on the complex plane (where RG certainly works). The realization of the RG to the non-trivial $\log^2 \epsilon$ level demonstrated in this paper supports this conjecture.

Appendix A. Some notation and useful formulas

The expression for the 2-loop amplitude (3.1) in the "period matrix" representation for the measure is [17] (we rescale A_M to absorb the usual zero mode factor $(\alpha')^{-D/2}$)

$$A_M = c_0 g^{-2} A_M[\Sigma_0] + c_1 \frac{4}{(2\pi^2)^{D/2}} \int_{F_1} d^2\tau \frac{|\Delta_{(1)}(\tau)|^{-16}}{(Im\tau)^{14}} A_M[\Sigma_1]$$

$$+c_2 g^2 \frac{4\pi}{(2\pi^2)^D} \int_{F_2} d^2\Omega_{11} d^2\Omega_{12} d^2\Omega_{22} \frac{|\Delta_{(2)}|^{-4}}{(det\ Im\Omega)^{13}} A_M[\Sigma_2] + ... \qquad (A.1)$$

where

$$\Delta_{(1)} \equiv \theta[^0_0](0|\tau)\theta[^1_0](0|\tau)\theta[^0_1](0|\tau) = -\frac{1}{\pi}\theta'[^1_1](0|\tau)$$
$$= -2f^3(q^2)q^{1/4} \ ; \ q \equiv e^{i\pi\tau}, \qquad (A.2)$$

$$f(q^2) \equiv \prod_{n=1}^{\infty}(1 - q^{2n}) \ ; \qquad (A.3)$$

$$d\mu_1 = \frac{4}{(2\pi^2)^{D/2}} d^2\tau \frac{|\Delta_{(1)}(\tau)|^{-16}}{(Im\tau)^{14}} = \frac{1}{2}\frac{1}{(2\pi)^D} d^2\tau \frac{|f(q^2)|^{-48}}{(Im\tau)^{14}} e^{4\pi\tau_2} \ ;$$

$$\Delta_{(2)} \equiv \prod_{even\ e} \theta[e](0|\Omega) \ . \qquad (A.4)$$

Here F_1 is the fundamental modular domain for the torus$=\{\tau \in C : Im\tau > 0, |\tau| > 1, |Re\tau| \leq \frac{1}{2}\}$ and F_2 is the fundamental modular domain for genus 2 Riemann surfaces. F_2 can be defined by the three conditions (Ω_{ij} is the period matrix)
(1)$|Re\Omega_{ij}| \leq \frac{1}{2}$.
(2)$|Im\Omega_{ij}|$ is reduced in the sense of Minkowski, and $0 \leq |2Im\Omega_{12}| \leq Im\Omega_{11} \leq Im\Omega_{22}$.
(3)$|det(C\Omega + D)| \geq 1$ for all $\begin{pmatrix} A & B \\ C & D \end{pmatrix} \in Sp(4, Z)$.

The constants c_n can be fixed by the correspondence with the effective field theory. As an example, consider the massless-exchange factorization of Z_2 in the limit $\Omega_{12} \to 0$. This fixes the ratio c_2/c_1^2. In fact, in this limit one has

$$\Delta_{(2)} \to -\frac{i\pi}{2}\Omega_{12}\Delta^4_{(1)}(\Omega_{11})\Delta^4_{(1)}(\Omega_{22})(1 + O(\Omega^2_{12}) + ...) \ , \qquad (A.5)$$

where we have used the Jacobi triple identity, eq. (A.2), and the expansion

$$\theta[e](0|\Omega) = \sum_{n=0}^{\infty} \frac{1}{n!}(\frac{\Omega_{12}}{2\pi i})^n \theta^{(n)}[e_1](0|\Omega_{11})\theta^{(n)}[e_2](0|\Omega_{22}) \ .$$

From the definition of F_2 one sees that for $\Omega_{12} \to 0$ the fundamental domain F_2 is simply given by two copies of F_1 ($\Omega_{11}, \Omega_{22} \in F_1$), with an extra factor $\frac{1}{2}$, coming from the second condition. The vacuum partition function is given by

$$Z_2 \equiv c_2 g^2 \frac{4\pi}{(2\pi^2)^D} \int_{F_2} \frac{d^2\Omega_{11} d^2\Omega_{12} d^2\Omega_{22}}{(det\ Im\Omega)^{13}} |\Delta_{(2)}|^{-4} \xrightarrow{\Omega_{12}\to 0}$$

$$\frac{1}{2} c_2 g^2 \frac{4\pi}{(2\pi^2)^D} (\frac{2}{\pi})^4 \int_{F_1} \frac{d^2\Omega_{11}}{(Im\Omega_{11})^{13}} |\Delta_{(1)}(\Omega_{11})|^{-16} \int_{F_1} \frac{d^2\Omega_{22}}{(Im\Omega_{22})^{13}} |\Delta_{(1)}(\Omega_{22})|^{-16}$$

$$\times \int_\epsilon^1 \frac{d^2\Omega_{12}}{|\Omega_{12}|^4} (1 + \frac{D}{4} \frac{|\Omega_{12}|^2}{(Im\Omega_{11})(Im\Omega_{22})} + ...) . \qquad (A.6)$$

Hence we obtain for the massless exchange contribution

$$Z_2 \to -\frac{c_2}{c_1^2} \frac{4}{\pi^2} g^2 D Z_1^2 \log \epsilon \ .$$

Now, using eq. (4.19a)

$$d_2' = -\frac{D d_1^2}{4 d_0} , \qquad (A.7)$$

we obtain

$$\frac{c_2}{c_1^2} = \frac{\pi^2}{16 d_0} . \qquad (A.8)$$

Let us now quote some relations involving massless vertex operators at zero momentum. The soft dilaton and (trace of) graviton vertex operators are given by[20,21,10]

$$\begin{aligned} V_d &\equiv \frac{1}{4\pi\alpha'} \int d^2\sigma \sqrt{h} h^{ab} (: \partial_a X^\mu \partial_b X^\mu : -\frac{1}{2}\alpha' R^{(2)}) \\ &= \frac{1}{4\pi\alpha'} \int d^2\sigma \sqrt{h} h^{ab} (\partial_a X^\mu \partial_b X^\mu + \frac{1}{4}\alpha'(D-2) R^{(2)}) \qquad (A.9) \\ V_g &= \frac{1}{4\pi\alpha'} \int d^2\sigma \sqrt{h} h^{ab} : \partial_a X^\mu \partial_b X^\mu : \end{aligned}$$

where the normal ordering is understood to be with respect to the propagator on the sphere (see eq.(2.17)).

Let us consider the computation of the graviton tadpole at genus n (we choose conformal coordinates)

$$< V_g >_n = c_n g^{2n-2} \int_{F_n} d\mu_n \frac{1}{\pi\alpha'} \int <: \partial X \cdot \bar\partial X :>_n \ . \qquad (A.10)$$

Using the general expression for the propagator one has (cf. (4.8))

$$<: \partial X \cdot \bar\partial X :>_n = -\frac{\pi D}{2} \alpha' \omega_a (\Omega_2^{-1})_{ab} \bar\omega_b \ . \qquad (A.11)$$

Now, using $\int \omega_a \wedge \bar{\omega}_b = -2i(\Omega_2)_{ab}$, one obtains

$$< V_g >_n = -\frac{nD}{2} Z_n = -\frac{D}{4}(2-\chi) Z_n \ . \tag{A.12}$$

Defining

$$\hat{V} \equiv V_g + \frac{1}{2} nD \ , \quad \chi = 2 - 2n \ , \tag{A.13}$$

one gets

$$< \hat{V} >_n = 0 \ . \tag{A.14}$$

Similarly, one can obtain

$$< V_g V_g >_1 = \frac{Z_1 D}{4}(D-2) \ . \tag{A.15}$$

References

1. C. Lovelace, Phys. Lett.B135, (1984) 75; Nucl.Phys. B273 (1986) 413.
2. E.S. Fradkin and A.A. Tseytlin, Phys.Lett. B158 (1985) 316; Nucl.Phys. B261 (1985) 1;
   P. Candelas, G. Horowitz, A. Strominger and E. Witten, Nucl. Phys. B256 (1985) 46.
3. C. Callan, D. Friedan, E.Martinec and M. Perry, Nucl.Phys.B262 (1985) 593.
4. W. Fischler and L. Susskind, Phys.Lett.B171 (1986) 383; Phys.Lett. B173 (1986) 262.
5. J.Shapiro, Phys.Rev.D5 (1972) 1945; D11 (1975) 2937;
   M. Ademollo, R. D'Auria, F. Gliozzi, E.Napolitano, S. Sciuto and P. di Vecchia, Nucl.Phys.B94 (1975) 221
6. A.A. Tseytlin, Int. J. Mod. Phys. A4 (1989) 3269;
   H. S. La and P. Nelson, Boston University preprints BUHEP-89-9; Phys. Rev. Lett. 63 (1989) 24.
7. A.A. Tseytlin: "*Sigma models and renormalization of string loops*", Lectures at the 1989 Trieste Spring School on Superstrings, ICTP preprint IC/89/90, to appear in Int. J. Mod. Phys.
8. J. Russo and A.A.Tseytlin, SISSA preprint 110/EP/89.
9. B. Fridling and A. Jevicki, Phys. Lett. B174 (1986) 75.
10. A.A. Tseytlin, Phys. Lett. B208 (1988) 228.
11. A.B. Zamolodchikov, JETP Lett. 43 (1986) 730;
    A.M. Polyakov, Phys. Scripta T15 (1987) 191;
    A.A. Tseytlin, Nucl. Phys. B276 (1986) 391; Phys. Lett. B194 (1987) 63;
    H. Osborn, Nucl. Phys.B308 (1988) 629; Phys. Lett. B214 (1988) 555.
12. S. Weinberg, in: Proc. DPF Meeting of APS (Eugene, Oregon, 1985) Ed. R. Hua (World Scientific, Singapor, 1986);

N. Seiberg, Phys. Lett. B187 (1987) 56.
13. A.A. Tseytlin, Phys. Lett. B178 (1986) 34;
H. Osborn, Nucl. Phys.B294 (1987) 595.
14. J. Polchinski, Nucl. Phys. B307 (1988) 61.
15. S. Weinberg, Phys. Lett. B187 (1987) 287.
16. N. Marcus, Phys. Lett. B219 (1989) 265.
17. A.A.Belavin, V.G. Knizhnik, A. Morozov and A. Perelomov, Phys.Lett.B177 (1986) 324;
G. Moore, Phys. Lett. B176 (1986) 369;
A. Morozov, Phys. Lett. B184 (1987) 171.
18. H. Ooguri and N. Sakai, Phys. Lett. B197 (1987); Nucl. Phys. B312 (1989) 435.
19. U. Ellwanger and M.G. Schmidt, Nucl.Phys. B314 (1989) 175;
U. Ellwanger, Nucl.Phys. B326 (1989) 254.
20. S. de Alwis, Phys. Lett. B168 (1986) 59;
J.Liu and J.Polchinski, Phys.Lett. B203 (1988) 39.
21. A.A. Tseytlin, Int. J. Mod. Phys.3A (1987) 365.

LOOP EFFECTS ON MASSIVE MODES IN SUPERSTRING THEORIES

AKIHIKO TSUCHIYA
Tokyo Institute of Technology
Oh-okayama, Meguro-ku, Tokyo, 152 Japan

ABSTRACT

We examine to what extent the picture of a splitting probability per unit length holds for total decay rates in string theories.

1. Introduction

The existence of the tower of the infinitely many massive states is certainly one of the features which characterize string theories. Since string theories have been developed as serious candidates for the unified theory of all interactions in the universe and hence our concerns have been mainly on massless modes or low-energy phenomena, it is not too much to say that we have not yet acquired a great amount of knowledge concerning the massive modes. In recent advances on the high energy scattering of strings and finite temperature effects, massive modes will play crucial roles, and it is desirable to fully investigate them. In this talk, I would like to review recent developments on some aspects of loop effects of such highly excited states, mainly on their mass shifts and decay rates. Our main goal is to study to what extent the old idea of regarding the open string loop coupling constant g as a density of splitting probability per unit length holds in string theories. The contents of my talk are based on the works of [1]-[7]. Similar issues are studied in [8]-[9] and by the Princeton group [10]-[14], and some of the crucial ideas are due to them.

2. Mass Shift Expression

The mass shifts and the decay rates of the string modes correspond to the real and imaginary parts of the two-point massive amplitudes as in the ordinary quantum field theories. I begin with the construction of the vertex operators.

There is technical difficulty in systematically constructing vertex operators of the highly excited modes for realistic (compactified) string models, so we restrict ourselves to discuss the superstring theories at the critical dimension. By standard superconformal techniques, it is not so difficult to make vertex operators of the open model describing the emission of a series of massive bosons on the Regge trajectories:

$$V = \frac{1}{\sqrt{n!m!}} S\mu_1 \cdots \mu_m \nu_1 \cdots \nu_n \Psi^{\mu_1} \cdots \Psi^{\mu_m} \partial X^{\nu_1} \cdots \partial X^{\nu_n} \exp(ikX)$$

$$2\ell = 2n + m + 1 \, . \tag{1}$$

Small m is the number of fermion fields, which should be odd by the GSO condition, and small ℓ is the mass level number. There are two SO(9) irreducible states at the first level, and the vertex operators for the bosons on the leading trajectory (m = 1) is the extension of the state. These have maximal symmetry of space-time indices. These are in the F2 picture, and at the fixed mass level, as the number of fermion fields increases and the number of boson fields decreases, the corresponding operators describe boson states on the lower trajectories. These series of vertex operators can cover all combinations of spin J and mass square M^2, but of course there are many particle states which cannot be represented by these operators. Only the operator of the leading trajectory is the unique one.

It was Yamamoto [1] who first calculated the two-point amplitudes of the leading and next-to leading trajectories. The corresponding mass shifts with definite coefficient are found to be the same for these two trajectories, but mass shifts have non-trivial dependences on ℓ:

$$\delta m^2 = \frac{g^2}{\alpha'} \frac{N}{2} \frac{\ell}{(2\pi)^{2\ell+8}} \int_0^1 \frac{dw}{w} \int_0^\tau d\nu \frac{1}{\tau^5} \psi^{2\ell} \left[\frac{\partial^2}{\partial \nu^2} \ln\psi \right]^{\ell-1} ,$$

$$\psi = -2\pi i \frac{\theta_1(\nu|\tau)}{\theta_1'(0|\tau)} \, . \tag{2}$$

If mass shifts were simply proportional to ℓ, they could be successfully absorbed into a slope parameter redefinition. This result suggests that this program does not work.

It may be rather curious that for the two trajectories, mass shifts have the same integral expression and hence mass degeneracies remain at all mass levels, because there are no principles which forbid mass splittings for highly excited states. But these degeneracies are only accidental, and for the lower trajectories we have different, very complicated integral expression. For example, the mass shifts of bosons on the "third" trajectory (m = 5) become

$$\delta m^2 = \frac{g^2}{\alpha'}\frac{N}{2}\frac{1}{(2\pi)^{2\ell+8}}\int_0^1 \frac{dw}{w}\int_0^\tau d\nu\,\frac{1}{\tau^5}\,\psi^{2\ell}$$

$$\times\ [(8\ell+1)P^2 + 4(8\ell-5)G_2P + 20(\ell-1)G_2^2 + 15(\ell-2)G_4] \qquad (3)$$

where P is the Pe-function and G_i are modular forms. We will also consider the decay rates for these states, which do not correspond to straight strings in a classical sense.

3. Singularities

String amplitudes as conventionally given are formal objects. They are, for example, real and divergent, whereas physically they should be complex and finite at least for the imaginary parts. One-loop two-point amplitudes are manifestly real since they are defined as $Tr(V^\dagger V)$, but they should have imaginary parts required from unitarity. We expect that the on-shell two-point functions thus calculated are on cuts as functions of external momenta. Therefore, proper analytic continuations are needed to extract correct answers from them. In this process, we will obtain imaginary parts, which are also expected to give the decay rates of the corresponding massive modes.

First, let us look closely at the planar two-point amplitude for the lightest massive states. A similar analysis can be applied to the nonorientable amplitude. In eq. (2), we set $\ell = 1$ and introduce an ultraviolet cut off $\Lambda(0<\Lambda<1)$ conventionally to avoid the dilaton tadpole divergence in the ultraviolet region.

We subtract the first term of the expansion of the theta function (which we denote A_2) from the mass shift expression and define A_1 as the rest:

$$\delta m^2 = A_1 + A_2 \ ,$$

$$A_2 = \frac{g^2}{\alpha'} \frac{N}{2} \frac{1}{(2\pi)^8} \int_0^{\Lambda} \frac{dw}{w} \int_0^{\tau} d\nu \, \frac{1}{\tau^5} \, e^{\pi i \nu k^2 \left(1 - \frac{\nu}{\tau}\right)} \tag{4}$$

One can see that A_1 is convergent in a certain momentum region which includes the on-shell value and analytic in it.

If we recall that ν and τ are purely imaginary numbers with positive imaginary parts and satisfy $Im\nu < Im\tau$, then the integral A_2 clearly diverges exponentially when $Rek^2 < 0$. One can think that A_2 is the contribution from the propagation of internal massless particles, which would become dominant in the infrared region.

Here we do an analytic continuation to search for the region of k^2 where both A_1 and A_2 converge and then analytically continue them back to the on-shell values. By a simple consideration, we find that the region of complex k^2 where both A_1 and A_2 converge is unique: $0 < Rek^2 < 1$. The value of Imk^2 is irrelevant to the fact whether A converges or not. For A_1, it is convergent and analytic in the region $Rek^2 < 1$, so we can integrate it on shell $k^2 = -2$ from the beginning. As for A_2, we transform variables as follows:

$$A_2 \sim \int_{f(\Lambda)}^{i\infty} d\tau \int_0^1 dy \, \frac{1}{\tau^4} \, e^{\pi i \tau k^2 y(1-y)} \ , \quad (\tau,\nu) \to (\tau,y), \quad y = \frac{\nu}{\tau} \tag{5}$$

$$\sim \int_0^1 dy y^3 (1-y)^3 \int_{g(\Lambda,k^2)}^{\infty} dt \, \frac{e^{-t}}{t^4} \ , \quad (\tau,y) \to (t,y), \quad t = -\pi i \tau k^2 y(1-y).$$

Repeating the partial integrations for the t integral, we find the following Ei-function appears:

$$Ei(-x) = -\int_x^\infty dt\, \frac{e^{-t}}{t} = \ln x + \gamma - x + \sum_{r=2}^{\infty} \frac{(-x)^r}{r\cdot r!}\,. \tag{6}$$

This expression contains the logarithmic function. In our case, this gives $\ln k^2$ having a cut in the k^2 plane which arises at the origin and runs along the real axis to $-\infty$. This term gives the imaginary part, which appears when we let k^2 go back to on-shell by analytically continuing $\ln k^2$ as $\ln k^2 \equiv \ln e^{i\pi}(-k^2)$.

When partially integrating A_2 in $0<\mathrm{Re}k^2<1$, there appear surface terms approaching zero exponentially when $\tau\to\infty$. These terms are the contributions from the boundary of the moduli space and are divergent if we integrate out the amplitude on-shell. All processes described above are equivalent to simply dropping such boundary terms, which may be precisely the roles of contact terms, if any.

For the validity of our naive analytic continuation, see ref. [4]. We assume it even for the case of highly excited states, at least for their imaginary parts.

It may be an interesting issue to numerically estimate the order of the mass shift of the first level as an example of the estimation of the loop effects in the string theories. Here we consider the singlet state for the Chan-Paton factor as an example. There is no substantial difficulty in incorporating the contributions from the nonorientable and the nonplanar parts. We find that for the singlet trajectories the infrared singularities are cancelled between the planar and nonplanar amplitudes, and come solely from nonorientable diagram. After the numerical estimation, we find the value of the mass shift to be

$$\delta m^2_{\ sing} = \frac{g^2}{\alpha'(2\pi)^8}\left(2.2952 - \frac{\pi^3}{7\cdot 6\cdot 5}\, i\right). \tag{7}$$

The real parts have calculational error $\pm 2\times10^{-4}$ in the parenthesis, and have the value

$$\mathrm{Re}[\delta m^2_{\ sing}] = 9.4 \times 10^{-7} \times \frac{g^2}{\alpha'}\ .$$

The mass at the tree level is $\frac{1}{\alpha'}$. Recall that g is the dimensionless loop expansion parameter and is related to the ten-dimensional gauge coupling constant g_{10} as $g = g_{10}(2\alpha')^{-3/2}$. In string theory, the determination of the value of g is a purely dynamical issue, the problem of determining the dilaton vacuum expectation value. However, if the superstring theory is a realistic unified theory, the order of the g^2 is estimated to be $\sim 10^{-1}$ from some phenomenological arguments. In that case, the real parts of the mass shifts for the first excited level are of the order $\sim 10^{-7} \times \frac{1}{\alpha'}$. Then we conclude that the mass shift with a precise normalization has turned out to be much smaller ($\sim 10^{-7}$) than the tree level one. If the string coupling constant is much larger, the string perturbation is meaningless from the beginning. Thus, our result of the estimation shows that the string one-loop approximation is quantitatively valid, at least in the region treated in the above.

4. Decay Rates

In the work of Mitchell et al., a minute description to evaluate the imaginary parts of the highly excited modes for the bosonic open string has been given. It is most convenient to express amplitudes in terms of dummy variables which are used in a parameter representation of the propagaters. Here we calculate the explicit values of the decay rates for the highly excited modes of the type-I superstring theory in the critical dimension.

We expand the integrand of eq. (2) as

$$\delta m^2 \sim \sum_{p,q,B} C^B_{pq} \int_0^1 \frac{dw}{w} \int_0^\tau d\nu \, \frac{1}{\tau^B} \, x_2^{-\ell} \, x_1^P \, x_2^q \, e^{\frac{2\pi i \ell \nu^2}{\tau}} \tag{8}$$

$$x_1 = e^{2\pi i(\tau-\nu)} , \quad x_2 = e^{2\pi i \nu} \qquad w = x_1 x_2 = e^{2\pi i \tau} \quad .$$

We set $y = \frac{\nu}{\tau}$ and rewrite mass shifts in $y(0<y<1)$ and τ as

$$\sum_{p,q,B} C^B_{pq} \int_0 dy \int_0^{i\infty} d\tau \, \frac{1}{\tau^B} \, e^{2\pi i \tau A(y)} \tag{9}$$

where

$$A(y) = \ell y^2 - (P-q+\ell)y + P \ . \tag{10}$$

Since τ is purely imaginary with positive imaginary parts, terms in eq. (9) diverge if $A(y)<0$ for $0<y<1$. The conditions of the divergence are easily found to be

$$0 < \frac{P+\ell-q}{2\ell} < 1 \quad , \quad (P+\ell-q)^2 - 4\ell P > 0 \quad . \tag{11}$$

From this equation, we can easily derive that $P<\ell$, $q<\ell$, and, more strongly, $P+q<\ell$ as necessary conditions. The physical picture is remarkably clear: in definitions of two-point amplitudes $\mathrm{Tr}(x_1^L o\ V\ X_2^L o\ V)$, the operator L_o is nothing but

$$\frac{p^2}{2} + \text{mass operator}$$

and hence by fixing p and q we derive the decay rate of the external mass mode into intermediate two states, each for level p and q. The concrete values of the imaginary parts can be obtained by partially integrating and deforming the integral contours or equivalently performing analytic continuations.

The total decay rate is related to $\mathrm{Im}(\delta m^2)$ as

$$\Gamma = -\mathrm{Im}(\delta m) = -\frac{\mathrm{Im}(\delta m^2)}{2m} = -\frac{1}{\sqrt{8\ell}}\,\mathrm{Im}(\delta m^2) \quad . \tag{12}$$

It is now well known that if it is possible to regard the open string coupling constant as a splitting probability per unit length, the total decay rate Γ should be proportional to the square root of the mass level number. This means that the imaginary parts of the eq. (2) except for the prefactor ℓ:

$$\mathrm{Im}\,\frac{1}{(2\pi)^{2\ell}}\int_0^1 \frac{dw}{w}\int_0^\tau d\nu\,\frac{1}{\tau^5}\,\psi^{2\ell}\left(\frac{\partial^2}{\partial\nu^2}\ln\psi\right)^{\ell-1} \tag{13}$$

should be almost independent of ℓ at least for the sufficiently large mass states. At first sight, this is not plausible, as I have already noted for the possibility of absorbing mass shifts (real parts) into the Regge slope parameter redefinition. To see the answer, we performed computer analysis, and the result of the explicit values of the decay rates for the charged states of the type-I SO(32) superstring theory is shown in Fig. 1. We have checked the results by hand up to the third mass level. To our surprise, the decay rate values do not fluctuate at the lower mass levels, and almost all values are beautifully on

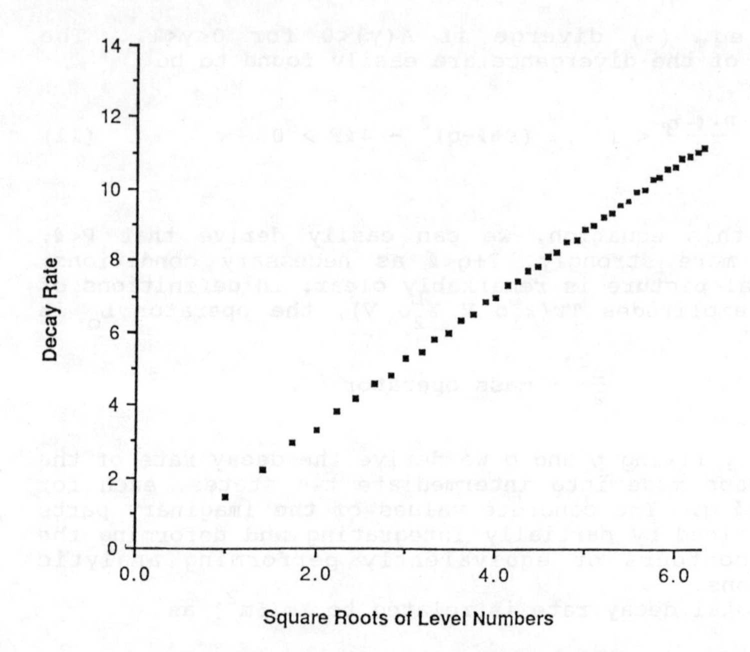

Fig. 1. Decay rates of the charged states in SO(32) type-I superstring in units of $\frac{g^2}{\alpha'} \frac{1}{2(2\pi)^8}$.

a straight line. We have to emphasize that there are no calculational errors in this result since we used computer analysis only for expanding theta functions. Eq. (13) has the value ~ 0.157, which is quite impressive since the number of the divergent terms in the expansion of theta functions which contribute to the imaginary parts of the mass shifts becomes large very soon as the level number ℓ increases.

Independently, Mitchell et al. showed the value 0.157 to be $1/2\pi$ by using some approximation. The significance of our computer analysis, if any, is to have confirmed that the validity of the splitting probability per unit length holds for almost all massive states contrary to our naive expectation. Even at the first level, whose imaginary part is

$$\frac{\pi^3}{7\cdot 6\cdot 5} \ (= 0.148) \ ,$$

this approximation is not so bad. It seems certain that the excited states rapidly approach their asymptotic region and behave classically as for their decay processes. Our result for the charged states leads to

$$\Gamma = \frac{g^2}{\alpha'} \frac{1}{2(2\pi)^8} 1.77 \; \ell^{1/2} \; . \tag{14}$$

How about for the bosons on lower trajectories? As for the boson states on the next-to leading trajectory, the mass shift expression is the same as that for the leading trajectory, so the answer is trivial. For the third trajectory, whose mass shifts are given in eq. (3), the results are shown in Fig. 2. Though these states do not

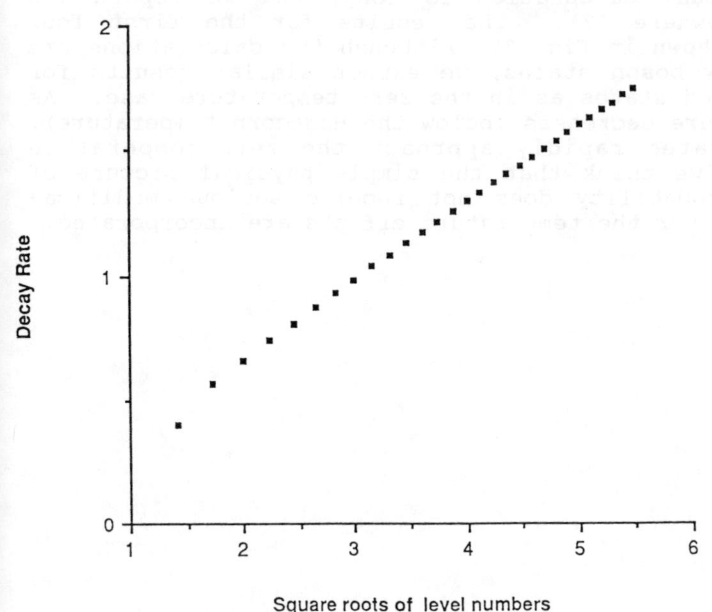

Fig. 2. Decay rates of the bosons on the "third" trajectory (U(1) sector).

correspond to the straight strings as I noted and there is no reason to have decay rates proportional to $\ell^{1/2}$, these are also approximately on a straight line. This suggests that the idea of the splitting probability per unit length holds well beyond our first naive expectation.

5. Thermal Effects on Decay Rates

It is a widely believed conjecture that as the Hagedorn temperature approaches most of the energy of the system goes into one long string and a phase transition occurs. It may be interesting to investigate the behavior of the mass shifts and the decay rates at finite temperature. Our interest here is how temperature effects deform the splitting probability rates.

We consider two-point amplitudes in their rest frame. Following the standard procedure, the static mass correction is obtained by compactifying the time direction and replacing the continuous energy E with discrete values.

The actual calculation is long, and we report its details elsewhere [7]. The results for the first four levels are shown in Fig. 3. Although the calculations are for some low boson states, we expect similar results for highly excited states as in the zero temperature case. As the temperature decreases (below the Hagedorn temperature), the decay rates rapidly approach the zero-temperature values. So we think that the simple physical picture of splitting probability does not receive serious modifications even after the temperature effects are incorporated.

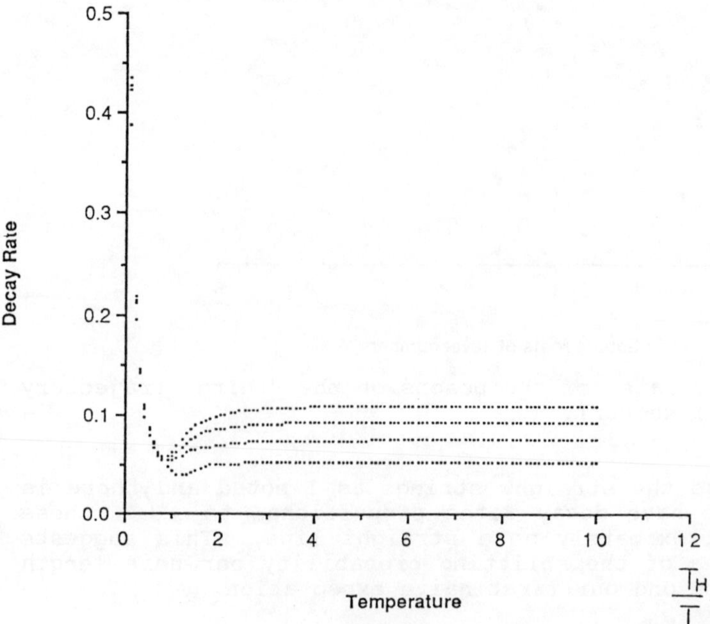

Fig. 3. Temperature dependence of decay rates. Shown are the first four states on the leading trajectory.

Note added:

Recently, we found that the decay rates for the states given by the vertex eq. (1) have the form

$$\Gamma = \frac{g^2}{\alpha'} \frac{N}{2(2\pi)^8} \frac{1}{\sqrt{8}} \left(A\ell^{\frac{1}{2}} + B\ell^{-\frac{1}{2}}\right)$$

where A and B are calculable constants. For example, (A,B) = $(1/2\pi, 0)$ for the leading and the next-to leading trajectories, and $(5/2\pi,\ 2/\pi)$ for the third trajectory. Therefore, all trajectories represented by eq. (1) have decay rates per unit length when the mass level number is sufficiently large.

It is also possible to give a simple summary of results at finite temperature. All values of the decay rates will be unaffected by temperature effects, so long as we consider the system below the Hagedron temperature.

References

[1] H. Yamamoto, Prog. Theor. Phys. 79 (1988) 189.
[2] K. Amano and A. Tsuchiya, Phys. Rev. D39 (1989) 565.
[3] A. Tsuchiya, Phys. Lett. B214 (1988) 35.
[4] K. Amano, to appear in Nucl. Phys.
[5] T. Nishioka, A. Tsuchiya and H. Yamamoto, to appear in Mod. Phys. Lett.
[6] H. Okada and A. Tsuchiya, TIT-HEP 149.
[7] H. Okada and A. Tsuchiya, in preparation.
[8] N. Marcus, Phys. Lett. B219 (1989) 265.
[9] Dai and J. Polchinski, Phys. Lett. B220 (1989) 387.
[10] B. Sundborg, Nucl. Phys. B306 (1988) 545.
[11] B. Sundborg, Nucl. Phys. B319 (1989) 415.
[12] B. Sundborg, Princeton preprint PUPT-1131 (1989).
[13] D. Mitchell, N. Turok and R. Wilkinson, Nucl. Phys. B315 (1989) 1; Imperial/TP/89/23.
[14] D. Mitchell, B. Sundborg and N. Turok, Princeton preprint PUPT-1186 (1989).

Finiteness Constraints on Open String Theory

Paul H. Cox*
Department of Physics, Texas A&I University
Kingsville, TX 78363

and

L. Clavelli and B. Harms
Department of Physics and Astr., University of Alabama
Tuscaloosa, AL 35487

ABSTRACT

Some recent results on open string theory in fewer than ten dimensions are presented and commented on. Finiteness constraints are found to be restrictive but difficult. No models in four dimensions were identified in our work.

1. Introduction

The rebirth of string theory in physics has probably attracted more interest among theoretical physicists than any other development in recent times. However, those involved directly in comparing experimental results to theory have frequently expressed the criticism that there have been as yet no phenomenological successes from pure string theory. This talk will present some recent results in the theory of Type I (open) strings which the Alabama-Texas A&I group has obtained, with some comments on the above-mentioned criticisms.

Although most of the milestones[1] of string theory in the 1960-1970's era, as well as the 1984 breakthroughs[2] that led to the current upsurge of interest involved, initially, advances in open string theory, most of the more recent effort has been concentrated in the theory of closed strings (Type II and heterotic). This has several reasons; for one, the mechanism for implementing internal symmetries in open string theories (Chan-Paton factors) is rather ad hoc and thus aesthetically less pleasing than the symmetry implementation in closed strings. The Marcus-Sagnotti idea[3] for a dynamical understanding of Chan-Paton factors has somewhat mitigated this problem. A more serious cause of the neglect of open string theories is the apparent collective judgement that closed strings are closer to phenomenology than are open string theories. While this evaluation may be correct, in view of the facts (1) that closed string theories are formed by pairing open string theories, and (2) the closeness to phenomenology seems to be intrinsically accompanied by enormous arbitrariness, lack of uniqueness, and absence of predictive power, one wonders whether further advances toward the ultimate string theory, the true 'theory of everything' (if there is one, as most of us trust), might well be provided by study of open strings.

*speaker

In open string theory the fundamental material entities are described by an analogy with bits of string with loose ends. Under interaction, relativistic open strings inevitably produce states which have a closed string structure; the converse, however, is not true. Although it has not been related to current experimental data, the mathematically consistent open string theory in 10 dimensions, as it was known by 1984, is uniquely successful in the following sense. It predicts the qualitative features of our physical universe, and uniquely singles out discrete parameter choices. More specifically, once one postulates that matter is fundamentally described as relativistic open strings with mathematically consistent interactions, the following predictions uniquely follow:

1. The existence of gravity.
2. The existence of non-Abelian gauge interactions.
3. The existence of anomaly-free chiral fermions.
4. Supersymmetry (SUSY).
5. Critical (maximum) space-time dimension (D=10).
6. Unique gauge group [SO(32), in D=10].

We consider the requirement of supersymmetry a phenomenological success although its experimental confirmation may have to await the Southern Super Collider.

Already in 1984 the closed string theories showed less uniqueness than the open strings, and since that time thousands of apparently equally consistent closed string theories have been found in fewer than 10 dimensions[4]. In open string theory the situation is much more tightly constrained. If one restricts one's attention to the "unmixed" fermionic construction in which, as in the D=10 theory, all anticommuting oscillators on boson lines are Neveu-Schwarz-like and all those on fermion lines are Ramond-like, then in lower dimensions a unique open string theory appears[5], this time in six dimensions, with the above features modified or extended by some new phenomenologically interesting results:

7. Family replication of fermions.
8. Spontaneous breakdown of space-time supersymmetry.
9. Unique prediction of a rank-4 Chan-Paton gauge group.
10. Appearance of Higgs-candidate scalar mesons, in gauge-group adjoints.
11. Higgs instability in perturbation theory, partly described by calculable milliweak dimensionless parameters.

To be fair one must note that (a) family replication and SUSY breaking occur also in many lower-dimension closed string theories, (b) this open string theory apparently loses space-time chirality, and (c) SUSY breaking is near the Planck scale so all the fermions acquire Planck-scale masses. In spite of these phenomenological drawbacks, the model merits some attention to its implementation of features 7 through 11 above, if only for its uniqueness within the specified limits. Of course this uniqueness would probably not continue once one allowed for "mixed" models; but all indications from our work so far indicate that there are still relatively very few consistent lower-dimensional models. (We do not discuss here the bosonic construction in which space is directly compactified. That method still seems to allow any number of Minkowski dimensions up to ten. In particular, we see no obstruction to reducing the dimension of the D=6 model to four by toroidal compactification of the other two dimensions, at the cost of additional ambiguities in the compactification radii. We also do not consider orbifold compactification schemes.[6])

2. Unmixed Models

To establish our notation, We outline the "unmixed" construction. It has been known since 1973[7] that the critical space-time dimension D of the bosonic string theory could be systematically reduced by adding some number d of anticommuting fields on the world sheet, according to the relation

$$2\, D + d = 52. \tag{1}$$

This relation leads, for instance, to the basic SO(44) invariance of the left-moving part of the four-dimensional heterotic string. The correspond relation for a superstring is

$$3\, D + d = 30. \tag{2}$$

A model may be indicated by giving its Virasoro superalgebra generato For an unmixed model, in a fermion sector, these are

$$F_M = \frac{1}{\sqrt{2i}} \oint \frac{dz}{2\pi i z}\, z^M \Big[P_\mu \Gamma^\mu + \Gamma_{D+1} \frac{f^{abc}}{6\sqrt{2C_2}} \Gamma^a \Gamma^b \Gamma^c \Big]; \tag{3a}$$

$$L_M = \frac{1}{2} \oint \frac{dz}{2\pi i z}\, z^M \Big[P_\mu P^\mu - \frac{1}{2} \dot{\Gamma}_\mu \Gamma^\mu + \frac{1}{2} \dot{\Gamma}^a \Gamma^a \Big]; \tag{3b}$$

where $\mu = 0$ to D–1, a = 1 to d. Closure of the superalgebra requires that f^{abc} be the structure constants of a d-dimensional semisimple Lie algebra, with $f^{abc} f^{ebc} = C_2 \delta^{ae}$ giving its Casimir operator; in this case

$$\Big\{ F_N, F_M \Big\} = 2\, L_{N+M} + \delta_{N+M,0} \Big[\frac{D}{2} N^2 + \frac{d}{12}(2\, N^2 + 1) \Big]. \tag{4}$$

The Γ zero modes give a Clifford algebra, which is acceptable if D, d are even. Thus the d-dimensional algebra $[SU(2)]^{10-D} = [SU(2)]^{d/3}$ is allowed, or, for D=4, SU(4)xSU(2) or SU(3)xSO(5). These give a superalgebra, but since $F_0{}^2 = L_0 + d/24$ is strictly positive the supersymmetry is broken, giving lowest fermion masses $\alpha' m^2 = d/24$; the Clifford algebra gives $2^{d/2}$ families, each in the Chan-Paton adjoint. The D=10 chiral projection $1+\Gamma^{11}$ here becomes $1+\Gamma^{D+1}\Gamma^{d+1}$ which leaves the fermion spectrum left-right symmetric (since Γ^{d+1} is equally ±1). The model has space-time scalars characterized by their fermion-line vertex:

$$V^a = ig\ e^{ikQ}\left[k\Gamma\ \Gamma^a + \Gamma_{D+1}\ \frac{f^{abc}}{2\sqrt{2C_2}}\ \Gamma^b\Gamma^c\right]; \tag{5}$$

they are in the Chan-Paton adjoint as well as carrying the explicit index for the d-dimension-group adjoint.

The main test we have applied for acceptability of lower-dimension models is finiteness: Unless amplitude calculations are finite except for usual singularities associated with physical states, the theory almost certainly will be undefinable. So, we check one-loop amplitudes for finiteness. For N external gauge bosons, standard methods reduce the amplitude to the form[8]

$$A_N = \int d\Omega\ \left(\frac{\tau}{i}\right)^{-D/2} \Big\{(\mathrm{Tr}\ \lambda^0)\left[U_B(\rho,w) + U_B(\rho,we^{2\pi i}) - U_F(\rho,w)\right]$$
$$- \xi\left[U_B(\rho,-w) + U_B(\rho,-we^{2\pi i}) - U_F(\rho,-w)\right]\Big\} \tag{6}$$

where

$$d\Omega = \frac{dw}{w}\prod_{i=2}^{N}\frac{d\rho_i}{\rho_i}\ ;\ \tau = \frac{\ln w}{2\pi i}\ ;\ \lambda^0 = \text{unit Chan-Paton matrix;}$$

ξ = +1 for orthogonal, -1 for symplectic, Chan-Paton group;
U = (modified) partition functions x vacuum expectation values of vertex expressions;

the first two U's (in each group) come from 1 and G in the G-parity (GSO) projected boson loop, the third U from the fermion loop (where the second term in the projector cancels); the terms with -w come from the Möbius loop diagram.

For w near 0, one finds expansions in the form

$$U_B(\rho,w) = w^{-1/2}c_{B1}(\rho) + c_{B2}(\rho) + w^{1/2}c_{B3}(\rho) + \ldots; \tag{7a}$$

$$U_F(\rho,w) = w^{d/24}\{c_{F1}(\rho) + w\ c_{F2}(\rho) + \ldots\}. \tag{7b}$$

The net result is that the integral is convergent in the $w \approx 0$ region for $D > 4$, marginally divergent for $D = 4$ (the ρ integrals give an extra factor of ln w), badly divergent for $D < 4$ (the divergence can be attributed to infrared behavior, which may be normal in $D = 4$ but would be a serious problem in $D < 4$). To investigate the $w \approx 1$ region, we apply Jacobi transforms:

$$\tau' = -1/\tau,\ w' = e^{4\pi^2/\ln w}\ \text{in the planar loop;} \tag{8a}$$
$$\tau'' = -1/4\tau,\ w'' = e^{\pi^2/\ln w}\ \text{in the Möbius loop.} \tag{8b}$$

Then

$$d\Omega = d\Omega'\tau^{N+1} = d\Omega''(2\tau)^{N+1}; \tag{9}$$

in the planar diagram terms

$$U_B(w) = \left(\frac{\tau}{i}\right)^{(D/2)-N-1} U_B(w'); \tag{10a}$$

$$U_B(we^{2\pi i}) = -\left(\frac{\tau}{i}\right)^{(D/2)-N-1} U_F(w'); \tag{10b}$$

$$U_F(w) = -\left(\frac{\tau}{i}\right)^{(D/2)-N-1} U_B(w' e^{2\pi i}); \tag{10c}$$

so the τ's cancel and the expression inside [] goes into itself with primed arguments.

But in the Möbius diagram terms,

$$U_B(-w) = e^{\pi i(10-D)/4} \left(\frac{2\tau}{i}\right)^{(D/2)-N-1} U_B(-w'' e^{2\pi i}); \tag{11a}$$

$$U_B(-we^{2\pi i}) = e^{-\pi i(10-D)/4} \left(\frac{2\tau}{i}\right)^{(D/2)-N-1} U_B(-w''); \tag{11b}$$

$$U_F(-w) = \left(\frac{2\tau}{i}\right)^{(D/2)-N-1} U_F(-w''); \tag{11c}$$

here factors of τ cancel leaving $2^{D/2}$, but the extra phases $e^{\pi i(10-D)/4}$ generally remain. In D = 10, those phases are absent and the divergence cancels if Tr $\lambda^0 = 2^{D/2}$, $\xi = 1$, hence SO(32) is selected. For D < 10, we have

$$A_N = \int d\Omega' \Big\{ (\mathrm{Tr}\ \lambda^0) \Big[U_B(w') + U_B(w' e^{2\pi i}) - U_F(w') \Big]$$

$$- \xi\ 2^{D/2} \Big[e^{-\pi i(D-10)/4} U_B(-w')$$

$$+ e^{\pi i(D-10)/4} U_B(-w' e^{2\pi i}) - U_F(-w') \Big] \Big\}. \tag{12}$$

The leading, divergent, terms in U_B cancel only if $e^{\pi i(D-10)/4}$ is real; that is, D = 10 mod 4. The w ≈ 0 divergence excludes D = 2, leaving only D = 6 as a candidate. The next-to-leading terms in U_B, also divergent, cancel only for Tr $\lambda^0 = 2^{D/2}$, $\xi = e^{\pi i(10-D)/4}$, which selects D = 6 with Sp(8) as Chan-Paton group.

This result, while interesting, has the following problems:

D = 6, not 4;

Fermions are too heavy (SUSY broken too high);

$2^{(d=12)/2}$ = 64 families: too many;

Sp(8) is rank 4 and contains SU(3)xSU(2)xU(1) but not with the right quantum numbers.

But it does give features 7 through 11 listed earlier; in particular there is Higgs instability, driven, in perturbation theory, by a calculable (small) wrong-sign 1-loop 2-point function or mass term:

$$\alpha' m^2 = - \frac{4g^2}{(2\pi)^5} \int \frac{dw}{w} \frac{\Theta_2^4}{\Theta_3^4} \left[1 - 4 \frac{\Theta_4^4}{\Theta_3^4} + 2 \frac{\Theta_4^2}{\Theta_2^2} \frac{\Theta_4^4 - \Theta_2^4}{\Theta_3^4} \right]$$

$$= - g^2 \,(1.499) \cdot 10^{-3}. \tag{13}$$

3. Mixed Models

The lack of phenomenological success with this unique unmixed model is one reason to look further. The proliferation of closed-string models is associated with altering boundary conditions on the fermionized modes; we can try the same. In place of Eq. (3a), we put, in a given sector (e. g. fermionic),

$$F = \frac{1}{\sqrt{2i}} \oint \frac{dz}{2\pi i z} z^M \left[P_\mu \Gamma^\mu + \Gamma_{D+1} \frac{f^{aij}}{C} \Gamma^a H^i H^j \right]; \tag{14}$$

or optionally include $\Gamma\Gamma\Gamma$ terms as well, as in Eq. (3); (in usual notation Γ (H) indicates a field with Ramond (Neveu-Schwarz) boundary conditions); L_N will include both Γ^a and H^i terms. Now closure of the algebra requires that all the f's give a Lie algebra, with Γ's and H's combining to give the adjoint, and that the Γ's alone give a subalgebra while the H's (spanning the quotient) define a symmetric space.

Such a model will have multiple 'sectors', each with a generalized GSO projector. Again there are constraints coming from finiteness. We begin by considering a general N-point 1-loop amplitude,

$$A_N \sim \sum_{\substack{\text{sectors}\\ \alpha}} \mathrm{Tr} \left(\sum_{\substack{\text{sectors}\\ \beta}} C^\alpha_\beta \, G^\alpha_\beta \right) \prod_1^N \left\{ (1 + \Omega) \, \Delta \, V(k, 1) \right\}, \tag{15}$$

where: V are vertex operators, Δ the propagator, and Ω the twist operator. α indexes the particles circulating in the loop. The β sum defines the generalized GSO projection for the sector α (a normalizing power of 2 is ignored here), using coefficients C^α_β with values 0, ±1 (if sector sums are suitably restricted 0 need not be used, but sometimes the appropriate restriction is not known *a priori*, or the restriction may be inconvenient) times fermion number operators G^α_β. α, β are d+D-2 component vectors, here as labels, with components either 0 or 1 (for complex fermion modes fractional values may be allowable; we have not pursued that possibility). (d+D-2 counts d internal modes plus fermionic partners of the coordinates, counting in light-cone gauge; by Lorentz symmetry the coordinate partners must all have the same values.) α^i = 1 (0) if in the sector the i-th mode has Ramond (Neveu-Schwarz) boundary conditions; β^i = 1 (0) if the number operator counts (omits) excitations of that mode (thus $G^{\alpha}_{\beta} = (-1)^{\alpha\cdot\beta}$).

The behavior of A_N near w = 0 is the same as for the unmixed models; that near w = 1 is more complicated, and provides constraints on the sector structure, which is already constrained in several ways.

First, in each contributing sector, the boundary conditions selected must obey the algebra closure requirements referred to above. Next, the GSO operation must define a projection and the amplitude must have the correct statistics, so

$$\left(\sum_{\beta} C^{\alpha}_{\beta}\, G^{\alpha}_{\beta}\right) \text{ must } = C^{\alpha}_{0} \prod (1 + C^{\alpha}_{b} G^{\alpha}_{b}) \tag{16}$$

where b runs through a 'basis' set, and $C^{\alpha}_{0} \prod 1 = C^{\alpha}_{0}$ must be +1 (−1, 0) if sector α is bosonic (fermionic, omitted). Here, the list of b's is a basis for the β's (i. e., for those with non-zero C's, as a subspace of the space of all such vectors,) in the usual sense with respect to component addition mod 2, or if each vector is identified with the set of its non-zero entries, then the operation is the set-theoretic symmetric difference.

For closed-string models, modular invariance through two loops leads to the requirement that the list of α's (sectors in the spectrum) be the same as the list of β's (terms in the GSO projectors); we have not established such a result for open strings (which do not have modular invariance; we have not explored their closed-string sectors) but it seems unlikely that the conditions we find could be satisfied without it, and we have only checked candidates with this property.

Investigating finiteness at w ≈ 1 requires Jacobi transforming A_N. In place of Eq. (6), we now have (indicating only the factors affecting the transformation structure)

$$A_N \sim \frac{dw}{w} \sum_{\alpha,\beta} C^{\alpha}_{\beta} \Big\{ (\mathrm{Tr}\ \lambda^0)\ (\Theta_1')^{-4} \Theta_2^{\alpha\cdot F/2} \Theta_4^{\beta\cdot F/2} \Theta_3^{(\alpha+\beta+F)\cdot F/2}\ T^{\alpha}_{\beta}(w)$$
$$- \xi\ (\Theta_1')^{-4} \Theta_2^{\alpha\cdot F/2} \Theta_4^{\beta\cdot F/2} \Theta_3^{(\alpha+\beta+F)\cdot F/2}\ T^{\alpha}_{\beta}(-w) \Big\} \tag{17}$$

where Θ are the widely known Jacobi functions, with argument τ understood in the first term, τ+1/2 in the second; F is the vector with all entries 1; vector addition (but not addition in dot products) is component-wise mod 2; and T denotes vacuum expectation values of vertex operator expressions.

After Jacobi transform, this is

$$A_N \sim \frac{dw}{w} \sum_{\alpha,\beta} C^{\alpha}_{\beta} \Big\{ (\mathrm{Tr}\ \lambda^0)\ (\Theta_1')^{-4} \Theta_4^{\alpha\cdot F/2} \Theta_2^{\beta\cdot F/2} \Theta_3^{(\alpha+\beta+F)\cdot F/2}\ T^{\beta}_{\alpha}(w)$$
$$- 2^{D/2}\ \xi\ (\Theta_1')^{-4} \Theta_2^{\alpha\cdot F/2} (e^{-\pi i/4}\ \Theta_3)^{\beta\cdot F/2}$$
$$(e^{\pi i/4}\ \Theta_4)^{(\alpha+\beta+F)\cdot F/2}\ T^{\alpha}_{\alpha+\beta+F}(-w) \Big\} \tag{18}$$

since because of their structure the T's transform similarly to Θ's.

Now relabeling $\alpha \leftrightarrow \beta$ in the first term and $\beta \leftrightarrow \beta+\alpha+F$ in the second (note that if the sums were understood to be limited so that the C's are all non-zero, then the two terms would no longer be assured of being summed over the same ranges, since the limitations might not match; in such a case term-by-term matching and cancellation of divergences would be a probably insurmountable problem)

$$A_N \sim \frac{dw}{w} \sum_{\alpha,\beta} \Big\{ (\mathrm{Tr}\,\lambda^0)(\Theta_1')^{-4} \Theta_2^{\alpha\cdot F/2} \Theta_4^{\beta\cdot F/2} \Theta_3^{(\alpha+\beta+F)\cdot F/2} C_\alpha^\beta T_\beta^\alpha(w)$$
$$- 2^{D/2} \xi\, (\Theta_1')^{-4} \Theta_2^{\alpha\cdot F/2} \Theta_4^{\beta\cdot F/2} \Theta_3^{(\alpha+\beta+F)\cdot F/2}$$
$$\frac{\epsilon_\beta}{\epsilon_{\alpha+\beta+F}} C^\alpha_{\alpha+\beta+F} T^\alpha_\beta(-w) \Big\} \tag{19}$$

where $\epsilon_A = (e^{\pi i/8})^{A\cdot A}$. For brevity we put $L = 2^{-D/2}\mathrm{Tr}\,\lambda_0$, dropping an overall $2^{D/2}$, and we now expand in w, keeping only potentially divergent terms:

$$A_N \sim \frac{dw}{w} \sum_{\alpha,\beta} \Big\{ L\, w^{(\frac{\alpha\cdot F}{16} - \frac{1}{2})} \{1 - \frac{\beta\cdot F}{2} w^{\frac{1}{2}} + \frac{(\alpha+\beta+F)\cdot F}{2} w^{\frac{1}{2}}\} C_\alpha^\beta T_\beta^\alpha(w)$$
$$- \xi\, w^{(\frac{\alpha\cdot F}{16} - \frac{1}{2})} \{1 - \frac{\beta\cdot F}{2} (-w)^{\frac{1}{2}} + \frac{(\alpha+\beta+F)\cdot F}{2} (-w)^{\frac{1}{2}}\}$$
$$\frac{\epsilon_\beta \epsilon_{\alpha+\beta}}{\epsilon_F} C^\alpha_{\alpha+\beta+F} T^\alpha_\beta(-w) \Big\}. \tag{20}$$

We now note that the leading term in T^α_β is independent of w, and of any β^i; thus, except when $\alpha = 0$ (the vector with all components 0), in Eq. 20 nonleading terms are convergent. Thus,

$$A_N \sim \frac{dw}{w} \Big[\sum_\alpha w^{(\frac{\alpha\cdot F}{16} - \frac{1}{2})} T^\alpha_\ell \sum_\beta \Big\{ L\, C^\beta_\alpha - \xi \frac{\epsilon_\beta \epsilon_{\alpha+\beta}}{\epsilon_F} C^\alpha_{\alpha+\beta+F} \Big\}$$
$$+ \sum_\beta \Big\{ L \frac{1}{\sqrt{w}} C^\beta_0 \Big[[1 - \frac{\beta\cdot F}{2}\sqrt{w} + \frac{(\beta+F)\cdot F}{2}\sqrt{w}]\, T^0_\beta(w) - T^0_\ell \Big]$$
$$- \xi \frac{1}{\sqrt{w}} C^0_{\beta+F} \frac{\epsilon_\beta \epsilon_\beta}{\epsilon_F} \Big[[1 - \frac{\beta\cdot F}{2}\sqrt{(-w)} +$$
$$\frac{(\beta+F)\cdot F}{2}\sqrt{(-w)}]\, T^0_\beta(-w) - T^0_\ell \Big] \Big\} \Big], \tag{21}$$

where T^α_ℓ is the leading term in T^α_β.

The leading ($\alpha = 0$) divergence may now be cancelable; if there are space-time fermions in the theory, then statistics gives $C^\beta_0 = \pm 1$ equally often and the sum on β will cancel the leading L term; the leading ξ

term may also be cancelable by suitable choices of $C^0{}_\beta$'s. The non-leading potentially divergent terms are not easily analyzed, and we have no general results.

We have explored a few cases, whose behavior may illustrate the general problem. Of course there is the unmixed case, where only 0 and F contribute to the α and β sums; the requirement of cancellation of the next-to-leading terms in $\sqrt{w}$ leads to the restrictions D= 6, d = 12, ξ = -1, Tr λ_0 = 8. Our next candidates, following leads suggested by heterotic models, were ones with four sectors,

$$0 = \{0^{D-2},\ 0^d\},$$

$$F = \{1^{D-2},\ 1^d\},$$

$$S = \{1^{D-2},\ (1,0,0)^{d/3}\},$$

$$S{+}F = \{0^{D-2},\ (0,1,1)^{d/3}\},$$

where the notation lists the Minkowski transverse modes, followed by internal modes, and exponents indicate repetition. Here, for reasonable GSO projections, sectors 0 and S have massless states, while F and S+F sectors have $\alpha' m^2 = d/24$ for their lowest-mass states; this pairing of sectors is associated with an unbroken space-time supersymmetry of the model (as in heterotic models, this supersymmetry is associated with the presence of sector S). The d-dimensional algebra is $SU(2)^{10-D}$ (note 10-D = d/3), with $U(1)^{10-D}$ as the subalgebra. The α, β sums thus give 16 terms; for external gauge mesons, seven vanish because of the trace with G^α_β. So for these candidates

$$A_N \sim (\mathrm{Tr}\ \lambda_0)[\ 9 \text{ terms }] - \xi\ [\ 9 \text{ terms},\ w \to -w\ ].$$

On Jacobi transforming, the first group of terms is merely rearranged, but in the second group there arise factors of $\exp(\pi i/4)$ (as in Eq. 18 above), so that some partition functions acquire phases u or u*, where $u = \exp((10-D)\pi i/4)$. The leading surviving terms then appear as

$$A_N \sim 2i \int \frac{dw}{w}\ \{[\mathrm{Tr}\lambda_0 - \xi\ 2^{D/2}\ \frac{u+u^*}{2}\ ]\ C + \ldots\}$$

where C is a combination of coefficients in w expansions, and ... represents safe powers of w. Thus finiteness requires a different relative normalization of the torus and Möbius terms than usual, since the Möbius terms will cancel partly among themselves; the finiteness condition is

$$\mathrm{Tr}\lambda_0 = 2^{D/2}\ \xi\ \mathrm{Re}\ u = 2^{D/2}\ \xi\ \cos\frac{(10-D)\pi}{4}.$$

The possible solutions, and their Chan-Paton groups, are:

D = 2	d = 24	SO(2)
D = 3	d = 21	SO(2)
D = 4	d = 18	No solution
D = 5	d = 15	Sp(4)
D = 6	d = 12	Sp(8)
D = 7	d = 9	Sp(8)
D = 8	d = 6	No solution
D = 9	d = 3	SO(16)
D = 10	d = 0	SO(32)

Note that D = 4 and 8 have no solutions; here the Möbius terms cancel themselves because of the phases. Features of the spectrum include (all in the Chan-Paton adjoint): O sector: scalar mesons (d/3 of them massless) and Chan-Paton gauge mesons. S sector: D-2 plus d/3 massless fermions. F, F+S sectors: massive states; all must provide representations of the Clifford algebra of 2d/3 zero modes, which gives a large 'family' replication.

4. Conclusion

Much remains to be investigated in open string models: we have barely mentioned the closed string sector; the finiteness constraints remain unsolved; and of course there are presumably many models still unidentified, including quite possibly the real world.

Note added: Since the workshop, we have recognized that the four-dimensional model described by Bern[9] corresponds in our terms to a 'truncation' of our mixed models: only the O and S sectors are retained.

Acknowledgements: One of us (PHC) wishes to thank the University of Alabama for hospitality while this work was developed. We also benefited from conversations with Z. Bern. The work was supported in part by Texas A&I University under their New Faculty Research Enhancement Program, and at the University of Alabama under DOE Grant no. DE-FG0584ER40141.

References

1. E. g., G. Veneziano, *Nuovo Cim.* 57A (1968) 190; Y. Nambu, Lectures at the Copenhagen Symposium, 1970; P. Ramond, *Phys. Rev.* D 3, (1971) 2415 ; Gliozzi, J. Scherk, and D. Olive, *Nucl. Phys.* B122 (1977) 253 .
2. M. B. Green and J. Schwarz, *Nucl. Phys.* B243 (1985) 285; *Phys. Lett.* 149B (1984) 117 .
3. N. Marcus and A. Sagnotti, *Phys. Lett.* 188B (1987) 58; L. Clavelli, P.H. Cox, and B. Harms, *Phys. Rev.* D 36 (1987) 3747 .

4. H. Kawai, D. C. Lewellen, and S.-H. H. Tye, *Phys. Rev. Lett.* 57 (1986) 1832 ; *Nucl. Phys.* B288 (1987) 1 ; I. Antoniadis, C. P. Bachas, and C. Kounas, *Nucl. Phys.* B289 (1987) 87 . For further references see S.-H. H. Tye, in Proceedings of the workshop "*Beyond the Standard Model*," held Nov. 18-20, 1988, Ames, IA.
5. L. Clavelli, P. H. Cox, and B. Harms, *Phys. Rev. Lett.* 61 (1988) 787 and *Phys. Rev. D* (in press).
6. J. A. Harvey and J. A. Minahan, *Phys. Lett.* 188B (1987) 44 ; G. Pradisi and A. Sagnotti, *Phys. Lett.* B216 (1989) 59; N. Ishibashi and T. Onogi, *Nucl. Phys.* B318 (1989) 239.
7. L. Clavelli and J. Shapiro, *Nucl. Phys.* B57 (1973) 490 , eq. 3.16.
8. For more details, see Ref. 5 and references cited therein.
9. D. Dunbar and Z. Bern, *Phys. Rev. Lett.* (1990) (in press) Z. Bern, these Proceedings.

CONSTRUCTION OF FOUR-DIMENSIONAL OPEN SUPERSTRINGS*

Zvi Bern
Theoretical Division, MS-P285, Los Alamos National Laboratory
Los Alamos, NM 87545, USA
and
David C. Dunbar
D.A.M.T.P., Liverpool University
Liverpool, UK

ABSTRACT

We present the first examples of space-time supersymmetric one-loop finite four-dimensional open superstrings.

The superstring explosion began with the Green-Schwarz SO(32) open superstring[1], but since then open superstrings have been neglected as potential theories of nature. However, there is no a priori reason for this. In the early days of the string revival, the $E_8 \times E_8$ heterotic string completely superceded the SO(32) open superstring because of the superior phenomenology. Since then the principle of modular invariance has led to a rich variety of four-dimensional heterotic strings[2–4], but there has been no analogous progress for open superstrings. The construction of open superstrings is technically more difficult since there is no simple symmetry principle, such as the modular invariance of closed strings, which guarantees much of the consistency. In this talk we present the first step in constructing phenomenologically sensible open superstrings: We have succeeded in constructing the first examples of one-loop finite four-dimensional open superstrings with space-time supersymmetry.

The basic rules for open string constructions have been given in a number of papers[5–9], but the only one-loop finite models which couple to modular invariant closed strings were found in dimensions $D = 2, 6, 10$. Additionally, in Cox's talk (in these proceedings) a number of other one-loop finite models have been given (which may not be coupled to modular invariant closed strings) in various dimensions not including $D = 4$. The source of the difficulty in $D = 4$ arises from the Jacobi transformation properties of the ϑ-functions describing the internal world-sheet fermionic degrees of freedom (see e.g. Cox's talk). Also, in $D = 4$ an apparent incompatibility between the allowed sets of basis vectors, as determined by the coupling to modular invariant closed strings and the twist operator was

* Talk presented by Z.B.

found[5]. In fact, this difficulty was so severe that there appeared to be no way to construct a four-dimensional open superstring within the fermionic formulation.

In this talk, we would like to present some recent work[10,11] which demonstrates how to get around this difficulty. The solution actually turns out to be technically quite simple: A particular modification of the standard twist operator allows for the construction in $D = 4$. We have constructed examples of four-dimensional open superstrings that are one-loop finite as well as space-time supersymmetric. Furthermore, the closed string poles which occur in the opens string amplitudes[12,13] correspond to the states of a modular invariant four-dimensional closed string.

In our construction of four-dimensional open superstrings, modular invariance is replaced with a series of principles:

1) Correct coupling between open and consistent modular invariant closed strings.
2) Physically sensible projections and correct space-time statistics.
3) Locality on the world-sheet.

The need for the second condition is completely analogous to the closed string situation with only one-loop modular invariance imposed[3]. The requirement of a physically sensible projection (PSP) is what leads to the modified twist operator in $D = 4$. Finally, locality on the world sheet can be used to relate the möbius to the Klein bottle amplitudes.

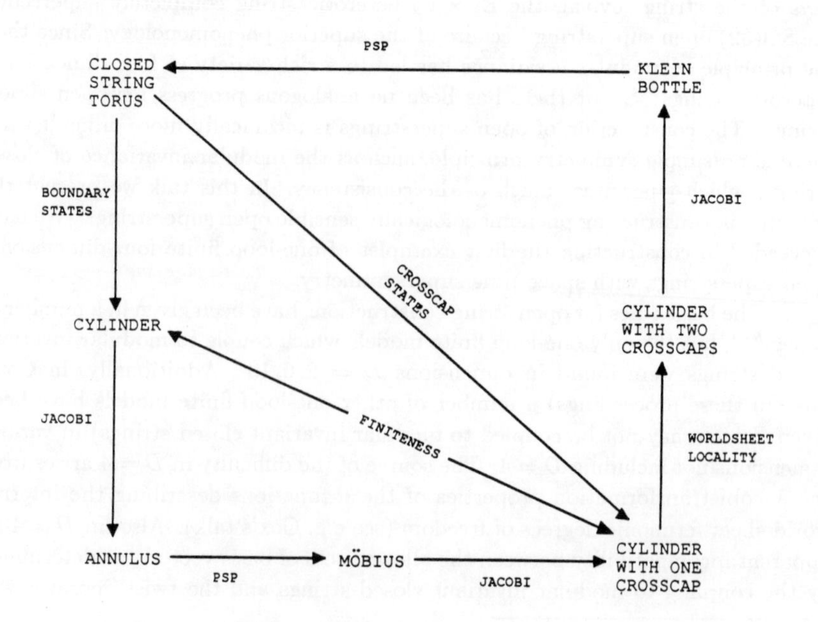

Fig. 1. Outline of our construction.

The outline of our construction is presented in fig. 1. The starting point is four-dimensional closed oriented superstrings in the fermionic formulation[14,3,4] whose constructions are quite well understood. For the unoriented superstrings we simply restrict attention to those models where all left-mover states can be identified with right-mover states. In order to relate the closed string to the cylinder we use the boundary states[13,15]. By using these boundary states we have developed a formalism which produces the annulus corresponding to an arbitrary given type II closed string[5,11]. Given the cylinder contribution, a Jacobi tranformation turns this into the annulus amplitude. Requiring a physically sensible projection (PSP) then dictates the form of the möbius amplitude, which by a Jacobi transformation yields a cylinder with one boundary and one crosscap. This cylinder with one crosscap and one boundary must also correspond to the result obtained from the closed string using crosscap and boundary states. The question of one-loop finiteness can most easily be investigated by comparing the cylinder with two boundaries to the cylinder with one boundary and one crosscap. Locality can then be used to argue that a cylinder with two crosscaps must necessarily also exist; the non-existence of such an amplitude would imply non-local correlations on the world-sheet since the crosscap would need to know that there is a boundary state and not a crosscap on the other side of the cylinder. Finally, the Jacobi transformation of the cylinder with two crosscaps yields the Klein bottle which for consistency must form a physically sensible projection with the torus.

For simplicity, in this talk we will only focus on two simple examples of four-dimensional superstrings, although we have developed a formalism for constructing open superstrings within the fermionic formulation[5,10,11,16], based on fig. 1. The hope is to develop four-dimensional open superstrings, for phenomenological purposes, to the same level of sophistication as the recent constructions of heterotic strings.

For our first simple example of a four-dimensional model we take the world sheet fermion's contribution to the partition function on the torus to be

$$Z^{\rm torus} = \frac{1}{2}(Z_{W_0}^{W_0} + Z_{\mathbf{0}}^{W_0}) + \frac{1}{2}(Z_{W_0}^{\mathbf{0}} - Z_{\mathbf{0}}^{\mathbf{0}})\,, \tag{1}$$

where we are following the notation of refs. [3,5]. The upper vectors specify the space boundary conditions and the lower vectors specify the time boundary conditions of the world sheet fermions on a torus. In this example, $W_0 = ((1/2)^{10}|(1/2)^{10})$ and $\mathbf{0} = (0^{10}|0^{10})$ respectively correspond to Neveu-Schwarz and Ramond boundary conditions on all twenty left- and right-mover complex world-sheet fermions. The complete model consists of tensoring the world-sheet fermionic contributions with the bosonic contributions and integrating over the modular parameter as described in, for example, refs. [17,3].

The consistency of this closed model follows from its modular invariance properties which can be easily checked using the transformations of $Z_W^{W'}$, which may be found, for example, in ref. [3]. (The standard bosonic contribution to the partition function is also modular invariant by itself.) Furthermore, the model can be truncated to a type I unoriented closed model[5], as follows from the left-right symmetry of the model.

The set of states of the closed string model which may couple to a boundary or crosscap state are the left-right-symmetric (LRS) ones, that is states which have the same left- and right-mover content[5,7]. Since both sectors of the model are LRS, both potentially may couple to the boundary and crosscap states. To compute this coupling we need to consider the restriction of the GSO projectors to LRS states,

$$\begin{aligned} \hat{P}(W_0)\Big|_{\rm LRS} &= \frac{1}{2}(1 + e^{2\pi i W_0 \cdot \hat{N}_{W_0}})\Big|_{\rm LRS} = 1 \\ \hat{P}(\mathbf{0})\Big|_{\rm LRS} &= \frac{1}{2}(1 - e^{2\pi i W_0 \cdot \hat{N}_{\mathbf{0}}})\Big|_{\rm LRS} = 0\,. \end{aligned} \tag{2}$$

The projectors collapse since, for LRS states, $N = (N_l|N_l)$ and so $\exp(2\pi i W_0 \cdot N)$ is unity for such states. Thus, the cylinder with two boundary states is (see ref. [5] for details)

$$\begin{aligned} Z^{BB}(\tau') &= \mathrm{tr}_{\,\rm LRS}[q^{\hat{H}^{\rm left}_{W_0}} \bar{q}^{\hat{H}^{\rm right}_{W_0}}\, \hat{P}(W_0)] \; + \mathrm{tr}_{\,\rm LRS}[q^{\hat{H}^{\rm left}_{\mathbf{0}}} \bar{q}^{\hat{H}^{\rm right}_{\mathbf{0}}}\, \hat{P}(\mathbf{0})] \\ &= \mathrm{tr}\,[|q|^{2\hat{H}^{\rm left}_{W_0}}] \\ &\equiv F^{U_0}_{U_0}(\tau')\,, \end{aligned} \tag{3}$$

where $\tau' = \ln|q|/i\pi$ and $U_0 = ((1/2)^{10})$ is a Neveu-Schwarz boundary condition vector of half the length of W_0 and $\hat{H}_{W_0}$ is the closed string hamiltonian for the world sheet fermions in the Neveu-Schwarz sector. By Jacobi transforming this, we obtain the annulus contribution to the open string partition function

$$Z^{\rm ann}(\tau) = F^{U_0}_{U_0}(\tau) = \mathrm{tr}\,[e^{2\pi i \tau \hat{H}^{\rm open}_{U_0}}\,] \equiv \mathrm{tr}\,[w^{\hat{H}^{\rm open}_{U_0}}\,]\,, \tag{4}$$

where $\tau = -1/\tau'$ and $\hat{H}^{\rm open}_{U_0}$ is the open string hamiltonian in the U_0 sector.

Given the annulus contribution, we can then construct the möbius contribution by requiring a physical sensible projection between the annulus and möbius contributions. This is equivalent to requiring the action of the twist operator[17,5] $\hat{\Omega}$ on open string states to satisfy $\hat{\Omega}^2 = 1$ when acting on a state. Since the open string U_0 sector, which does not have a GSO projector, contains states at both integer and half integer mass levels, the naive twist operator $\hat{\Omega} \sim e^{\pi i \hat{H}_{U_0}}$ is not sensible. In the presence of a GSO projector which would ensure that this twist operator is well defined, consistent open string models exist only in $D = 2, 6, 10$ as described in ref. [5,6] and not in $D = 4$. The models we present here do not contain such a GSO projector, although our second example will have a GSO projector act on some of the fermions, which is enough to ensure space-time supersymmetry and the absence of tachyons.

The solution to this difficulty is actually quite simple: use a GSO projector to separate the states into integer and half-integer mass levels before applying the twist operator and associate different phases with the two parts of the twist operator so as to ensure $\hat{\Omega}^2 = 1$. The explicit form for the twist operator is then

$$\hat{\Omega} = e^{\pi i d/24}\Big[\eta_1\Big(\frac{1+(-1)^{\hat{N}_{U_0}}}{2}\Big)e^{\pi i\hat{H}^{\rm open}_{U_0}} - i\eta_2\Big(\frac{1-(-1)^{\hat{N}_{U_0}}}{2}\Big)e^{\pi i\hat{H}^{\rm open}_{U_0}}\Big]\,, \tag{5}$$

where $\hat{N}$ is the usual fermion number operator and $d = 10$ for a four-dimensional model. The phase $e^{\pi i d/24}$ absorbs the phase in $e^{\pi i\hat{H}^{\rm open}_{U_0}}$ due to the zero-point energy of the hamiltonian $\hat{H}^{\rm open}_{U_0}$. For $\eta_1 = \eta_2$ this twist operator is equivalent to the one given by Clavelli and Shapiro 16 years ago[19]. The choice of the $\eta_i = \pm 1$ determines the Chan-Paton gauge group representation[18] of the various mass levels. (See ref. [17] for details.) (We are allowing for an independent choice of the Chan-Paton gauge group representation at integer and half-integer mass levels because these lie in different representations of the two-dimensional Kac-Moody symmetry.) The Chan-Paton representation of the massless gauge bosons corresponds to the adjoint representation of the gauge group, so for $\eta_2 = +1, -1$, which controls the representation of the massless states, the gauge group will respectively be a Sp or SO group.

With this choice of twist operator the möbius contribution is

$$\begin{aligned}Z^{\rm mob} &= e^{i\pi d/24}\Big(\frac{1}{2}\eta_1{\rm tr}\,[w^{\hat{H}^{\rm open}_{U_0}}(1+(-1)^{\hat{N}_{U_0}})\hat{\Omega}] - \frac{i}{2}\eta_2{\rm tr}\,[w^{\hat{H}^{\rm open}_{U_0}}(1-(-1)^{\hat{N}_{U_0}})\hat{\Omega}]\Big)\\ &= e^{i\pi d/24}\Big(\frac{1}{2}\eta_1\big(F^{U_0}_{U_0}(\tau+1/2) + F^{U_0}_{\mathbf{0}}(\tau+1/2)\big)\\ &\qquad\qquad - \frac{i}{2}\eta_2\big(F^{U_0}_{U_0}(\tau+1/2) - F^{U_0}_{\mathbf{0}}(\tau+1/2)\big)\Big)\,.\end{aligned} \tag{6}$$

The Jacobi transformation properties of the möbius contributions to the partition function are given by

$$F^U_V(\tau+1/2) = e^{4\pi i(U-U_0)\cdot(V-U_0)}e^{2\pi i(V\cdot V-d/6)}e^{\pi i(U\cdot U-d/6)}F^{\overline{U}}_{\overline{V-U}}(\tau'/4+1/2)\,, \tag{7}$$

where U and V consist of Neveu-Schwarz and Ramond open string boundary conditions and the overbar on the boundary condition vector indicated that it should be evaluated mod 1. By Jacobi transforming the möbius partition function (6) (after the usual rescaling[1] $\tau'/4 \to \tau'$) we obtain the cylinder with one boundary and one crosscap ($D = 4, d = 10$)

$$\begin{aligned}Z^{BC} = e^{i\pi d/24}\Big(&\frac{1}{2}\eta_2\big(F^{U_0}_{U_0}(\tau'+1/2) + F^{U_0}_{\mathbf{0}}(\tau'+1/2)\big)\\ &- \frac{i}{2}\eta_1\big(F^{U_0}_{U_0}(\tau'+1/2) - F^{U_0}_{\mathbf{0}}(\tau'+1/2)\big)\Big)\,,\end{aligned} \tag{8}$$

which contains the same closed string states as the cylinder with two boundary states (3).

The potential divergences of the amplitudes in eqs. (3) and (8) are determined by the leading and next to leading terms in the Taylor expansion, in $e^{2\pi i\tau'}$, corresponding to the tachyon and massless scalars propagating into the vacuum. In the Jacobi transformed möbius contribution (8), the coefficient of the tachyon singularity is $2^2\eta_2 M$ while for the massless scalar the coefficient is $2^2\eta_1 M$, where we are including the factor $2^{D/2}M$ arising from the bosonic and Chan-Paton contributions[17,5]. Similarly for the Jacobi transformed annulus (9), the contribution to both these singularities is M^2 for gauge groups $\mathrm{Sp}(M)$ or $\mathrm{SO}(M)$. Thus, for cancellation of both tachyon and massless scalar divergences

$$M + \eta_2 2^2 = 0\,, \qquad\qquad M + \eta_1 2^2 = 0\,, \tag{9}$$

so the gauge group is SO(4). (Of course, the presence of tachyons means this model should not be taken seriously except as an illustrative example.)

Finally, we comment that it is not difficult to carry out the remaining steps in fig. 1 so as to show that the Klein bottle amplitude, as obtained from the implied definition of the crosscap states in eq. (8), forms a PSP with the torus amplitude.

As a second example of a $D = 4$ open superstring we present a model whose low energy spectrum corresponds to $N = 4$ supergravity coupled to $N = 4$ super-Yang-Mills. The torus contribution to the partition function for this model is

$$Z^{\text{torus}} = Z^{D=10} \times Z^{d=6}\,, \tag{10}$$

where $Z^{D=10}$ is the torus contribution to the partition function of the standard type I superstring[1] in ten space-time dimensions ($d = 4$) and $Z^{d=6}$ is the partition function for the previous model (1), but with $d = 6$. Again the modular invariance of this model is easy to check, while the world sheet supersymmetry follows from the possible grouping of internal world sheet fermions into triplets satisfying the required conditions[3,5]. As for the previous model, truncation to a type I model follows from its left-right symmetry.

Computing the cylinder contribution as before, we obtain

$$Z^{BB} = \frac{1}{2}\Big[F^{1/2}_{1/2}(\tau')^4 - F^{1/2}_{0}(\tau')^4 - F^{0}_{1/2}(\tau')^4 + F^{0}_{0}(\tau')^4\Big]F^{1/2}_{1/2}(\tau')^6\,. \tag{11}$$

Hence, after the Jacobi transformation, we obtain the annulus partition function

$$\begin{aligned} Z^{\text{ann}} &= \frac{1}{2}\Big[F^{1/2}_{1/2}(\tau)^4 - F^{1/2}_{0}(\tau)^4 - F^{0}_{1/2}(\tau)^4 + F^{0}_{0}(\tau)^4\Big]F^{1/2}_{1/2}(\tau)^6 \\ &= \frac{1}{2}\Big(F^{U_0}_{U_0}(\tau) - F^{U_0}_{\overline{U_0+U_1}}(\tau)\Big) - \frac{1}{2}\Big(F^{\overline{U_0+U_1}}_{U_0}(\tau) - F^{\overline{U_0+U_1}}_{\overline{U_0+U_1}}(\tau)\Big)\,, \end{aligned} \tag{12}$$

where the first factor in brackets is the world sheet fermion contribution to the partition function for the standard $D = 10$ ($d = 4$) type I superstring. The boundary condition vector $U_1 = (1/2, (0,0,1/2)^3)$, where we have explicitly displayed the grouping of the internal fermions into world sheet supersymmetric triplets.

Since the model factorises, the twist operator also factorises as $\hat{\Omega} = \hat{\Omega}_4\hat{\Omega}_6$. The $\hat{\Omega}_4$ will be simply $\exp(\pi i\hat{H}^4)$ (up to a phase) which is well defined because of the projection of states in this subspace keeps only integer mass levels. For $\hat{\Omega}_6$, as in the first example, we must split the action of $\hat{\Omega}_6$ into its action upon the states with even and odd fermion numbers. The full twist operator, in the U_0 sector, is

$$\hat{\Omega}_{U_0} = -ie^{\pi id/24}e^{\pi i\hat{H}^4}\Big[\eta_1\Big(\frac{1+(-1)^{\hat{N}^6}}{2}\Big)e^{\pi i\hat{H}^6} - i\eta_2\Big(\frac{1-(-1)^{\hat{N}^6}}{2}\Big)e^{\pi i\hat{H}^6}\Big], \qquad (13)$$

where $\hat{H}^4$ is the hamiltonian for the $d = 4$ subspace of four complex fermions and $\hat{H}^6$, $\hat{N}^6$ the hamiltonian and number operator for the remaining six. The overall phase $-i$ is necessary because the projection in eq. (12) keeps only states with odd fermion numbers in the $d = 4$ subspace. The twist operator in the $\overline{U_0 + U_1}$ sector will be just as above but with η_i replaced by η_i'. Using this twist operator we obtain the möbius amplitude

$$\begin{aligned} Z^{\text{mob}} = &-\frac{i}{4}e^{\pi i10/24}\Big((F^{1/2}_{1/2})^4 - (F^{1/2}_0)^4\Big) \\ &\qquad\times\Big(\eta_1((F^{1/2}_{1/2})^6 + (F^{1/2}_0)^6) - i\eta_2((F^{1/2}_{1/2})^6 - (F^{1/2}_0)^6)\Big) \\ &+\frac{i}{4}e^{\pi i10/24}\Big((F^0_{1/2})^4 - (F^0_0)^4\Big) \\ &\qquad\times\Big(\eta_1'((F^{1/2}_{1/2})^6 + (F^{1/2}_0)^6) - i\eta_2'((F^{1/2}_{1/2})^6 - (F^{1/2}_0)^6)\Big). \end{aligned} \qquad (14)$$

This Jacobi transforms into the cylinder with crosscap

$$\begin{aligned} Z^{BC} = &-\frac{i}{4}e^{\pi i10/24}\Big((F^{1/2}_{1/2})^4 - (F^{1/2}_0)^4\Big) \\ &\qquad\times\Big(-\eta_2((F^{1/2}_{1/2})^6 + (F^{1/2}_0)^6) + i\eta_1((F^{1/2}_{1/2})^6 - (F^{1/2}_0)^6)\Big) \\ &+\frac{i}{4}e^{\pi i10/24}\Big((F^0_{1/2})^4 - (F^0_0)^4\Big) \\ &\qquad\times\Big(-\eta_2'((F^{1/2}_{1/2})^6 + (F^{1/2}_0)^6) + i\eta_1'((F^{1/2}_{1/2})^6 - (F^{1/2}_0)^6)\Big), \end{aligned} \qquad (15)$$

where the argument of the F's is $(\tau'+1/2)$ as in eq. (8). The closed string tachyon is projected out, as necessary, and the only potential infinities are due to closed string massless states from either the U_0 or $\overline{U_0 + U_1}$ sectors. Comparing with the cylinder (11) we obtain the following conditions for cancelling these infinities

$$M - \eta_2 2^2 = 0\,, \qquad M - \eta_2' 2^2 = 0 \qquad (16)$$

(again using the relative weights of the möbius and annulus amplitudes[5]). Hence, the divergences will cancel provided $\eta_2 = \eta'_2 = 1$ and $M = 4$. For a consistent supersymmetric model both the gauge bosons in the U_0 sector and the gauginos in the $\overline{U_0 + U_1}$ sector must be in the same representation so $\eta_1 = \eta'_1$. Choosing $\eta_1 = -1$ gives the Chan-Paton gauge group SO(4), while $\eta_1 = +1$ gives Sp(4) as the gauge group. Although we have only shown cancellation in the vacuum amplitudes the cancellation in more complicated amplitudes will follow as in ref. [5].

Further models can be constructed by the addition of extra boundary condition basis vectors. In ref. [11] examples of open-closed string models with $N = 4, 2, 1$ space-time supergravity were presented.

Much work remains to be done before open strings are as well developed as the heterotic string for phenomenological purposes. Besides the straightforward extension of our work to the other standard formalisms for the internal degrees of freedom of string theories such as lattices and orbifolds[2], there are a number of avenues which should be investigated. For example, there has been a suggestion to have projectors act non-trivially on the Chan-Paton factors which potentially would lead to a much richer structure for the gauge group[9,7]. There is also the question of trying to construct chiral models, perhaps through the use of higher level Kac-Moody algebras[20]. Another important topic is the investigation of higher loop amplitudes[21]. Finally, within any particular formalism, it may be possible to classify the set of all one-loop finite open string models since the number of internal degrees of freedom is significantly smaller than for the four-dimensional heterotic string.

In conclusion, we have shown that it is possible to construct a variety of four-dimensional one-loop finite open superstring theories and are hopeful that such constructions will play a role in string theory phenomenology. In particular we have constructed models which are space-time supersymmetric and where the supergravity of the massless states is $N = 4, 2$ or 1.

This work was supported in part by the SERC and in part by the US Department of Energy.

References

1. M.B. Green and J.H. Schwarz, *Phys. Lett.* **151B** (1985) 21.
2. L. Dixon, J. Harvey, C. Vafa, and E. Witten, *Nucl. Phys.* **B261** (1985) 678; *Nucl. Phys.* **B274** (1986) 285; K.S. Narain, *Phys. Lett.* **169B** (1986) 41; K.S. Narain, M.H. Sarmadi and C. Vafa, *Nucl. Phys.* **B288** (1987) 551; W. Lerche, D. Lüst and A.N. Schellekens, *Nucl. Phys.* **B287** (1987) 477; D. Bailin, D.C. Dunbar and A. Love, to appear Nucl. Phys. **B**.
3. H. Kawai, D.C. Lewellen and S.-H.H. Tye, *Phys. Rev. Lett.* **57** (1986) 1832,

Nucl. Phys. **B288** (1987) 1.
4. I. Antoniadis, C.P. Bachas and C. Kounnas, *Nucl. Phys.* **B289**(1987) 87.
5. Z. Bern and D.C. Dunbar, *Phys. Lett.* **203B** (1988) 109; D.C. Dunbar, *Nucl. Phys.* **B319** (1989) 72; Z. Bern and D.C. Dunbar, *Nucl. Phys.* **B319** (1989) 104.
6. L. Clavelli, P.H. Cox, B. Harms and A. Stern, University of Alabama preprint UAHEP-882; L. Clavelli, P.H. Cox and B. Harms, *Phys. Rev. Lett.* **61** (1988) 787; L. Clavelli, P.H. Cox, P. Elmfors and B. Harms, preprint UAHEP-891; P.H. Cox, these proceedings.
7. N. Ishibashi and T. Onogi, *Nucl. Phys.* **B 318** (1989) 239; N. Ishibashi, *Mod. Phys. Lett.* **A4** (1989) 251.
8. G. Pradisi and A. Sagnotti, *Phys. Lett.* **216B** (1989) 59.
9. J.A. Harvey and J.A. Minahan, *Phys. Lett.* **188B** (1987) 44.
10. Z. Bern and D.C. Dunbar, preprint LA-UR-89-3213, to appear in *Phys. Rev. Lett.*
11. Z. Bern and D.C. Dunbar, preprint LA-UR-89-3653, LTH-89-246.
12. D.J. Gross, A. Neveu, J. Scherk and J.H. Schwarz, *Phys. Rev.* **D2** (1970) 697; E. Cremmer and J. Scherk, *Nucl. Phys.* **B50** (1972) 222;
13. M. Ademollo et. al., *Nucl. Phys.* **B94** (1975) 221.
14. K. Bardakci and M.B. Halpern, *Phys. Rev.* **D3** (1971) 2493.
15. C.G. Callan, C. Lovelace, C.R. Nappi and S.A. Yost, *Nucl. Phys.* **B293** (1987) 83; J. Polchinski and Y. Cai, *Nucl. Phys.* **B296** (1988) 91.
16. Z. Bern and D.C. Dunbar, in preparation.
17. J.H. Schwarz, *Phys. Rep.* **89** (1982) 223.
18. J.E. Paton and Chan Hang-Mo, *Nucl. Phys.* **B10** (1969) 519. N. Marcus and A. Sagnotti, *Phys. Lett.* **119B** (1982) 97.
19. L. Clavelli and J.A. Shapiro, *Nucl. Phys.* **B57** (1973) 490.
20. D.C. Lewellen, preprint SLAC-PUB-5023.
21. M. Bianchi and A. Sagnotti, *Phys. Lett.* **211B** (1988) 407; preprint ROM2F-89/15.

NONPERTURBATIVE VACUUM STRUCTURE AND THE OPEN BOSONIC STRING

V. Alan Kostelecký*

Physics Department, Indiana University
Bloomington, IN 47405, U.S.A.

and

Stuart Samuel

Physics Department, The City College of New York,
New York, NY 10031, U.S.A.

The nonperturbative vacuum structure of the open bosonic string is explored using covariant string field theory and functional methods. A candidate nonperturbative vacuum is identified by examining successive level truncations of the string field. Numerical vacuum expectation values of all fields to level two are presented. Perturbation theory in this vacuum is shown to exist and is used to determine low-lying levels of the mass spectrum. We show that string couplings run at tree level. One consequence is that the number of states associated with a given set of particle fields is substantially smaller in the nonperturbative vacuum than in the canonical vacuum. Another is that the open bosonic string is asymptotically free. These effects are generic in string theory.

1. Introduction

Superstring theory [1] may eventually provide a consistent theory of quantum gravity, complete with finite and anomaly-free [2] models of the other known basic interactions and observed fundamental particles. However, superstrings or heterotic strings [3] in their canonical ten-dimensional vacua do not seem to provide a description of the physical universe. The dimensionalities of spacetime and of the gauge groups are too large and the particles appearing do not match reality. To obtain a realistic theory based on one of these models six of the dimensions

---

* Speaker

must be rendered unobservable, for example, by compactification on a Calabi-Yau manifold [4]. Furthermore, the large gauge symmetry must be reduced, at least at low energies, to the standard-model group.

One attractive possibility would be for both the compactification and the gauge-group breaking to occur spontaneously. If, for instance, the canonical vacuum is unstable, the string theory might naturally exist in another vacuum that retains desirable aspects of string theory and that is physically realistic. The structure of the vacuum may therefore play an essential role in the viability of string models.

One approach to the study of vacuum structure in particle physics uses functional integrals and field theory. Candidate vacua of interest may be found by examining extrema of the effective potential. Vacua other than the true one may decay by barrier penetration [5,6].

A similar approach may be taken to the investigation of string vacuum structure. For the open bosonic string, a covariant field theory exists [7] and has been extensively studied [8-35]. The theory has been reformulated in a Fock-space representation in terms of particle fields. Perturbation theory about the canonical vacuum correctly reproduces the Koba-Nielsen tree-level amplitudes and generates off-shell extensions. The N-point one-loop and higher-loop amplitudes have been determined both on- and off-shell. Section 2 presents some useful background material concerning string field theory.

Here, functional methods are applied to the covariant string field theory. The aim is to explore the vacuum structure and physics of the open bosonic string. An attractive feature of this theory is its relative simplicity: neither fermions nor gravity are present at the semiclassical level.

An interesting aspect of the theory is that the canonical 26-dimensional vacuum is unstable because the spectrum includes a tachyon. The presence of a negative-sign mass-squared term in the quadratic part of the action makes pair-production energetically favorable. An analogous situation exists in the standard electroweak model in its naive vacuum, where the Higgs scalar fields have negative mass squared. The instability is eliminated when the scalars acquire vacuum expectation values. Similarly, the appearance of a tachyon in the spectrum of the open bosonic string may merely be a sign that the vacuum has been incorrectly selected. String interactions may cause the tachyon field ϕ to acquire an expectation value resulting in a different, stable vacuum.

One method adopted here is to investigate the possibility of a stable vacuum in the theory by constructing the static effective potential for ϕ. The lowest-order nontrivial interaction in the static tachyon potential has previously been calculated [34]. It was also demonstrated that interactions involving arbitrary powers of ϕ are generically equally important. This means that the vacuum structure in the open bosonic string is inherently nonperturbative.

Nonperturbative information about the effective potential can be obtained in a level-truncation scheme. The method is discussed in section 3 and applied in section 4 to the identification of a candidate nonperturbative vacuum. Physics in this vacuum is developed in section 5. A summary and discussion is provided in section 6. The notational conventions of ref. [36] are followed throughout.

2. Background

The string vibrational modes are particle states, with higher modes representing heavier particles. Using this Fock-space representation, the string field Ψ may be expanded as a linear combination of ordinary particle fields with coefficients being solutions at the first-quantized level [10-17].

The expansion begins

$$\begin{aligned}\Psi = \big(\phi + A_\mu \alpha^\mu_{-1} + i\alpha b_{-1} c_0 + \frac{1}{\sqrt{2}} i B_\mu \alpha^\mu_{-2} + \frac{1}{\sqrt{2}} B_{\mu\nu} \alpha^\mu_{-1} \alpha^\nu_{-1} \\ + \beta_0 b_{-2} c_0 + \beta_1 b_{-1} c_{-1} + i k_\mu b_{-1} c_0 \alpha^\mu_{-1} + \dots \big) \mid 0 \rangle .\end{aligned} \tag{2.1}$$

Here, ϕ is the tachyon, A_μ is the massless vector, B_μ is a vector at the first massive level, and $B_{\mu\nu}$ is a symmetric two tensor at the first massive level. Note that these descriptions apply only in the canonical vacuum. The fields α, β_0, β_1, and k_μ are auxiliary. The first-quantized string vacuum is $|0\rangle = c_1|\Omega\rangle$, where $|\Omega\rangle$ is the $Sl(2,R)$-invariant vacuum [37]. All fields are real; the reality condition [7] $\Psi^\ddagger = \Psi^\dagger$ is implemented by factors of i determined from Appendix C of ref. [21]. The coefficients of $1/\sqrt{2}$ establish standard normalizations for kinetic energy terms. The level of a state is given by its mass squared in units of $1/\alpha'$ measured from the tachyon level. For instance, the tachyon ϕ is at level 0 and A_μ is at level 1.

The action of the covariant string field theory is a Chern-Simons form [7]:

$$S = \frac{1}{2\alpha'} \int \Psi * Q\Psi + \frac{g}{3} \int \Psi * \Psi * \Psi \quad . \tag{2.2}$$

This action is invariant under the gauge transformations

$$\delta\Psi = \frac{1}{\sqrt{2\alpha'}} Q\Lambda + \sqrt{\frac{\alpha'}{2}} g(\Psi * \Lambda - \Lambda * \Psi) \ . \tag{2.3}$$

The functional integral is well defined once a gauge is chosen. We select the Siegel-Feynman gauge [38] $b_0\Psi = 0$. This eliminates terms containing c_0 in Ψ. Quantization in this gauge is considered in refs. [26,27].

When the representation (2.1) of Ψ is substituted into the action (2.2), the particle-field action is found. The interaction part may be expanded as $\mathcal{L}_{int} = \sum_n \mathcal{L}^{(n)}$, where the superscript is called the order; it denotes the sum of the level numbers of the fields. Terms in the action to order six involving level-two fields may be found in ref. [39]. All particle fields f in interaction terms appear as fields smeared over a distance $\sqrt{\alpha'}$:

$$\tilde{f} = exp\big[\alpha' \ ln(3\sqrt{3}/4)\partial_\mu \partial^\mu\big] \ f \ . \tag{2.4}$$

For example, $\mathcal{L}^{(0)} = \kappa g \tilde{\phi}^3$, where g is the on-shell three-tachyon coupling at tree-level, and $\kappa = 3^3\sqrt{3}/2^7$.

3. Methodology

First, we present a string-field framework for the analysis of nonperturbative vacuum structure and the associated excitations. Let Ψ_B be a background string field consisting of expectation values satisfying the equations of motion:

$$Q\Psi_B + g\alpha'\Psi_B * \Psi_B = 0 \ . \tag{3.1}$$

Define a new BRST operator for the background vacuum by

$$Q_B\Psi = Q\Psi + g\alpha'(\Psi_B * \Psi - (-1)^{g(\Psi)}\Psi * \Psi_B) \ , \tag{3.2}$$

where Ψ is an arbitrary string field and $g(\Psi)$ is its grading or ghost number, e.g., $g(\Lambda) = 0$ for a gauge-transformation field and $g(\Psi) = 1$ for the string field. The operator Q_B has all the properties of a BRST operator needed to define a string field theory. Eq. (3.1) implies nilpotency: $Q_B Q_B = 0$. It is also distributive across the star product: $Q_B(\Psi_1 * \Psi_2) = Q_B\Psi_1 * \Psi_2 + (-1)^{g(\Psi_1)}\Psi_1 * Q_B\Psi_2$. Finally, Q_B satisfies $\int Q_B\Psi_1 * \Psi_2 = (-1)^{g(\Psi_1)+1} \int \Psi_1 * Q_B\Psi_2$.

Perform the shift $\Psi = \Psi_B + \Psi$, where Ψ represents the string field fluctuations in the new vacuum, and substitute into the action to find

$$A = \frac{1}{2\alpha'} \int \Psi_B * Q \Psi_B + \frac{g}{3} \int \Psi_B * \Psi_B * \Psi_B + \frac{1}{2\alpha'} \int \Psi * Q_B \Psi + \frac{g}{3} \int \Psi * \Psi * \Psi \ . \quad (3.3)$$

The first two terms in Eq. (3.3) represent the contribution of the background solution to the cosmological constant. The remaining terms form the action for the new field Ψ. Eq. (3.3) is invariant under

$$\delta \Psi = \frac{1}{\sqrt{2\alpha'}} Q_B \Psi + g \sqrt{\frac{\alpha'}{2}} (\Psi * \Lambda - \Lambda * \Psi) \ . \quad (3.4)$$

The structure of the nonperturbative version of the theory is the same as the perturbative version. Consequently, the framework and analysis of the BRST perturbation series carries over to the background case. If fluctuations are small so that the interaction term in (3.3) can be treated as a perturbation then the spectrum of states is given by the cohomology of Q_B, i.e., the spectrum consists of states Ψ_S such that $Q_B \Psi_S = 0$ but $\Psi \neq Q_B \Lambda$. It is natural to call such states the physical on-shell states of the nonperturbative vacuum. Corrections to the masses and to scattering processes can be obtained by including perturbatively the effects of the interaction terms.

Here, we use level truncation of the string field Ψ to determine Ψ_B and to obtain solutions of $Q_B \Psi = 0$. Several types of level truncation are possible: the expansion in Ψ can be limited to certain fields and all lagrangian terms can be considered, or the truncation can be performed in the order n of the interaction lagrangian $\mathcal{L}_{int} = \sum_n \mathcal{L}^{(n)}$. A combination is used here: first, Ψ is truncated and then successive terms in $\mathcal{L}_{int}$ are analysed.

Level truncation is a plausible approach because the couplings in $\mathcal{L}^{(n)}$ are proportional to a factor of $(4/3\sqrt{3})^n \approx exp(-0.26n)$, i.e., they decrease exponentially as n increases [39]. If $4/3\sqrt{3} \approx 0.77$ were a small number then level truncation would be an excellent approximation. For the true value ≈ 0.77, the situation is less clear *a priori* but higher-order corrections to lower-order results provide insight.

The level-truncation method may be tested by calculating in successive orders the tree-level ϕ^4 contribution to the static tachyon effective potential and comparing with the known exact result [34]. Without Paton-Chan factors, only even levels in the truncation are relevant because the s- and t-channel contributions of odd-level fields to the four-point amplitude cancel. Note also that the momentum dependence in the interaction lagrangian is irrelevant to the static tachyon potential. Saturating the intermediate line with fields to level two generates 72% of the exact result. An additional 12% is provided by fields at level four. The net contribution to the static four-point tachyon amplitude to level four is therefore 84%. This suggests convergence of the level-truncation scheme. Numerical results presented below provide further evidence in favor of this.

4. Nonperturbative Vacuum Structure

An effective potential for a subset of fields may in principle be constructed by suitable integrations over other fields in the functional integral. Since the instability of the canonical vacuum originates with the tachyon field, it is appropriate to consider the effective tachyon potential. Extrema of effective actions for other sets of fields are necessarily extrema of the tachyon potential. Any fields integrated over in the functional integral become functionals of ϕ. A nonzero vacuum expectation value $\langle\phi\rangle$ then implies expectation values for other fields, obtained by the substitution of $\langle\phi\rangle$ for ϕ in the equations of motion.

The effective tachyon potential is a complicated nonlocal function for ϕ. Since the structure of the vacuum is determined by zero-momentum states, derivatives of ϕ can be disregarded. An exact treatment of the resulting potential is difficult. Instead, dominant contributions to ϕ^n for all powers of n can be found via the level-truncation method. This method is systematic; as higher levels are added, the potential is more accurate.

Consider first the lowest-order truncation, which limits the particle-field content of Ψ to ϕ. This is called the level-zero truncation; it is relevant for comparison with higher-order calculations and for a qualitative understanding. Setting $\tilde{\phi} = \phi$ for the static situation, the order-zero terms in the static lagrangian are

$$V_{eff}(\phi) = -\frac{\phi^2}{2\alpha'} + g\kappa\phi^3 + \dots \quad . \tag{4.1}$$

The canonical vacuum at $\phi = 0$ is a local maximum and hence unstable. However, a local minimum in V_{eff} occurs at

$$\langle\phi\rangle = \frac{1}{3\kappa g\alpha'} \approx \frac{0.91}{g\alpha'} \quad . \tag{4.2}$$

Equation (4.2) shows explicitly that $\langle\phi\rangle$ is of order $1/g$ and is therefore nonperturbative. The cosmological constant Λ acquires a contribution $\Lambda^{-1} = -2\cdot 3^3\kappa^2 g^2\alpha'^3$ at the new minimum.

The next-order effect involves level-two fields. Consider the expansion terms $\mathcal{L}^{(2)}$ and $\mathcal{L}^{(4)}$. This is called the order-four truncation. The extra terms have the form ϕ-ϕ-B and ϕ-B-B, where B is a generic level-two field. Integrating over these fields in the functional integral produces the comb approximation restricted to level-two fields. Each vertex of a comb diagram has at least one external leg. The relevant diagrams have tachyons as external states and level-two fields on internal lines. The resulting effective potential can be expressed analytically because the integrations are all quadratic.

In fact, only β_1 and $B := 1/\sqrt{26}\eta_{\mu\nu}B^{\mu\nu}$ contribute to the static potential. The coefficient $1/\sqrt{26}$ provides a conventional kinetic-energy normalization for B. The static effective potential for ϕ is

$$\begin{aligned} V_{eff}(\phi) = &-\frac{1}{2\alpha'}\phi^2 + \frac{3^3\sqrt{3}}{2^7}g\phi^3 \\ &+ \frac{3^4 17}{2^{13}}g^2\phi^4\left(\frac{53\sqrt{3}}{2^2 3^2 17}g\phi - 1\right)\left(1+\frac{131\sqrt{3}}{2^4 3^2}g\phi + \frac{97}{2^6 3^3}g^2\phi^2\right)^{-1} . \end{aligned} \tag{4.3}$$

The local minimum found at order zero in Eq. (4.2) persists. The value of $\langle\phi\rangle$ grows about 20% to $\langle\phi\rangle \approx 1.083/g\alpha'$ and the contribution to the cosmological constant becomes $\Lambda \approx -0.192/g^2\alpha'^3$. The associated vacuum values for the level-two fields are $\langle B_{\mu\nu}\rangle = \eta_{\mu\nu}\langle B\rangle/\sqrt{26}$, $\langle B\rangle \approx 0.37/g\alpha'$, and $\langle\beta_1\rangle \approx -0.35/g\alpha'$.

There is a second local mimimum of V_{eff}, with $\langle\phi\rangle \approx -6.98/g\alpha'$. However, this corresponds to a large value of $|g\alpha'\langle\phi\rangle|$. The shift of ϕ by its vacuum value then generates large terms in the action. Higher-order contributions therefore may overwhelm this minimum. This possibility is a consequence of our use of semiclassical techniques for which results are most reliable when expectation values are small. In fact, as discussed below, incorporation of $\mathcal{L}^{(6)}$ eliminates the minimum of V_{eff} for negative ϕ. Nonetheless, the nontrivial vacuum structure found at this order suggests the possibility of interesting complications in a more exact treatment.

Finally, consider the next-order effects, which arise from the incorporation of all ϕ and B terms in $\mathcal{L}_{int}$ up to and including $\mathcal{L}^{(6)}$. This is the order-six truncation. The appearance of trilinear terms in level-two fields excludes an analytical treatment. However, a numerical analysis is feasible [39]. No minimum of $V_{eff}(\phi)$ exists other than the one with $\langle\phi\rangle \approx 1.09$. This means the candidate vacuum at $\langle\phi\rangle \approx -6.98/g\alpha'$ found at order four is unreliable. The numerical values of the vacuum values and the contribution to the cosmological constant are found to be

$$\langle\phi\rangle \approx 1.088/g\alpha' \ , \quad \langle B\rangle \approx 0.404/g\alpha' \ , \quad \langle\beta_1\rangle \approx -0.380/g\alpha' \ , \quad \Lambda \approx -0.194/g^2\alpha'^3 \tag{4.4}$$

The small difference between orders four and six is indicative of convergence of the level-truncation scheme.

5. Physics in the Nonperturbative Vacuum

Physics in the candidate nonperturbative vacuum may be explored by shifting fields by their vacuum values and considering small oscillations. This section considers the possibility of tensor-induced spontaneous Lorentz-symmetry breaking, determines the spectrum of states in successive truncation orders, and establishes the perturbative stability of the vacuum.

There is a natural mechanism in string theories for spontaneous breaking of the higher-dimensional Lorentz symmetry [35]. As mentioned in the introduction, a realization of such breaking is especially attractive because the physically observed Lorentz group is $O(3,1)$ rather than the higher-dimensional Lorentz groups $O(9,1)$ and $O(25,1)$ associated with the canonical vacua of superstrings and the bosonic string, respectively. Phenomenological consequences of such breaking are discussed in refs. [40,41].

The mechanism stems from the existence of couplings of the form ST_MT^M in the covariant string field theory, where S and T_M denote generic scalar and tensor fields. If some scalars S acquire vacuum values of the appropriate signs then one or more tensor fields might obtain a negative mass squared. Any such tensor would then be analogous to a Higgs field in a gauge theory: it would in turn acquire a vacuum value, inducing spontaneous Lorentz-symmetry breakdown. For instance, there is an interaction of the form $\phi A_\mu A^\mu$ present in $\mathcal{L}^{(2)}$. If $\langle\phi\rangle$ develops a negative value then A^μ necessarily gains a negative mass squared because there is no contribution to the mass from the kinetic terms in the lagrangian density. The resulting expectation value for A_μ reduces the Lorentz group from $O(25,1)$ to $O(24,1)$. However, if $\langle\phi\rangle$ develops a positive value then Lorentz-symmetry breaking is unlikely.

In this context, then, it is interesting to examine the possibility of tensor-induced spontaneous Lorentz-symmetry breaking in the candidate nonperturbative vacuum. The order-four truncation does contain a minimum of the tachyon potential with a negative expectation for ϕ, but we have seen that this minimum is unreliable. The shift of ϕ, $B_{\mu\nu}$ and β_1 by the expectation values in the order six truncation as given in Eq. (3.4) generates only positive mass-squared terms for A_μ, B_μ and $B_{\mu\nu}$. There is therefore no Lorentz-symmetry breaking present in the theory at this order.

Another interesting issue is whether breaking of the internal symmetry or of the string gauge group occurs. This question is best addressed in a more general framework, rather than within the Siegel-Feynman gauge adopted here. In fact, it can be shown that part of the string gauge group does break but that the internal symmetry is not affected. See ref. [42].

Let us next consider general features of the spectrum of states in the candidate nonperturbative vacuum. The spectrum is found by considering small fluctuations about the shifted fields. The quadratic terms in the shifted action yield a quadratic form whose inverse is the propagator. A pole in the momentum-space propagator at some value of $p^2 = -p_0^2$ represents the mass squared of a state.

There are limits on the reliability of states found in the truncation scheme. The point is that timelike directions are involved in determining the spectrum. Masses large on the scale of the truncation involve large values of p_0. However, as may be seen directly from Eq. (2.4). the effective coupling runs in momentum space even though the analysis is at tree level. This is a direct consequence of the extended nature of the string. At large values of p_0 the effective coupling $g(p_0) = g\ exp\left[3\alpha' p_0^2 ln(3\sqrt{3}/4)\right]$ is also large. Perturbation theory is therefore unreliable, and any poles found at scales beyond the truncation scale cannot be trusted. Evidently, the spectrum of the free theory may be substantially changed at high mass levels once interactions are present.

The decomposition of fields into Lorentz-irreducible multiplets simplifies the analysis of the quadratic form. Distinct irreducible multiplets do not mix, thereby block-diagonalizing the quadratic form. For example, the field A_μ contains a scalar A^L and a vector A_μ^T, where $A_\mu = A_\mu^T + \partial_\mu A^L/\sqrt{|\partial^2|}$ with $A^L = \partial^\lambda A_\lambda/\sqrt{|\partial^2|}$. Other Lorentz multiplets are similarly obtained [39]. To level-two fields, the irreducible representations include one two-tensor, $b_{\mu\nu}^{TT}$, three vectors, A_μ^T, B_μ^T and b_μ^{TL}, and six scalars ϕ, A^L, B^L, B, b^{LL}, and β_1.

As a qualitative illustration of the effects in a nonperturbative vacuum and as a basis for comparison with higher-level calculations, consider first the order-two truncation. This involves the particle fields ϕ, A_μ and the terms in $\mathcal{L}_{int}$ up to and including $\mathcal{L}^{(2)}$.

The value of $\langle\phi\rangle$ is provided in Eq. (3.3). Shifting ϕ by this expectation value and studying small oscillations shows that there is no mixing between ϕ and A^L. The quadratic form for the field ϕ in euclidean momentum space is

$$p^2 - \frac{1}{\alpha'} + \frac{2}{\alpha'}\, exp\left[-2\alpha' p^2\, ln(3\sqrt{3}/4)\right] \quad . \tag{5.1}$$

Setting $p^2 = -p_0^2$ and examining the resulting equation demonstrates that there is no pole associated with the tachyon field. This results from the running of the coupling with momentum stemming from exponentials of p^2: a transcendental term is present in the equation determining the poles of the Minkowski-space propagator. If ϕ instead of $\tilde{\phi}$ appeared in Eq. (5.1), the field ϕ would have a pole at a mass squared value of $1/\alpha'$. The running of the coupling has substantially affected the spectrum.

A similar analysis reveals that there is no pole for the transverse vector field. In contrast, the scalar A^L does have a pole, at $m_{A^L} = 1/\sqrt{\alpha'}$. Note that the absence of the transcendental factor $exp\left[-\alpha' p^2\, ln(3\sqrt{3}/4)\right]$ would shift the pole in A^L to $p^2 = -2^5/43\alpha' \approx -0.74/\alpha'$ and would generate a pole in A^T_μ at $-2^5/3^3\alpha' \approx -1.19/\alpha'$.

Consider next the order-four truncation. The A_μ fields are found to decouple from the others. There is no pole for A^T_μ. The state associated with the field A^L is shifted by about 17% from $m = 1/\sqrt{\alpha'}$ to $m_{A^L} \approx 1.17/\sqrt{\alpha'}$.

There is no pole for the tensor field, $b^{TT}_{\mu\nu}$. The vectors B^T_μ and b^{TL}_μ mix, producing a matrix quadratic form when $p^2 = -p_0^2$. There is only one pole in the propagator in Minkowski momentum space at $m_{vec} \approx 1.45/\sqrt{\alpha'}$. The scalars ϕ, β_1, B, b^{LL} and B^L all mix, generating a five-by-five matrix quadratic form. In Minkowski momentum space, the propagator has only two poles, at $m_1 \approx 1.14/\sqrt{\alpha'}$ and $m_2 \approx 1.44/\sqrt{\alpha'}$.

At order six, the qualitative results are the same and numerical values change by less than 1% [39].

Summarizing, in the candidate nonperturbative vacuum we find three scalar poles at $m_{A^L} \approx 1.17/\sqrt{\alpha'}$, $m_1 \approx 1.14/\sqrt{\alpha'}$, and $m_2 \approx 1.44/\sqrt{\alpha'}$, and one vector pole at $m_{vec} \approx 1.45/\sqrt{\alpha'}$. Numerically, higher-order terms leave essentially unchanged the lower-order results.

Let us address the question of the quantum-mechanical stability of the candidate nonperturbative vacuum. Consider first perturbative stability. The question is whether a meaningful perturbation theory exists in the functional integral representation. A perturbative analysis is possible if all propagators in euclidean momentum space are everywhere finite. This is equivalent to the requirement of the nonvanishing of the determinant of the quadratic form. A numerical examination of these determinants for each particle field in each truncation order described above shows that they are nonvanishing everywhere in euclidean momentum space. In contrast, a similar analysis shows that perturbation theory does not exist for the canonical bosonic-string vacuum, as expected, because the tachyon propagator diverges at $p^2 = 1/\alpha'$.

Next, consider nonperturbative stability. The issue is whether the candidate vacuum can penetrate the potential-energy barrier. As a general rule, a suppression factor arises for each degree of freedom. String field theory is like a particle field theory with an infinite number of fields. For this reason any barrier penetration should be suppressed relative to the particle case. It may be that a local minimum in the effective potential is nonperturbatively stable. An alternative possibility is that the lifetime for decay may be greater than the age of the universe. In any event, this issue deserves further study.

6. Discussion

We have developed a level-truncation scheme for extracting nonperturbative information in covariant string field theory using functional-integral methods. The method is systematic in that successive orders approach a point of convergence. The effective potential has been evaluated in different truncation orders. A candidate nonperturbative vacuum has been identified and the vacuum expectation values of all fields through truncation-level two have been analytically or numerically obtained, along with the contribution to the cosmological constant produced at tree-level. The physics about the candidate vacuum was analyzed. It was shown to be perturbatively stable. The spectrum of states in the new vacuum was numerically calculated by considering small fluctuations about the vacuum values. Higher-level corrections to these results should occur, but the numerical analyses indicate that the consequent changes are small.

The approach taken here of constructing the effective potential is generally applicable to the identification of candidate vacua in string theories for which a field theory exists. From the involved form of the effective potential found in section 4, it appears that the vacuum structure of a general string theory might be quite complicated. For the open bosonic string, some of the interesting behavior arises for values of the tachyon field for which higher-level corrections are significant. These may cause any local minima to disappear or could induce additional complicated functional dependence of the effective potential.

In the nonperturbative vacuum, the spectrum of states is substantially different from that in the canonical vacuum. Instead of ten states associated with the fields to level two, only four exist. The spectrum includes no symmetric level-one two tensor, only one vector, and only three scalars. The vector and two of the three scalars are formed from linear combinations of fields at different levels. The inclusion of higher-level fields is likely to induce further mixing. However, the mixing should decrease as level number increases because the couplings decrease. The number of entries in the mass-mixing matrix grows quadratically but the magnitude of off-diagonal entries decreases exponentially. We anticipate analogous behavior in other string theories in nonperturbative vacua. Substantial changes in the mass spectrum should occur along with mixing of fields at different levels, especially for lower levels.

In the nonperturbative vacuum, certain poles expected in the propagators are absent. This effect originates in the tree-level running of the couplings. In momentum space, these include a factor $exp\left[-c\alpha' p^2 ln(3\sqrt{3}/4)\right]$, as can be inferred directly from Eq. (2.4). The running coupling is a physical consequence of the extended nature of the string, which smears interactions over the scale $\sqrt{\alpha'}$ and generates good ultraviolet behavior. The results should therefore be generic to string models. Any tree-level string field theory in a nonperturbative vacuum should have features typically found at the loop level in particle field theories. For example, scattering amplitudes require wavefunction renormalization. Also, propagator poles are renormalized and disappear in some cases.

The absence of certain poles in propagators may have significant impact on a string model purporting to be phenomenologically viable. Whenever fields acquire vacuum values, factors of $exp(-c\alpha' p^2)$ enter the quadratic form. The propagator poles in Minkowski space are then fixed by a transcendental equation in p_0^2 rather than by a polynomial, which can eliminate many states. Naive perturbation theory is therefore unsuitable for determining the number and type of states in a nonperturbative string vacuum.

Generally, the number of degrees of freedom in a particle field theory is unaffected by the appearance of expectation values. In the electroweak model, for example, three naively massless gauge bosons absorb three scalars but thereby gain mass, preserving the total number of degrees of freedom. In string theory, the running of the coupling causes some poles to acquire an imaginary component. For instance, we have seen in section 5 that at order two the ϕ propagator has no real pole. However, a complex pole exists at $p_0^2 \approx (0.24 - 1.9i)/\alpha'$. Apparently, the string condensate does not support propagation of certain modes beyond the Planck scale. The existence of the tachyonic degree of freedom for zero expectation values appears simply to indicate the unsuitability of the canonical vacuum.

Since for short spatial distances or large spacelike momenta the effective couplings are small, any string field theory should incorporate tree-level asymptotic freedom. One consequence is perturbative calculability: at distances smaller than the string scale $\sqrt{\alpha'}$, a perturbative analysis is possible in any vacuum. Another issue is the problem of establishing high-level mass spectrum associated with strong coupling, as discussed in section 5.

The tree-level spectrum in the nonperturbative vacuum probably consists exclusively of massive states. This implies the absence of infrared divergences. It is therefore possible that the open bosonic string in the nonperturbative vacuum is a finite 26-dimensional theory. However, loops introduce closed-string poles, including the closed-string tachyon. This destabilizes the vacuum once more. A study of the combined effective potential is necessary to establish the existence of a stable nonperturbative vacuum. If one exists and if the resulting spectrum is also massive, then a finite 26-dimensional theory ensues.

To conclude, we have identified a candidate nonperturbative vacuum in the open bosonic string. The spectrum of states in this vacuum is substantially different from the usual case. The effects we have presented are a consequence of the extended nature of strings. They illustrate anew the importance of nonperturbative effects in string theories.

Acknowledgments

This work was supported in part by the United States Department of Energy under contracts DE-AC02-84ER40125 and DE-AC02-83ER40107.

References

1. J. H. Schwarz, ed., Superstrings, Vols. I and II, (World Scientific, Singapore, 1985); M. B. Green and D. J. Gross, eds., Unified String Theories, (World Scientific, Singapore, 1986); M. B. Green, J. H. Schwarz and E. Witten, Superstring Theory, Vols. I and II, (Cambridge University Press, 1987).
2. M. B. Green and J. H. Schwarz, Phys. Lett. 149B (1984) 117; Phys. Lett. 151B (1985) 21; Nucl. Phys. B225 (1985) 93.

3. D. J. Gross, J. A. Harvey, E. J. Martinec and R. Rohm, Nucl. Phys. B256 (1985) 253; Nucl. Phys. B267 (1986) 75.
4. P. Candelas, G. T. Horowitz, A. Strominger and E. Witten, Nucl. Phys. B258 (1985) 46.
5. P. Frampton, Phys. Rev. D10 (1977) 2922.
6. S. Coleman, Phys. Rev. D10 (1977) 2929.
7. E. Witten, Nucl. Phys. B268 (1986) 253.
8. S. Giddings, Nucl. Phys. B278 (1986) 242.
9. S. Giddings and E. Martinec, Nucl. Phys. B278 (1986) 91.
10. D. Gross and A. Jevicki, Nucl. Phys. B283 (1987) 1.
11. E. Cremmer, A. Schwimmer and C. Thorn, Phys. Lett. 179B (1986) 57.
12. S. Samuel, Phys. Lett. 181B (1986) 255.
13. N. Ohta, Phys. Rev. D34 (1986) 3785.
14. D. Gross and A. Jevicki, Nucl. Phys. B287 (1987) 225.
15. C. Thorn, in *Proceedings of the XXIIIth International Conference of High-Energy Physics, Berkeley, 1986.*
16. E. Cremmer, in *Proceedings of the Paris-Meudon Colloquium, 1986.*
17. S. Samuel, in *Proceedings of the 18th International GIFT Conference: Strings and Superstrings, El Escorial, Spain, 1987.*
18. S. Samuel, Phys. Lett. 181B (1986) 249.
19. D. Gross and A. Jevicki, Nucl. Phys. B293 (1988) 29.
20. K. Suehiro, Nucl. Phys. B296 (1988) 333.
21. S. Samuel, Nucl. Phys. B296 (1988) 187.
22. K. Itoh, K. Ogawa and K. Suehiro, Nucl. Phys. B289 (1987) 127.
23. H. Kunitomo and K. Suehiro, Nucl. Phys. B289 (1987) 157.
24. S. Giddings, E. Martinec and E. Witten, Phys. Lett. 176B (1986) 362.
25. O. Lechtenfeld and S. Samuel, Nucl. Phys. B308 (1988) 361.
26. C. Thorn, Nucl. Phys. B287 (1987) 61.
27. M. Bochicchio, Phys. Lett. 188B (1987) 330; Phys. Lett. 193B (1987) 31.
28. J. H. Sloan, Nucl. Phys. B302 (1988) 349.
29. S. Samuel, Nucl. Phys. B308 (1988) 285.
30. S. Samuel, Nucl. Phys. B308 (1988) 317.
31. D. Freedman, S. Giddings, J. Shapiro and C. Thorn, Nucl. Phys. B298 (1988) 253.
32. R. Bluhm and S. Samuel, Nucl. Phys. B323 (1989) 337.
33. R. Bluhm and S. Samuel, Nucl. Phys. B325 (1989) 275.

34. V. A. Kostelecký and S. Samuel, Phys. Lett. B207 (1988) 169.
35. V. A. Kostelecký and S. Samuel, Phys. Rev. D39 (1989) 683.
36. V. A. Kostelecký, O. Lechtenfeld, W. Lerche, S. Samuel and S. Watamura, Nucl. Phys. B288 (1987) 173.
37. D. Friedan, E. Martinec and S. Shenker, Nucl. Phys. B271 (1986) 93.
38. W. Siegel, Phys. Lett. 142B (1984) 276; 151B (1985) 391, 396.
39. V.A. Kostelecký and S. Samuel, Nucl. Phys. B, in press.
40. V. A. Kostelecký and S. Samuel, Phys. Rev. Lett. 63 (1989) 224.
41. V. A. Kostelecký and S. Samuel, Phys. Rev. D40 (1989) 1886.
42. V.A. Kostelecký and S. Samuel, preprint IUHET 174 (October 1989).

THERMAL EFFECTS IN PARTICLE AND STRING THEORIES

by

B.-S. SKAGERSTAM

Institute of Theoretical Physics
Chalmers University of Technology
S-412 96 GÖTEBORG
Sweden

ABSTRACT

Thermal mass shifts and their physical interpretation are illustrated using quantum field theory of elementary particles and are extended to the theory of strings. A one-loop string calculation is outlined which illustrates a decrease in the effective number of space-time dimensions at high temperatures for open strings.

1. Introduction

1.1 *Thermal Mass Shifts in Particle Theory.* In the presence of a thermal background of particles, physical parameters like masses and coupling constants become temperature (T)-dependent. As was discussed a long time ago (see e.g. Refs.1), in a scattering process a particle can absorb particles from and emit particles to the heat-bath in such a way that the corresponding amplitudes can interfer quantum mechanically with the T=0 vacuum amplitudes. This interference leads, as we discuss below, to temperature-dependent renormalized physical parameters.

The net effect of a thermal background can conveniently be decribed, at least on the one-loop level to which order we essentially will restrict ourselves, by making use of $T\neq 0$ real-time propagators[2]. In such a theoretical framework, the $T\neq 0$ contributions are explicitly separated from the T=0 vacuum contributions. For a charged scalar particle with four-momentum p, the temperature-dependent part of the self-energy, i.e. $\Sigma_T(p)$, will then not be a Lorentz invariant function of energy (p_0) and momentum ($\mathbf{p}$) but rather an O(3)-symmetric function i.e. $\Sigma_T(p) \equiv \Sigma_T(p_0, |\mathbf{p}|)$ due to the presence of a preferred frame of reference namely the heat-bath.

The relationship between p_0 and $\mathbf{p}$ is determined by solving for the equation

$$p^2 - m^2 - \Sigma_T(p_0, |\mathbf{p}|) = 0 \quad . \tag{1.1}$$

Within a reasonable accuracy (≈ 10 percent) one finds, by a numerical investigation, that the solution to Eq.(1.1) can be described by the following dispersion relation

$$p_0 = \sqrt{|\mathbf{p}|^2 + m^2(T)} \quad , \tag{1.2}$$

where

$$m(T) = m + \delta m(T) \quad , \tag{1.3}$$

and where the thermal mass shift $\delta m(T)$ is given by

$$\frac{\delta m(T)}{m} = \frac{\alpha\pi}{3}\left(\frac{T}{m}\right)^2 \tag{1.4}$$

for electrically charged particles and independent of spin[3].

Several things can be noted. **(i)** The mass shift $\delta m(T)$ is completely different in structure from the temperature dependence of the particle masses in the standard model induced by a T-dependence of the Higgs field[4,5]. **(ii)** For $m \leq p_0 \leq m + \delta m(T)$ and for $|\mathbf{p}|$ sufficiently small, the momentum $\mathbf{p}$ will be a complex variable. Such modes will therefore not propagate in the medium. **(iii)** $\delta m(T) = \Sigma_T(p_0 = m, |\mathbf{p}| = 0)$ can be calculated on-shell i.e. for $p^2 = m^2$. **(iv)** $\delta m(T)$ does not depend on Planck's constant. It should therefore be possible to understand the thermal mass shift $\delta m(T)$ in terms of a purely classical argument. This is indeed possible in a remarkably straightforward way. As was discussed by Kibble[6], a charged particle is refracted by an intense electromagnetic wave. The net effect can be described by replacing the mass of the particle by an effective mass, m_{eff}, given by

$$m_{eff}^2 = m^2 + \frac{e^2}{\omega^2}\,\mathcal{E} \quad , \tag{1.5}$$

where $\mathcal{E}$ and ω corresponds to the energy density and the characteristic angular frequency of the external electromagnetic wave. Eq. (1.5) can also be obtained by making use of the known solution for a Dirac electron in a plane wave[7]. Now taking a thermal average of m_{eff} in Eq. (1.5) with a Planck blackbody frequency distribution with temperature T, one is immediately lead to Eq. (1.4). **(v)** $\delta m(T)$ is a dynamical mass in the sense that it actually enters into dynamical equations of motion for the particle in question. This feature, which actually is somewhat difficult to see in Kibble's approach, has been discussed in great detail by Donoghue and collaborators[8] and others[1,3]. **(vi)** Certain experimentally studied thermally induced energy-shifts of highly excited Rydberg levels[9] can be interpreted entirely due to the change of the mass of the electron by an amount $\delta m(T)$. Low lying states will be much less effected by a thermal background of photons as compared to highly excited levels for which the electron is quasi-free. Barton[10] has recently discussed this point in great detail and we will not discuss this issue further here. **(vii)** Since $\delta m(T)/m = O(1)$ for temperatures relevant for the big bang nucleosynthesis, it is interesting to investigate $T \neq 0$ radiative corrections to the corresponding dominant weak interaction transition rates. It has been found[11] that such corrections are in general too small to be observable. A similar result is true for $T \neq 0$ radiative corrections to baryon-number generation in GUTS[12].

1.2 *Corrections to the Anomalous Magnetic Moment.* It is a question of principal interest to what extent purely statistical effects, as discussed above, influence the anomalous magnetic moment of elementary particles. The correction to the electron anomalous magnetic moment, $\delta a(T)$, due to the presence of a thermal background of photons, has been studied by many groups (Peressutti and Skagerstam in Refs.3 and Refs.8,13) with the result

$$\delta a(T) = -\frac{2\pi\alpha}{9}\left(\frac{T}{m}\right)^2 . \tag{1.6}$$

Here the Bohr magneton is finite temperature renormalized, i.e. the anomalous magnetic moment is given in units of $e/2(m + \delta m(T))$. $\delta a(T)$ is of order 10^{-12}, the present experimental accuracy[14], at $T \approx 10^5$ K which is far beyond any temperature relevant to present day high-precision experiments.

In passing we would also like to mention that recently the finite density correction due to exchange processes of a fermion with identical particles in the heat-bath has been calculated[15]. Apart from an exponentially small correction, it is actually possible to write down the corresponding correction, $\delta a(T,\mu)$, in a very simple form, i.e.

$$\delta a(T,\mu) = -\frac{2\alpha}{3\pi}\int_1^\infty \frac{dx}{x^2}\sqrt{\frac{x+1}{x-1}}\,(1 - 6x^2 + x^3)\,\frac{1}{1 + \exp(\frac{m}{T}(x - \frac{\mu}{m}))} , \tag{1.7}$$

where μ is the chemical potential and where the Bohr magneton unit now is given by $e/2(m + \delta m(T,\mu))$. The mass-shift $\delta m(T,\mu)$ can be calculated on the one-loop level in a straightforward manner and is of the approximative form[16]

$$\delta m(T,\mu) \approx \delta m(T) + \frac{\alpha}{2\pi}\left(\left(\frac{\mu}{m} - 2\right)\sqrt{\left(\frac{\mu}{m}\right)^2 - 1} - 3\ln\left(\frac{\mu}{m} + \sqrt{\left(\frac{\mu}{m}\right)^2 - 1}\right)\right) \quad . \quad (1.8)$$

For high temperatures, i.e. $T >> m$, Eq. (1.7) renormalizes $\delta a(T)$ in Eq. (1.6) to the value $\delta a(T) = -\frac{\pi\alpha}{3}\left(\frac{T}{m}\right)^2$, which agrees with previously obtained high-temperature expansion results by Fujimoto and Yee[13] and by Donoghue et. al.[8]. At small densities n and for T=0, it easily follows from Eq. (1.7), using the ideal-gas relationship between μ and n, that $\delta a(0,\mu)$ becomes positive and has the form

$$\delta a(0,\mu) = \frac{16\alpha}{3\pi}\,\frac{(3\pi^2 n)^{1/3}}{m} \quad . \qquad (1.9)$$

An estimate of n appropiate to e.g. the Wesley and Rich experiment[17] leads to $\delta a(0,\mu) \approx 5\cdot 10^{-11}$, which is larger than the muonic, hadronic or electroweak correction to the electron anomalous magnetic moment. The n-dependence was actually studied by Wesley and Rich. Their experimental accuracy was, at the time of their experiment, $\pm 3.5\cdot 10^{-9}$. The trend of similar experiments since then have, however, been towards smaller densities and in the experiment by Dehmelt et. al.[18] one measures the anomalous magnetic moment of a single electron. In such a situation our considerations should not apply.

1.3 *Thermal Gravitons and the Classical Jeans Instability.* If the thermal background consists of gravitons one expects a gravitational instability. The thermally induced mass for external gravitons, $m_g^2(T)$, which corresponds to the well-known plasma frequency in QED for photons, was first calculated in the context of quantum field theory by Gross, Perry and Yaffe[19] in 1981 with the result $m_g^2(T) = -4m_J^2$, where $m_J^2 = 4\pi G\rho$ is the classical Jeans mass (ρ the energy density of the thermal gravitons and G Newton's constant). $m_g^2(T)$ is tachyonic as expected. Ambiguities in the calculational procedure was discussed by Kikuchi, Moriya and Tsukahara[20]. These authors pointed out, among other things, that $m_g^2(T)$ depends on how one actually defines the gravitational fluctuations and suggested a scheme-independent definition and obtained the result $m_g^2(T) = -3m_J^2$. Recently Gribosky, Donoghue and Holstein[21] pointed out that the flat space-time background used in Refs. 19 and 20 is inconsistent with the presence of a thermal heat bath. Using a self-consistent approach and eliminating a potentially divergent tadpole contribution, these authors obtained the result $m_g^2(T) = -2m_J^2$.

Here we would like to suggest that it could be of great interest to recalculate $m_g^2(T)$, at least on the one-loop order, in a string theory containing gravitons. A string theory gives us directly the proper definition of the gravitational field by conformal invariance. The effect of the thermal background could perhaps be described in terms of a Fischler-Susskind mechanism[22]. We have not yet any complete results to report on such a calculation. At T=0 a similar issue has been addressed in Ref. 23.

For external photons one finds[24] similarly that $m_\gamma^2(T) = 0$ and for external massive scalars $m_s^2(T) = m^2(1 - 4\pi G T^2)$.

2. Thermal Mass Shifts in Open String Theories

2.1 *Universality of the Hagedorn Temperature.* It turns out to be convenient to consider lower-dimensional type I open superstrings as a theoretical laboratory when studying thermal mass shifts in string theory. This is primarily due to the observation that such string models contain massless scalar mesons at the tree-level. The covariant NSR-formulation[25] is, furthermore, of basic importance when extending the mass-shift calculations to $T\neq0$, which should be clear when comparing with the calculational rules in $T\neq0$ particle theory as briefly discussed in the introduction.

The lower-dimensional open string theories in D space-time dimensions are constructed by adding a set of d world-sheet fermions (Refs. 26, also see the contributions by P. Cox[27] and Z. Bern[28] for more technical details and further references). Conformal invariance, or the factorizability in the graviton sector, then leads to d=30-3D. The space-time boson ($F_B(w)$) and fermion ($F_F(w)$) partition functions have the generic structure

$$F_B(w) = \frac{1}{\sqrt{w}} \prod_{n=1}^{\infty} \left(\frac{1+w^{n-1/2}}{1-w^n}\right)^{D-2} \prod_{n=1}^{\infty} (1+w^{n-1/2})^d , \tag{2.1}$$

and

$$F_F(w) = w^{d/24} 2^{(D-2)/2} \prod_{n=1}^{\infty} \left(\frac{1+w^n}{1-w^n}\right)^{D-2} 2^{d/2} \prod_{n=1}^{\infty} (1+w^n)^d \ . \tag{2.2}$$

We will especially consider the case D=6. The extension of our considerations to other dimensions is in principle straightforward[29] except for D=4 in which case, however, recent progress have been made[28].

In D=10 the G-projected bosonic partition function and $F_F(w)$ are, of course, equal. In D=6 this is not true. It nevertheless follows that the asymptotic forms of the corresponding boson and fermion level-densities, i.e. ρ_B and ρ_F, are equal. To show this one considers in a standard manner the expansion ($F(w) = F_F(w)$ or $F_B(w)$)

$$F(w) = \sum_{n=1}^{\infty} d(n) w^n \tag{2.3}$$

and evaluates d(n) using a saddle point technique. We find that[30]

$$\rho_B(m) \approx \rho_F(m) \approx \rho_0 \frac{e^{m\beta_H}}{m^{D-1/2}} \quad , \tag{2.4}$$

where ρ_0 is a constant and $\beta_H = 1/T_H = \pi\sqrt{8\alpha'}$ is the inverse of the Hagedorn temperature for D=10. This apparent universality of T_H, which easily can be shown[31] to be true for all subcritical models so far constructed, is similar to the universality of $\rho(m)$ in the case of closed string theories. As a consequence of modular invariance it is known[32] that asymptotically

$$\rho(m) = \frac{e^{m\beta_H}}{m} \tag{2.5}$$

for closed strings independent of the structure of the compact target space-manifold of the string, a result which can be understood in terms of a very simple classical model of high-density string gas[33].

The open string gas partition function Z is given by:

$$\ln Z = \frac{V}{(2\pi)^{D-1}} \int_0^{\infty} d^{D-1}\mathbf{k} \, \{\rho_F(m) \ln(1 + e^{-\beta\omega}) - \rho_B(m) \ln(1 - e^{-\beta\omega})\} \ , \tag{2.6}$$

where $\omega = \sqrt{m^2 + |\mathbf{k}|^2}$. In Eq. (2.6) the momentum integrals can now be carried out using the asymptotic forms in Eq. (2.4) with the following result for the energy density, $\varepsilon(T)$, for temperatures close to T_H

$$\varepsilon(T) = -\frac{1}{V}\frac{\partial}{\partial\beta}\ln Z \propto \frac{1}{(1-T/T_H)^2} \quad , \tag{2.7}$$

i.e. T_H is a limiting temperature for the D=6 open string gas. In the next section we will verify that the two-point function at $T\neq 0$ is, nevertheless, finite for all temperatures, i.e. also for $T \geq T_H$.

2.1 *One-loop Mass Shifts at Finite Temperature.* In $D < 10$ there are massless scalars at the tree level with, in the case when D=6, an Sp(8) Chan-Paton gauge symmetry and a global $SU(2)^4$ symmetry among the 12-fold replicated Sp(8) adjoint representation. For the planar loop, the T=0 mass shift δm^2 is of the form:

$$\delta m^2 = ig^2 \int \frac{d^Dp}{(2\pi)^D} \langle p,0 \mid Tr\{\Delta V^a(k_1,1)\, \Delta V^b(k_2,1)\} \mid p,0\rangle \tag{2.8}$$

apart from Sp(8) Chan-Paton factors. g is the open string coupling constant. $V^a(k,z)$ is the vertex operator for emission of a scalar meson with momentum k and with a $SU(2)^4$ index a. Δ is the open string propagator. Here k_1, k_2 are such that $k_1+k_2 = k$. The limit k, $k_1 \cdot k_2 \to 0$ is taken after the momentum-integrations have been performed following the regularization prescription of Neveu and Scherk[34]. To Eq. (2.7) one should add the Möbius strip contribution. At T=0 and for D=6 it was then found that[35]

$$\frac{\alpha' \delta m^2}{g^2} = -1.49 \cdot 10^{-3} \tag{2.9}$$

For k_1, k_2 on-shell and no external (space-)momentum, it is now a rather straightforward extension of the the considerations in Section 1 to find δm^2 at $T\neq 0$. The $T\neq 0$ real-time field-theoretical techniques can e.g. be extended to string theory in a somewhat abstract manner[36] in order to derive the calculation rule that in a zero-mode energy integral one should simply make the replacement

$$\int_{-\infty}^{\infty} \frac{dE}{2\pi} f(E) \quad \rightarrow \quad \left(\frac{1}{-i\beta}\right) \sum_{n=-\infty}^{\infty} f(E) \quad , \tag{2.10}$$

where in the sum $E = \frac{2\pi n}{-i\beta}$ $\left(= \frac{\pi(2n+1)}{-i\beta}\right)$ for bosons (fermions) circulating in the loop-diagram. Eq. (2.10) can also be derived as follows. If a function f vanishes sufficiently fast at infinity and has no other singularities than simple particle poles, corresponding to $E=\omega_k$, with residues $\mathrm{Resf}(\omega_k)$, Eq. (2.10) follows from the following simple integral-representation

$$\sum_{E=\omega_k} \frac{1}{i} \frac{\mathrm{sign}(E)}{e^{\beta|E|} \pm 1} \mathrm{Resf}(E) = \int_{-i\infty+\varepsilon}^{i\infty+\varepsilon} \frac{d\,E}{2\pi} \frac{f(E) + f(-E)}{e^{\beta E} \pm 1} =$$

$$= \left(\frac{1}{-i\beta}\right) \sum_{n=-\infty}^{\infty} f(E) \quad - \quad \int_{-i\infty}^{i\infty} \frac{dE}{2\pi} f(E) \quad , \tag{2.11}$$

where + (-) correponds to fermions (bosons). Eq. (2.11) represents a sum over the contributions from the particle poles, weighted with the appropiate statistical factors, which in turn corresponds to particles emitted or absorbed by the heat-bath in the picture of thermal processes described in the introduction. The last integral in Eq. (2.11) corresponds to the (Wick-rotated) T=0 contribution.

If we write the T=0 mass shift δm^2 in the form

$$\frac{\alpha'\delta m^2}{g^2} = i \int_0^1 \frac{dw}{w} (-\ln(w)/\pi)^{-(D-1)/2} \int_{-\infty}^{\infty} dE\, e^{-\alpha' E^2 \ln w} \int_w^1 \frac{d\rho}{\rho} \Psi(\rho,w)^{2\alpha' k_1\cdot k_2}$$

$$\cdot\{I_B(\rho,w) + I_F(\rho,w)\} \quad , \tag{2.12}$$

then $I_B(\rho,w)$ ($I_F(\rho,w)$) represents the contribution from bosons (fermions) circulating in the loop. $\Psi(\rho,w)$ comes from the correlation of generalized plane waves in a standard manner[24]. In Eq. (2.12) we now replace the zero-mode energy integral with the prescription in Eq. (2.9). Similarly as for the T=0 mass shift[34], the integrals in the expression for the T≠0 mass shift, $\delta m^2(T)$, can be simplified and the remaining parametric w-integral can easily be done numerically[30]. $\delta m^2(T)$ is found to be finite for all temperatures and, for the D=6 dimensional model, also negative. For sufficiently high temperatures it also becomes linearly dependent on the temperature, as can be verified analytically. For supersymmetric models $\delta m^2(T)$ becomes positive except in the case D=3, in which $\delta m^2(T)$ is exponentially small and negative. For D=7 and D=9, $\delta m^2(T)$ becomes divergent for T slightly larger than the Hagedorn temperature[31].

3. Conclusions and Final Comments

We have briefly indicated how one, in a rather straightforward manner, can extend one-loop string calculations to finite temperature and we have also presented the results of one such explicit calculation. For open strings, supersymmetric or not, $\delta m^2(T)$ exhibits a linear dependence on T for sufficiently large temperatures (except for D≠7,9). Comparing this behaviour with particle theory, in which case $\delta m^2(T) \propto T^{D-2}$, one may interpret this feature by saying that the string T≠0 gas corresponds to an effective D=3-dimensional particle theory. (For T≠0 closed strings Atick and Witten[37] have argued that the effective dimension is D=2). It is interesting to notice that the D=3 behaviour we observe has been observed using a completely different reasoning by Gross and Manez[38]. They have noticed that in high energy scattering of open strings, the polarization of all particles must lie in the plane of scattering.

We have restricted ourselves to an evaluation of the real part of the two-point function at T≠0. The imaginary part at T≠0 for the type I open supersymmetric string in D=10 space-time dimensions has been discussed by Tsuchiya[39] at this conference. The imaginary part of the two-point function is found to vary slowly as a function of the temperature and it is, as $\delta m^2(T)$, also linear in T for sufficiently large temperatures.

4. Acknowledgment

It is a pleasure to thank the organizers for the opportunity to present some results of a fruitful collaboration with L. Clavelli, B. Harms, P. Elmfors and A. Stern. We have also reported some work done in collaboration with G. Peressutti and P. Salomonson. We are also grateful to Z. Bern, L. Clavelli, P. Elmfors, H. Konno and A. Tsuchiya for discussions on some of their recent work in progress and G. Barton for informing us about his work prior to publication. We appreciate a discussion with I. Lindgren on high-precision QED experiments and B. Harms for useful comments on the manuscript. This research is supported by the Swedish National Science Research Council under contract No. 8244-103.

5. References

1. E. P. Tyron, *Phys. Rev. Lett.* **32** (1974) 1139 D. Eimerl; *Phys. Rev.* **D12** (1975) 427.
2. L. Dolan and R. Jackiw, *Phys. Rev.* **D9** (1974) 3320. For a recent discussion and review see e.g. N. P. Landsmann and Ch. G. van Weert, *Phys. Rep.* **145** (1987) 1.
3. F. Englert, *Bull. Class. Sci. Acad. R. Bel.* **45** (1959) 782; V. V. Klimov, *Sov. Phys. JETP* **55** (1982) 199; G. Peressutti and B.-S.Skagerstam, *Phys. Lett.* **110B** (1982) 406; D. A. Weldon, *Phys. Rev.* **D26** (1982) 2789.
4. D. A. Kirzhnits and A. D. Linde, *Phys. Lett.* **42B** (1972) 471; S. Weinberg, *Phys. Rev.* **D9** (1974) 3357; A. D. Linde, *Rep. Prog. Phys.* **42** (1979) 389.
5. M.B.Kisslinger and P.D. Morley, *Phys. Rev.* **D13** (1976) 2765; *Phys. Rep.* **51C** (1979) 63
6. L.S. Brown and T.W.B. Kibble, *Phys. Rev.* **133** (1964) A705; T.W.B. Kibble, *Phys. Rev. Lett.* **16** (1966) 1054.
7. D.A. Volkov, *Z. Physik* **94** (1935) 25. Also see C. Itzyskon and J.-B. Zuber, *Quantum Field Theory* (McGraw-Hill, 1980).
8. J.F. Donoghue, B. R. Holstein and R. W. Robinett, *Phys. Rev.* **D30** (1984) 2561; *Ann. Phys.* **164** (1985) 233.
9. L. Hollberg and J. L. Hall, *Phys. Rev. Lett.* **53** (1984) 230.
10. G. Barton, "*On the Finite-Temperature Quantum Electrodynamics of Free Electrons and Photons*", preprint University of Sussex, 1989, and references cited therein.
11. A. E. I. Johansson, G. Peressutti and B.-S. Skagerstam, *Phys. Lett.* **117B** (1982) 171; J.-L. Cambier, J. R. Primack and M. Sher, *Nucl. Phys.* **B209** (1982) 372 and (E) **B222** (1983) 517; D.A. Dicus, E.W. Kolb, A.M. Gleeson, E.C. G. Sudarshan, V.L. Teplitz and M.S. Turner, *Phys. Rev.* **D26** (1982) 2694; W. Keil, *Phys. Rev.* **D38** (1988) 152.
12. K. Takahashi, *Phys. Rev.* **D29** (1984) 632.
13. Y. Fujimoto and J.H. Yee, *Phys. Lett.* **114B** (1982) 359; A. E. I. Johansson, G. Peressutti and B.-S. Skagerstam, *Nucl. Phys.* **B278** (1986) 324.

14. For a review of recent theoretical and experimental achievments see e.g. T. Kinoshita, talk presented at the *Symposium on the Hydrogen Atom*, Scuola Normale Superiore, Pisa, Italy, June, 1988 (CERN-TH. 5097/88).
15. P. Elmfors, G. Peressutti and B.-S. Skagerstam, to appear.
16. E. J. Levinson and D. H. Boal, *Phys. Rev.* **D31** (1985) 3280; T. Toimela, *Nucl. Phys.* **B273** (1986) 719.
17. J. C. Wesley and A. Rich, *Phys. Rev.* **A4** (1971) 1341 and *Rev. Mod. Phys.* **44** (1972) 250.
18. P. B. Schwinberg, R.S. Van Dyck Jr and H. G. Dehmelt, *Phys. Rev. Lett.* **47** (1981) 1679.
19. D. Gross, M. J. Perry and L.G. Yaffe, *Phys. Rev.* **D25** (1982) 330.
20. Y. Kikuchi, T. Moriya and H. Tsukahara, *Phys. Rev.* **D29** (1984) 2220.
21. P.S. Gribosky, J.F. Donoghue and B. R. Holstein, Ann. Phys. (N.Y.) **190** (1989) 149 , and B. R. Holstein in the *Proceedings from Workshop on Thermal Field Theories and Their Applications*, Ohio 1988, Eds. K.L Kowalski, N.P. Landsman and Ch.G.van Weert, *Physica* **A158** (1989) 387.
22. C. Lovelace, *Nucl. Phys.* **B273** (1986) 413; W. Fischler and L. Susskind, *Phys. Lett.* **173B** (1986) 262.
23. J. A. Minahan, "*Calculation of the One-Loop Gravitation Mass-Shift in Bosonic String Theory*", preprint (UFIFT-HEP-89-2), University of Florida, Gainesville, 1989; N. Marcus, *Phys. Lett.* **B219** (1989) 265.
24. A. E. I. Johansson, G. Peressutti and B.-S. Skagerstam, *Mod. Phys. Lett.* **A3** (1988) 667; D. A. Johnston, *Nucl. Phys.* **B297** (1988) 721.
25. M. B. Green, J. H. Schwartz and E. Witten, *Superstring Theory* (Cambridge University Press, Cambridge, 1987).
26. L. Clavelli, P. H. Cox, B. Harms and A. Stern ,"*Lower-Dimensional Type I String Theories* ", preprint, UAHEP882, University of Alabama, 1988; L. Clavelli, P. H. Cox and B. Harms, *Phys. Rev. Lett.* **61** (1988) 787; Z. Bern and D. Dunbar, *Nucl. Phys.* **B319** (1989) 104.
27. P. Cox, these Proceedings.
28. Z. Bern, these Proceedings; Z. Bern and D. C. Dunbar, "*Open Strings In Four Dimensions*", preprint (LA-UR-89-3213), Los Alamos National Laboratory, 1989.
29. L. Clavelli, P. H. Cox, P. Elmfors and B. Harms, Phys. Rev. **D40** (1989) 4078.

30. L. Clavelli, P. Elmfors, B. Harms, B.-S. Skagerstam and A. Stern, *Int. J. Mod. Phys.* **A5** (1990) 175.
31. L.Clavelli and P. Elmfors, *Thermal Properties of Open Superstring Models*, preprint, UAHEP898, University of Alabama and ITP 89-64, Chalmers University of Technology, 1989.
32. I. Antoniades, J. Ellis and D.V. Nanopoulos, *Phys. Lett.* **B199** (1987) 402; M. Axenides, S.D. Ellis and C. Kounnas, *Phys. Rev.* **D37** (1988) 2965; P. Salomonson and B.-S. Skagerstam, in the *Proceedings from Workshop on Thermal Field Theories and Their Applications*, Ohio 1988, Eds. K.L Kowalski, N.P. Landsman and Ch.G.van Weert, *Physica* **A158** (1989) 499; R. Brandenberger and C. Vafa, *Nucl. Phys.* **B316** (1989) 391; F. Englert and J. Orloff, " *Universality of the Closed String Phase Transition*", preprint, ULB-TH-89/08, Université Libre de Bruxelles, 1989.
33. P. Salomonson and B.-S. Skagerstam, *Nucl. Phys.* **B268** (1986) 349.
34. A. Neveu and J. Scherk, *Nucl. Phys.* **B36** (1972) 317.
35. L. Clavelli, P.H. Cox and B. Harms, "*Higgs Instability in Lower Dimensional Type I String Theory* ", preprint, UAHEP 889, University of Alabama 1989.
36. Y. Leblanc, Phys. Rev. **D39** (1989) 1139 and in the *Proceedings from Workshop on Thermal Field Theories and Their Applications*, Ohio 1988, Eds. K.L Kowalski, N.P. Landsman and Ch.G.van Weert, *Physica* **A158** (1989) 536.
37. J.J. Atick and E. Witten, *Nucl. Phys.* **B310** (1988) 291.
38. D. Gross and Manez, *Nucl. Phys.* **B326** (1989) 73.
39. A. Tsuchiya, these Proceedings.

The C Theorem, Vertex Condensates and New Phases of the Coulomb gas

D. Boyanovsky
Department of Physics and Astronomy
University of Pittsburgh
Pittsburgh, P.A. 15260

and

R.Holman
Department of Physics
Carnegie Mellon University
Pittsburgh,P.A. 15213

Abstract

We give an alternative proof of the c-theorem within the context of perturbation theory with particular attention to the renormalization aspects. The c-function is related to the thermodynamic entropy by analyzing the corresponding field theory at finite temperature. We also study a generalized Coulomb gas of electric charges and magnetic monopoles in two dimensions. This theory is a simple model for bosonic string theories perturbed by vertex operators. We find a new non-trivial self-dual fixed point with a continuous spectrum of anomalous dimensions and $c > 1$. This fixed point describes a new universality class with condensates of vertex operators and excitations with fractional statistics.

1 The C-theorem, Entropy and the Density of States

Zamolodchikov's c-theorem[1] is one of the more important results in conformal field theory since it allows us to study two dimensional systems away from criticality[2-4]. If we introduce operators $\mathcal{O}_i$ and conjugate couplings g_i that move an initially conformal theory away from criticality, then the theorem states that there exists a function $c(g_i)$ such that: a) $c(g_i)$ is non-negative and non-increasing along infrared renormalization group flows. b) the stationary points of $c(g_i)$ correspond to the critical fixed points of the field theory, and c) at the fixed points, $c(g_i^*)$ is the conformal anomaly. The original proof[1] makes use of positivity arguments for the correlation functions of the trace and light-cone components of the energy momentum (E.M) tensor. This proof does not, however, address some important questions regarding the renormalization of the theory. In particular, conservation of the E.M tensor implies that it is free of multiplicative renormalizations (no anomalous dimension). However, one subtraction is required to render it finite, and since this subtraction is scheme dependent, we must now question some aspects of the proof. Since the subtractions are arbitrary, can we be sure that the required positivity properties of the correlation functions still hold? Another question is: is the theorem scheme dependent? Another point to be dealt with concerns the question of renormalizability. The theory will be strictly renormalizable with a finite number of operators when these operators are marginal. However we are interested in operators that move the theory away from criticality towards an infrared stable fixed point. The renormalization procedure must regularize the infrared (for relevant operators) or ultraviolet (for irrelevant operators) divergences and provide a prescription to handle possible new operators that appear in the process. There are several possibilities one of which is the Wilsonian approach[5]. Here one allows all possible operators consistent with the symmetries to appear in the action, thereby effectively dealing with a very large coupling constant space. We can also introduce a controlling parameter just like the ϵ expansion of critical phenomena[6] to keep the irrelevant operators at bay, allowing us to keep renormalizability within this expansion with a finite number of relevant operators. Though the first approach is very appealing, it is hard to perform explicit calculations within it. The second one provides us with a controlled perturbative expansion that allows for the study of the superrenormalizable and non-renormalizable regimes while keeping a finite number of operators. Here we provide an alternative proof[7] of the c- theorem addressing these issues. We use the second approach, with $\epsilon_i = dim[\mathcal{O}_i] - 2$.

The main ingredient is the result from conformal invariance that predicts that the finite temperature contribution to the free energy density in a conformally invariant theory is given by[8,9]:

$$\mathcal{F}(T) = \mathcal{F}(0) - \frac{\pi c T^2}{6} \tag{1}$$

where c is the conformal anomaly. This expression is valid in the critical (fixed point) theory in flat space. The crucial observation is that the free energy density, like the stress tensor does *not* acquire an anomalous dimension under renormalization and scales with dimension two. This fact allows us to write the finite temperature free energy density **even away from the critical point** as:[7,9]

$$\mathcal{F}(g_B, T) = \mathcal{F}(g_B, 0) - \frac{\pi c(g_B, T, a) T^2}{6} \tag{2}$$

where $c(g_B, T, a)$ is a dimensionless function of the bare couplings (g_B) and a is a cut-off introduced to regularize the theory.

It is a well known result of finite temperature field theory, that the counterterms and subtractions required to renormalize the theory are the same as those in the $T = 0$ theory. This stems from the fact that the short distance behaviour of the theory is insensitive to temperature. Since at finite temperatures the correlation functions fall-off exponentially at large distances ,the finite temperature contribution to the correlation functions is infrared finite.

This analysis then shows that all the required subtractions in the free energy density are absorbed in the zero temperature contribution $\mathcal{F}(g_B, 0)$. Furthermore, the function c does not acquire anomalous dimensions, therefore it is not multiplicatively renormalized. Introducing a **dimensionless** renormalized coupling λ_R, at $T = 0$ and a renormalization mass scale μ (where we now concentrate on the case of only one coupling for simplicity) the function $c(g_B, T, a)$ is **finite** in terms of λ_R. All the cut-off dependence is absorbed in the definition of the renormalized coupling (at $T = 0$) and the function c now depends on the renormalized coupling, temperature and the renormalization scale. Being a dimensionless function, we may write:

$$c(g_B, T, a) = c(\lambda_R, \frac{\mu}{T}) \tag{3}$$

Since c is R.G. invariant, it obeys an homogeneous R.G. equation whose solution is [7] $c = c(\lambda(t))$ where $\lambda(t)$ is the running coupling constant obeying $\partial\lambda(t)/\partial t = -\beta(\lambda(t))$, $\lambda(t = 0) = \lambda_R$ and $t = \ln(\mu/2\pi T)$. Assuming that the operator $\mathcal{O}$ is primary, it scales with definite conformal weight at the fixed points $\lambda = 0$, $\lambda = \lambda^*$. Having been subtracted in the plane (by $\mathcal{F}(T = 0)$) its expectation value at $T \neq 0$ vanishes at the fixed points. From eq.(1), c coincides with the conformal anomaly at the fixed points and as argued above, $\partial c/\partial\lambda = 0$ at the fixed points, which suggests that $\partial c/\partial\lambda = 0$ may be related to the beta function.

In ref.(7) we analyze the perturbative series for $\partial c/\partial\lambda$ and β in a **double expansion** in λ_R and ϵ and show to **all orders in the double expansion** that:

$$\begin{aligned} \frac{\partial c}{\partial\lambda} &= c_0(\epsilon)\beta(\lambda)F(\lambda) \\ c_0(\epsilon) &= 6\pi^2 + O(\epsilon) + ... \\ F(\lambda) &= 1 + A(\epsilon)\lambda + B(\epsilon)\lambda^2 + ... \end{aligned} \tag{4}$$

We first note that the $c_0(\epsilon)$ and $F(\lambda)$ are positive definite within the double expansion since we cannot balance a term of $O(\epsilon, \lambda)$ against a term of $0(1)$ in perturbation theory. From the R.G. equation obeyed by c we find[7]

$$\frac{\partial c}{\partial t} = -\beta^2[c_0(\epsilon)F(\lambda)] \tag{5}$$

and the term in square bracket is positive definite. Unitarity determines the sign of the two point function from which $c_0(\epsilon)$ is obtained[7]. The above equation shows that the function c decreases along infrared flows and as argued above it coincides with the conformal anomaly at the fixed points.

From the expression for $F(\lambda)$ given above, we see that it is possible to find a reparametrization $g(\lambda) = \lambda[1 + g_1\lambda + g_2\lambda^2 + ...]$ in which

$$\frac{\partial c(g)}{\partial g} = c_0(\epsilon)\beta(g) \tag{6}$$

However this reparametrization **defines a particular scheme** and the above relationship between the coupling derivative of the c-function and the β function is valid only in *this scheme.* The beta and F functions above certainly depend on the renormalization scheme used. However the c-function is **universal** since it determines the finite-size scaling corrections to the free energy density[9,7].

Although we have only studied the case of one coupling, there is no loss of generality in doing this. The reason is as follows. If we consider the case of many coupling constants, the beta functions define a set of vector fields in the space of couplings whose integral curves are the R.G. trajectories. Given two fixed points, we can always choose the integral curve that joins them and along this particular trajectory our analysis holds. In the next section we will apply this analysis to a specific theory.

The study of the c-theorem by means of finite temperature field theory, offers the following interesting payoff. The free energy density at finite temperature, given by eq.(1) can also be written as $\mathcal{F} = \langle H \rangle - T\mathcal{S}$ with $\mathcal{S}$ the entropy given by:

$$\mathcal{S} = \frac{\pi c T}{3}$$

As a function of energy, it is given by

$$S = (\frac{2\pi c V E}{3})^{\frac{1}{2}} \tag{7}$$

In the thermodynamic limit the microcanonical density of states $\rho(E)$ is (up to power-law corrections) $\rho(E) \approx e^{\mathcal{S}(E)}$. For any fixed point theory in D space-time dimensions

$$\begin{aligned} \mathcal{F}(T) &= \mathcal{F}(0) - \alpha T^D \\ \mathcal{S}(T) &= D\alpha T^{D-1} \end{aligned} \tag{8}$$

The constant α plays in D-dimensions the same role as c in $D = 2$ in the sense that it "counts" the degrees of freedom. The entropy is a monotonic increasing function of temperature (and energy) and this fact may yield a c-theorem in D dimensions[10].

2 The Coulomb gas: Vertex operators and Parafermionic Condensates

The Coulomb gas is a fascinating model of 2-D statistical mechanics since it is conjectured to describe the critical behavior of a wide variety of models[11]. Our motivation for studying this model is then the possibility of studying many different conformal field theories by means of analyzing this model. The details of our investigation may be found in ref.(12). The partition function of the theory reads:

$$\begin{aligned} Z[K, N, y_0, y_N] \; = \; & Z_{sw} \sum_{n,m} y_0(m(\vec{r})) y_N(n(\vec{r})) \exp[\sum_{\vec{r},\vec{r}'} \{ -\frac{N^2}{2K} n(\vec{r}) G(\vec{r} - \vec{r}') n(\vec{r}') \\ & -2\pi^2 K m(\vec{r}) G(\vec{r} - \vec{r}') m(\vec{r}') - iNn(\vec{r}) \Theta(\vec{r} - \vec{r}') m(\vec{r}') \}] \end{aligned} \tag{9}$$

where, Z_{sw} is a gaussian partition function, G and Θ are the two dimensional Green function and angle respectively (in Euclidean space). The variables $n(\vec{r})$ and $m(\vec{r})$ are integer valued fields representing magnetic monopoles and electric charges respectively[11,12] and $y_0(m), y_N(n)$ are the fugacities for the corresponding charges. Notice that whereas both magnetic and electric charges interact amongst themselves via the coulomb potential, they interact with each other via the Aharonov-Bohm potential given by the angle between the particles. After some straightforward manipulations, keeping only the fugacities for the $0, \pm 1$ charges, this partition function can be written in terms of vertex operators $V_{a,b}$ as:

$$
\begin{aligned}
Z &\simeq \int [\mathcal{D}\phi] \exp[-\int d^2r\{\frac{1}{2}(\partial_\mu\phi)^2 + \frac{Y_+}{16\pi a^2}V_{1,0} + \frac{Y_-}{16\pi a^2}V_{-1,0} \\
&+ \frac{G_+}{16\pi a^2}V_{0,1} + \frac{G_-}{16\pi a^2}V_{0,-1}\}]
\end{aligned} \tag{10}
$$

where

$$V_{a,b} = \exp[i(a\beta\phi + b\gamma\tilde{\phi})] \tag{11}$$

and

$$\beta = 2\pi\sqrt{K}; \gamma = \frac{2\pi N}{\beta}$$

The dual field $\tilde{\phi}$ is related to the field ϕ by $i\partial_\mu\phi = \epsilon_{\mu\nu}\partial_\nu\tilde{\phi}$. The couplings $Y_\pm, G_\pm$ are proportional to the fugacities and a is an ultraviolet cutoff. This theory has a duality: $Y_\pm \leftrightarrow G_\pm$, $\phi \leftrightarrow \tilde{\phi}$.

The vertex operators $V_{a,b}$ are the parafermions of Kadanoff and Fradkin[13]. The operators $V_{\pm1,0}$, $V_{0,\pm1}$ create a unit of electric and magnetic charges of strength β and γ respectively and obey $V_{a,b}(\vec{\mathbf{r}})V_{a,b}(\vec{\mathbf{r}'}) = e^{2\pi i Nab}V_{a,b}(\vec{\mathbf{r}'})V_{a,b}(\vec{\mathbf{r}})$. For general a,b these operators create excitations with *fractional statistics* with spin Nab. In the gaussian theory we have:

$$
\begin{aligned}
dim[V_{\pm1,0}] &= \frac{\beta^2}{4\pi};\ dim[V_{0,\pm1}] = \frac{\gamma^2}{4\pi} \\
dim[Y_\pm] &= 2\delta;\ dim[G_\pm] = 2\tilde{\delta} \\
\delta &= \frac{\beta^2}{8\pi} - 1;\ \tilde{\delta} = \frac{\gamma^2}{8\pi} - 1
\end{aligned} \tag{12}
$$

From these equations we see that the interaction is **renormalizable** when the vertex operators are marginal corresponding to $\beta^2 = 8\pi, N = 4$ ($\delta = 0, \tilde{\delta} = 0$). At this value of β the vertex operators $V_{\pm1,0}$ create on-shell tachyons in string theory language. We want to study the theory in the super and non-renormalizable regimes to search for nontrivial infrared stable fixed points. In order to perform a controlled expansion and renormalization we use a multiple expansion in the couplings $Y_\pm, G_\pm$ and $\delta, \tilde{\delta}$. This is equivalent to perform an expansion in β^2 around 8π and N about 4. In ref.(12) we proved that within this expansion the theory is multiplicatively renormalizable and that the free energy requires one subtraction. The renormalization procedure used in ref.(12) is non-standard so we highlight the main ingredients (see ref.(12) for details). The first step is to normal order the vertex operators in the interaction picture of a free massless boson. Next, expand the exponential in the partition function as a power series in terms of the vertex operators **without expanding the vertex operators themselves in power series in the fields.** Inside the path integral we can replace the product of vertex operators by their **operator product expansion** (O.P.E). We isolate the divergent terms and carry out the integrals using I.R and U.V cutoffs. The divergences (logarithmic in the cutoffs) can be removed order by order by wave-function, coupling constant renormalization and a subtraction for the free energy. Once the renormalization constants are obtained the rest is standard. The β functions (in terms of the renormalized couplings) are given by:

$$\begin{aligned} \beta_{Y_\pm} &= Y_\pm[2\delta - G_+G_-/64] \\ \beta_{G_\pm} &= G_\pm[2\tilde{\delta} - Y_+Y_-/64] \\ \beta_{\beta^2} &= \frac{\beta^2}{32}[(1+\delta)Y_+Y_- - (1+\tilde{\delta})G_+G_-] \end{aligned} \quad (13)$$

Defining the **asymmetry** parameters $\alpha = Y_-/Y_+, \tilde{\alpha} = G_-/G_+$, we can use eq.(13) to show that $\alpha\ \tilde{\alpha}$ are **R.G. invariant.** The asymmetries act as superselection rules and split the theory up in sectors. Defining $Y_+ = Y, G_+ = G$ only three independent β functions remain. We can find a non-trivial self-dual fixed point at: $\delta^* = \tilde{\delta}^*$, $\alpha Y^{2*} = \tilde{\alpha} G^{2*} = 128\delta^*$ with:

$$\begin{aligned} \beta^{2*} &= 2\pi N; \delta^* = \frac{\epsilon}{4} \\ Y^* &= \sqrt{\frac{32\epsilon}{\alpha}}; G^* = \sqrt{\frac{32\epsilon}{\tilde{\alpha}}} \end{aligned}$$

where $\epsilon = N - 4$. Linearizing the flow equations near the fixed point and diagonalizing the matrix of anomalous dimensions $\gamma_{ij} = \partial_i\beta_j\ |_*$, we find that for $\epsilon > 0$ ($N > 4$) there are two relevant eigenvalues and one irrelevant, while for $\epsilon < 0$ there is one irrelevant eigenvalue and a pair of **complex conjugate** eigenvalues. The appearance of these complex eigenvalues signals a breakdown of unitarity for the case $N < 4$. Using the β functions we can show that there exists a **fixed line** joining the gaussian fixed point to the non-trivial fixed point. It is defined by:

$$\begin{aligned} Y &= \sqrt{\frac{\alpha}{\tilde{\alpha}}}G \\ \delta^* &= \tilde{\delta}^* = \frac{\epsilon}{4} \end{aligned} \quad (14)$$

This fixed line is the **only** integral curve of the vector fields in coupling space connecting the two fixed points *and* lying entirely within the perturbative regime. Flows along this line converge to free field theory in the I.R. for $\epsilon > 0$ and to the nontrivial fixed point for $\epsilon < 0$. Now consider, the unitary case ($\epsilon > 0$). The new fixed point is unstable in the I.R. along this integral curve. In this case the new fixed point is multicritical and separates four phases with very rich structure. Along the fixed line the fixed point separates two phases: the first corresponds to an almost free field theory (with logarithmic corrections to scaling) where the excitations are just gaussian fluctuations (spin waves) coexisting with bound pairs of electric charges and magnetic monopoles. On the other side of the fixed point along the fixed line ($Y > Y^*, G > G^*$), is a plasma phase where electric charges and magnetic monopoles are unbound. This is a strong coupling phase where the fugacities are very large, the system is topologically disordered and the plasma is not neutral when the asymmetry parameters are not equal to 1. This is a massive phase, with screening of the long range Coulomb interactions. The Debye screening length is approximately given by $\xi \simeq \mu^{-1}(Y - Y^*)^{-1/\epsilon}$. This phase may be thought of as being a "condensate" of the vertex operators that create these excitations. Since in this phase both electric and magnetic charges are present we conjecture that there are parafermionic excitations with fractional statistics.

At the non-trivial fixed point the long distance properties are described by a non-neutral, dilute plasma of electric charges and magnetic monopoles. It is dilute since the fugacities are $O(\sqrt{\epsilon})$. Viewed as a String Theory the fixed point theory corresponds to a "condensate" of tachyonic modes (slightly off-shell because the anomalous dimensions are close to 2) and the corresponding duals . Since at the fixed point both electric and magnetic charges are present we conjecture that there will be excitations with fractional statistics. In our paper[12] we show that at the new fixed point the spectrum of anomalous dimensions is **continuous.**

If the fixed point theory is perturbed in a direction perpendicular to the fixed line, the fixed point separates two phases one with a dense plasma of electric charges where magnetic charges are tightly bound and another with a dense plasma of magnetic charges in which electric charges are bound. These two phases are massive and dual to each other.

To compute the conformal anomaly at the new fixed point, we used the fact that it is connected to gaussian via an R.G. flow along the fixed line and then applied the results of the previous section. Since the operator that moves the theory along the fixed line has dimension $\epsilon/2$ at the trivial fixed point and $-\epsilon$ at the new fixed point and since the coupling derivative of the c-function defined in the previous section is proportional to the β function to lowest order in the couplings and ϵ, we find:

$$c^* = 1 + \frac{3}{8}\epsilon^2 \tag{15}$$

This result is consistent with Zamolodchikov's theorem for the unitary case $\epsilon > 0$ because the new fixed point is *unstable* in the infrared and the I.R. trajectories along the fixed line lead towards the trivial fixed point whose conformal anomaly must be smaller. For $\epsilon < 0$ the new fixed point is attractive along the fixed line and this value for c seems to contradict the theorem. However, for $\epsilon < 0$ the theory is *not unitary* and the theorem does not apply in this case. We also showed in our paper that the fixed point theory is conformally invariant due to the existence of a traceless energy momentum tensor. Therefore we conclude that the new fixed point describes a **new conformal field theory** with $c > 1$, with a continuous spectrum of anomalous dimension and whose ground state seems to be a kind of "condensate" of off-shell tachyonic modes and their duals. We point out that the presence of the dual operators is *crucial* for the existence of the new fixed point. We conjecture that the new fixed point theory may describe critical phenomena of thin film layers of superfluid HeII on subtrate potentials with N-fold symmetry.

We are now investigating theories in which the vertex operators are of the Frenkel-Kac type in terms of vectors in the root lattice of semi-simple Lie groups and incorporating their duals. We are also studying the case when the vertex operators create other (massive) modes. We hope that a better knowledge of these theories may shed light on possible new conformal field theories with a very rich phase structure and interesting excitation spectrum.

Acknowledgments:

The work of R.H. was supported in part by DOE grant DE-AC02- 76ER03066.

References

1. A. B. Zamolodchikov, JETP Lett. **43** (1986) 730; Sov. J. Nucl. Phys. **46** (1987) 1090.
2. J. L. Cardy, Les Houches Lectures on Fields, Strings and Statistical Mechanics (1988); Phys. Rev. Lett. **60** (1988) 2709, Phys. Lett. **215B** (1988) 749.
3. C. Vafa, Phys. Lett. **212B** (1988) 28.
4. A. A. Tseytlin, Phys. Lett. **194B** (1987) 63; N. E. Mavromatos and J. L. Miramontes, Phys. Lett. **212B** (1988) 33.
5. T. Banks and E. Martinec, Nucl. Phys. **294** (1987) 733.
6. See for example: D. J. Amit "Field Theory, the Renormalization Group and Critical Phenomena", McGraw-Hill (1978)
7. D. Boyanovsky and R. Holman, Phys. Rev. D. **40**, 1964 (1989).
8. H. W. Blote, J. L. Cardy and M. P. Nightingale, Phys. Rev. Lett. **56** (1986) 742; I. Affleck, Phys. Rev. Lett. **56** (1986) 746.
9. A. W. W. Ludwig and J. L. Cardy, Nucl. Phys. **285** (FS19) (1987) 687.
10. D. Boyanovsky and R. Holman in progress

11. L. P. Kadanoff, J. Phys. A: Math. and Gen. **11**, (1978) 1399; B. Nienhuis, J. Stat. Phys. **34** (1984), 371.

12. D. Boyanovsky and R. Holman to appear in Nucl. Phys. (FS) (1990).

13. E. Fradkin and L. P. Kadanoff Nucl. Phys. **B170** (FS 1) (1980) 1.

STATUS OF P-ADIC STRINGS

Paul H. Frampton
Department of Physics and Astronomy
University of North Carolina
Chapel Hill, NC 27599-3255

ABSTRACT

A review of the status of p-adic strings, as of November 1989, is provided. After a short elementary introduction to p-adic numbers, a number of aspects of p-adic bosonic strings are covered: N-tachyon tree amplitudes, effective field theory, p-adic sigma models, stability of the vacuum and p-adic worldsheet. Finally, there is a brief discussion of p-adic supersymmetry and p-adic superstrings.

I wish I could give you a physical motivation to study p-adic strings but there is at least a mathematical motivation. We shall use number theory, the purest of pure mathematics, where other approaches to string theory, such as conformal field theory and string field theory, use much more boring and conventional mathematics such as complex analysis and topology!

This talk will be in three sections: (I) P-Adic Numbers, the elementary ABC thereof, (II) Bosonic String, - this will be the bulk of this talk, (III) Superstring.

1. Introduction to P-Adic Numbers [1]

One completion of the rational numbers $Q = \{x/y\}$, x, y being integers, is the real number field R. The process by which Q is completed to R uses Cauchy sequences $q_1, q_2, \ldots, q_n, \ldots$ and the conventional norm $|q_n - q_{n+1}| < \varepsilon$.

In order that the number field which results from completion of Q be very useful in mathematics, it is essential that the norm satisfy three basic requirements, namely:

$$|x| = 0 \quad \text{iff} \quad x = 0 \tag{1a}$$
$$|xy| = |x|\,|y| \tag{1b}$$
$$|x+y| \leq |x| + |y| \tag{1c}$$

There exists an infinite set of such norms, one for each prime number p. The definition of the p-adic norm is

$$\left|\frac{x}{y}\right|_p = \left|\frac{x'}{y'}\, p^n\right|_p = p^{-n} \tag{2}$$

where x', y' are prime with respect to p. Using the norm (2), Q may be completed to Q_p, a p-adic number field.

A central theorem in number theory, credited to Ostrowski, is that every nontrivial norm on Q, satisfying (1a), (1b) and (1c) is equivalent to either a p-adic norm for some p or the usual absolute value.

Having defined Q_p, we first discuss integration over Q_p. Integrals become much simpler than those over R. As a normalization we choose

$$\int_{|x|_p \le 1} dx = 1 \tag{3}$$

We may change p-adic variable by $d(ax) = |a|_p\, dx$. This leads to

$$\int_{|x|_p \le p^{-k}} dx = p^{-k} \tag{4}$$

$$\int_{|x|_p = p^{-k}} = p^{-k}\,(1-p^{-1}) \tag{5}$$

It is useful to introduce a p-adic counterpart of the exponential function in order to define p-adic Fourier transforms. We wish to have

$$X(x+y) = X(x)\, X(y) \tag{6a}$$

$$|X(x)| = 1 \tag{6b}$$

and can define

$$X(x) = \exp\,(2\pi i\{x\}) \tag{7}$$

where $\{x\}$ is the p-adic rational part of x obtained by expanding

$$x = p^k \sum_{n=0}^{\infty} A_n\, p^n \tag{8}$$

with $a_o \neq 0$, $0 \le a_n \le (p-1)$, and k = integer. If $k \ge 0$, $X(x) = 1$. If $k < 0$, $\{x\}$ is defined by truncation of the series in (8) at $n = -k - 1$. One can show, using the rules already given, that:

$$\int_{|x|_p \leq p^{-k}} X(x)\, dx = \begin{cases} p^{-k} & k \geq 0 \\ 0 & k < 0 \end{cases} \qquad (9a \,\&\, 9b)$$

A multiplicative character satisfies

$$\Pi(xy) = \Pi(x)\,\Pi(y) \qquad (10)$$

For example

$$\Pi(x) = |x|_p^s \qquad (11)$$

where s is a complex variable.

This $\Pi(x)$ is useful in defining the p-adic gamma function $\Gamma_p(s)$. Recall that the usual formula over a real variable t is

$$\Gamma(z) = \int_0^\infty dt \quad t^{z-1} e^{-t} \qquad (12)$$

The appropriate generalization useful for p-adic strings is

$$\Gamma_p(z) = \int_{Q_p} X(x)\, |x|_p^{z-1}\, dx \qquad (13)$$

This can be straightforwardly computed, using Eq. 9, to give

$$\Gamma_p(z) = \frac{1 - p^{z-1}}{1 - p^{-z}} \qquad (14)$$

The poles of $\Gamma_p(x)$ are at

$$Z = \frac{2\pi in}{\ln p} \qquad (15)$$

where n = integer. All the residua are equal to

$$\frac{1 - p^{-1}}{\ln p.}$$

These are some elementary properties of p-adic numbers and we now proceed to p-adic strings.Whether p-adic strings are useful can be summarized by multiple choice questions:

1) Superstrings are the theory of
   (a) Everything (b) Gravity (c) Nothing

2) In the understanding of superstrings, p-adic strings are
   (a) Crucial (b) Useful (c) Irrelevant

In either case (a) is "WOW", (b) is "wow" and (c) is "YUK".

2. Bosonic Strings

At the tree level, we reconsider the Veneziano 4-tachyon amplitude by using the p-adic integral [2]

$$A_4^{(p)} = \int_{Q_p} dx\, |x|_p^{\alpha}\, |1 - x|_p^{\beta} \tag{16}$$

with $\alpha = -\alpha_s - 1$, $\beta = -\alpha_t - 1$. Dividing this integral into three regions $|x|_p > 1$ (u term), $|x|_p < 1$ (s term), $|x|_p = 1$ (t term and contact term) and using the integration rules given above we find

$$A_4^{(p)} = \sum_{x = s,t,u} \frac{(1 - p^{-1})\, p^{\alpha_x}}{1 - p^{\alpha_x}} + (1 - 2p^{-1}) \tag{17}$$

This can be rewritten, using $\alpha_s + \alpha_t + \alpha_u = -1$ as

$$A_4^{(p)} = \prod_{x = s,t,u} \left(\frac{1 - p^{-\alpha_x - 1}}{1 - p^{\alpha_x}} \right) \tag{18}$$

By the use of the product representation, Eq. 18, and the well-known formula

$$\zeta(z) = \prod_p (1 - p^{-z})^{-1} \tag{19}$$

one finds formally [3]

$$A_4^{-1} = \prod_p A_4^{(p)} \tag{20}$$

This is called an adelic formula, in which A_4 is the usual Veneziano amplitude (formally $p \to \infty$).

The analysis extends rather neatly [4,5] to the N-tachyon amplitude where one deduces Feynman rules:

$$\text{Propagator } \Pi = \frac{(1 - p^{-1})\, p^{\alpha}}{1 - p^{\alpha}} \tag{21a}$$

$$\text{Vertex} \quad V^m = \prod_{n=2}^{m-2} (1 - np^{-1}) \quad (m \geq 4) \tag{21b}$$

$$V^3 = 1$$

One draws all tree diagrams with arbitrary polynomial vertices and the amplitudes add to $A_N^{(p)}$ It is then natural to ask whether the adelic formula, Eq. 20, extends to $N \geq 5$?

I used the word "formally" prior to Eq. 20 for a reason. The relevant product formula is

$$\prod_x \prod_p \left(\frac{1 - p^{-1-\alpha_x}}{1 \;\; p^{\alpha_x}} \right) = \prod_x \frac{\zeta(-\alpha_x)}{\zeta(1 + \alpha_x)} \tag{22}$$

The numerator product in Eq. 22 is convergent for Re $\alpha_x > 0$ while the denominator requires Re $\alpha_x < -1$, clearly inconsistent. One method [6] for regularization is to replace p by $p^{\}}$, take the infinite product then let $\} \rightarrow 1$. This validates the adelic formula for the kinematic region in the vicinity of all $a_x \approx -\frac{1}{3}$. But there exist other kinematic regions for N=4 where Eq. 20 is not valid.

For the case N = 5, it has been shown [7] that in the neighborhood of all $\alpha_{ij} \approx -1/2$ ($\Sigma\alpha_{ij} = -5$ with ten terms) the generalization of Eq. 20 does not work. It is an open question [6] whether in other kinematic regions there is an analytic continuation such that the adelic formula holds for $N \geq 5$. The technical difficulty involved in addressing this question is that for any infinite product over primes where the factors are anything significantly more complicated than in Eq. 19 the analytic continuation is unknown. It is, however, a general property of such infinite products that there are several different analytic continuations for different regions separated by essential boundaries [5].

What about string loops? In one attempt to p-adicise these [8] the one-loop vacuum (N = 0) diagram was re-written from the form

$$A_0 = \int_0^{\infty} \frac{dt}{[\eta(it)]^{24}} \tag{23}$$

where with q = exp(-2pt) we have

$$\eta(it)^{24} = q \prod_{n=1}^{\infty} (1 - q^n)^{24} \tag{24}$$

In the p-adic form

$$A_0^{(p)} = \int_{|q|_p < 1} \frac{dq}{|q|_p^2 \prod (1 - q^n)_p^{24}} \tag{25a}$$

$$= \int_{|q|_p < 1} \frac{dq}{|q|_p^2} = -1 \tag{25b}$$

This is a simple finite answer with no adelic property. For N > 0 there develop double sums [8] e.g.

$$A_2^{(p)}(1 \text{ loop}) = -\frac{2}{p^2} + \frac{(p-1)^3}{p^3} \sum_{k=1}^{\infty} k \sum_{l=1}^{k-1} p^{\frac{l^2 - lk + k}{k}} \tag{26}$$

Again we find a simple formula but with no adelic property. This is just one approach; in general, string loops are "p-adic terra incognita".

Returning to the tree amplitudes, for fixed p there is a nonlocal effective lagrangian [9] corresponding to the Feynman rules in (21) above:

$$L = \frac{1}{2}\phi\left(\frac{1 - p^{1 + \frac{1}{2}\Box}}{1 - p^{-1}}\right)\phi \qquad - \sum_{m=3}^{\infty} V^m(p)\frac{\phi^m}{m!} \tag{27}$$

The classical field equation is

$$p^{\frac{1}{2}\Box}(1 - \phi) = (1 + \phi)^{1/p} \tag{28}$$

Eq. 28 suggests redefining $\chi^p = (1 + \phi)$ giving

$$p^{-\frac{1}{2}\Box}\chi = \chi^p \tag{29}$$

The corresponding potential $V(\chi)$ is

$$V(\chi) = \frac{1}{2}\chi^2 - \frac{1}{p+1}\chi^{p+1} \tag{30}$$

which, for odd p, has the appearance in the Figure 1.

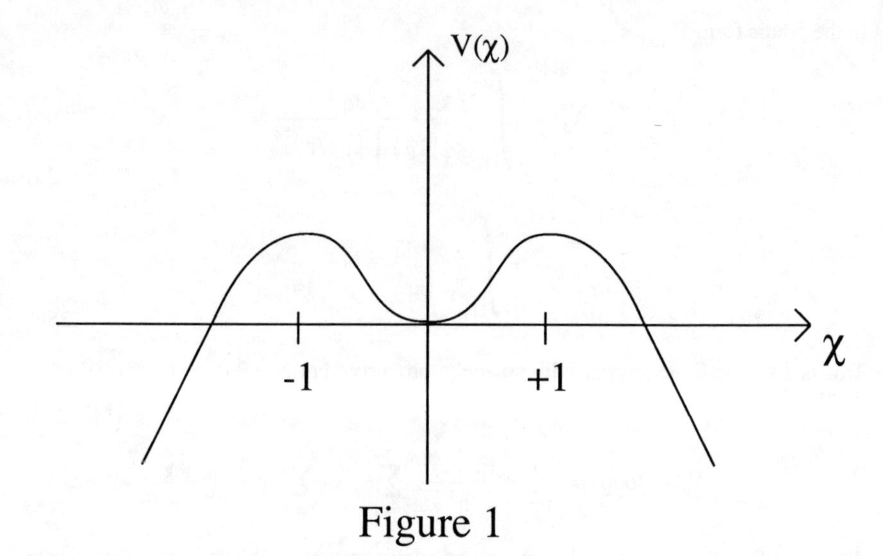

Figure 1

There is the tachyonic vacuum at $\chi = +1$ and a local minimum at $\chi = 0$. More on the $\chi = 0$ vacuum will be discussed later.

In a flat background, we may write a p-adic sigma model with free action [10]

$$S_0 = -\frac{1}{2}\int_{Q_p} d\sigma\, X^\mu(\sigma)\, \Delta\, X_\mu(\sigma) \tag{31}$$

where the p-adic Laplacian is the non-local form

$$\Delta f(x) = C_p \int_{Q_p} \frac{dy\, f(y)}{|x-y|_p^2} \tag{32}$$

where C_p is a p-dependent constant. The Green's function is

$$\Delta\,(-\ln|x-y|_p) = -\,\delta_p(x-y) \tag{33}$$

We may add a tachyon background to the action by writing

$$S = S_0 - g \int_{Q_p} \phi[X^\mu(\sigma)]d\sigma \tag{34}$$

To regularize calculations, and develop a renormalization group we introduce, for a large integer k, the cut-off Green's function

$$G(\sigma) = \{ -\ln(|\sigma|_p m) \quad |\sigma|_p \geq p^{-k} \tag{35a}$$

$$\{ -\ln(p^{-k} m) \quad |\sigma|_p < p^{-k} \tag{35b}$$

Expanding the background

$$X^\mu(\sigma) = x^\mu + \xi^\mu \tag{36}$$

in ξ_μ we may find successive approximations by loop diagrams to the renormalization of ϕ. The β-function of such a generalized renormalization group is defined by

$$\beta = \frac{\Delta\phi}{\Delta k} \tag{37}$$

It has been shown [11] that setting $\beta = 0$ reproduces the spacetime field equation, Eq. 28 above. Thus the p-adic field theory is equally as consistent as a real field theory in this regard.

All of the above p-adic tree amplitudes, effective field theory and p-adic sigma model work equally well for the closed-string case. [12]

In a curved background a p-adic sigma model can give a new version of gravity. We define [13] a derivative

$$\partial f(x) = C_p \int_{Q_p} dy \, \frac{sg(x-y) \, f(y)}{|x-y|_p^{3/2}} \tag{38}$$

This satisfies

$$\partial(\partial f(x)) = \Delta f(x) \tag{39}$$

given by Eq. 32.

The $sg(x - y)$ function in Eq. 38 is taken to be $sg_p(x - y)$ for $p = 3 \pmod 4$ and $sg_{-1}(x - y)$ for $p = 2$. For $p = 5 \pmod 4$, it is an open question how to write Eq. 38.

As a gravitational action, for the allowed p,

$$S = \int_{Q_p} \frac{1}{2} G_{\mu\nu}[X(x)] \, \partial X^\mu(x) \, \partial X^\nu(x) \tag{40}$$

In a background

$$X^\mu = X_0^\mu + \xi^\mu \tag{41a}$$

$$g_{\mu\nu}(X) = g_{\mu\nu}(X_0) - \frac{1}{6}\xi^\rho\xi^\sigma R_{\rho\mu\nu\sigma} + \dots \tag{41b}$$

The propagator for ξ^μ is

$$< \xi^\mu\xi^\nu > (u) = \frac{1}{c|u|_k^{1+\varepsilon}} \, g^{\mu\nu} \tag{42}$$

in which k and ε signify infrared and ultraviolet regularization respectively. The corresponding β-function is [13]

$$\beta_{\mu\nu}(g) = R_{\mu\nu}(g) + 0(\alpha') \tag{43}$$

The one-loop β-function thus vanishes in an empty Einstein space.

One interesting question is the p-adic worldsheet. In the case of the real open string the interior of the worldsheet admits a homogeneous space description with an SL(2,R) invariant metric. The boundary of the worldsheet is R. The p-adic worldsheet may be taken to be a Bruhat-Tits tree (T_p), an infinite-dimensional discrete space. This is endowed with a GL(2,Q_p) invariant metric. The boundary is Q_p. By guessing a simple action, one can again rederive [14] the p-adic tree amplitude. $A_N^{(p)}$ agreeing with Eqs. (21) above; the same technique may be used to attempt string loop amplitudes [15].

Let me go back to the question of the local minimum in the potential, Eq. 30, where the propagator in χ is $p^{1/2\Box}$ and has no pole for finite k^2 and hence no particles perturbatively. Localized at $\chi = 0$ the field equation, Eq. 29, has solution [5,10]

$$\chi_0(r^2) = A(p) \exp\left[- B(p)\; r^2\right] \tag{44}$$

This exists as a solution in any spacetime dimension D; the same solution is obviously a static soliton solution for spatial dimension d = (D - 1). The energy of this soliton

$$E = \int d^d x \, (- L_p) \tag{45}$$

turns out to be positive, suggesting a positive-energy "particle" in a displaced vacuum which therefore might be tachyon-free.

In order to investigate this further, we consider a perturbation [16]

$$\delta\chi_0 = \varepsilon \exp\!\left(- B(p) r^2\right) \tag{46}$$

and calculate the corresponding shift in energy

$$\delta E = (\delta E)_1\,\varepsilon + (\delta E)_2\,\varepsilon^2 + O(\varepsilon^3) \tag{47}$$

Here $(\delta E)_1 = 0$ and

$$(\delta E)_2 = -\frac{1}{2}p^{2+d/2}\left[\frac{\pi}{(p+1)\,B(p)}\right]^{d/2} \tag{48}$$

which is always negative. We may consider more generally

$$\delta\chi_n = \varepsilon\, r^{2n} \exp\left(-B(p)r^2\right) \tag{49}$$

For n= 1, $(\delta E)_2$ is proportional to (2p-D), negative for D > 2p at any fixed p. For n > 1, the number of negative modes increases with D for fixed p. Note that

$$|\chi_0(x = 0)| > 1$$

for all p, d implying that the center of the soliton is outside the potential barrier.

The instanton solution is an interpolation between $\chi = 0$ for all $\underline{x}$ at $t = -\infty$ and at $t = 0$, a state appearing as shown in Figure 2.

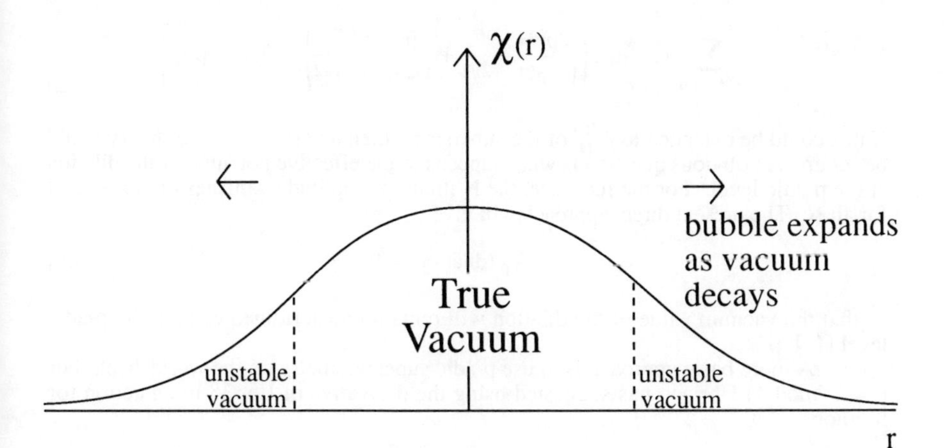

Figure 2

The conclusion is that the soliton does not represent a particle localized and confined in the proximity of the $\chi = 0$ minimum. Instead, the instanton solution signifies tunnelling in Minkowski space from the $\chi = 0$ minimum to the unboundedly-low energy vacuum as $\chi \to \infty$.

III. P-Adic Supersymmetry and P-Adic Superstrings

Let me close with some brief remarks about p-adic supersymmetry and its possible application to superstrings; this is, in my opinion, the most exciting area of p-adic strings today. The combination of p-adic number fields and supersymmetry is not always an easy one.

Taking a direct approach to superstrings the amplitude for four massless particles in the type II string is

$$A_4 = C \prod_{x = s,t,u} \frac{\Gamma(-x/8)}{\Gamma(1 + x/8)} K \tag{50}$$

where C is a constant and K is a kinematic factor. This may be rewritten

$$A_4 = C' \left(s^2 + t^2 + u^2\right) K \text{ (Virasoro form)} \tag{51}$$

The Virasoro amplitude can be integrated over a quadratically-extended p-adic number field to find a p-adic amplitude [12]

$$A_4^{(p)} = -\frac{64}{3} c(s^2 + t^2 + u^2)$$

$$\sum_{x = s,t,u} (1 - p^{-1}) \left(\frac{2p^{(\alpha_x - 2)/2}}{1 - p^{(\alpha_x - 2)/2}} + \frac{p^{(\alpha_x - 4)/2}}{1 - p^{(\alpha_x - 4)/2}} \right) + 3(1 - 2p^{-1}) \Big] \tag{52}$$

If this could be extended to $A_N^{(p)}$ of the superstring then an effective field theory could be set up. An obvious question is what happens to the effective potential of the dilation at the p-adic level. For the real case, the N-dilation amplitude vanishes on mass shell for all N. The present direct approach will give

$$V_p \text{ (dilation)} = 0 \tag{53}$$

so that the vacuum value of the dilation will remain undetermined even at the p-adic level [17].

A more basic approach is to use p-adic supersymmetry on the worldsheet. For p = 3 (mod 4) Ubriaco has suggested using the derivative of Eq. 38 in an action for fermions

$$S_F = -\frac{1}{2} \int_{Q_p} dx \, \psi(x) \, \partial\psi(x) \tag{54}$$

One may combine S_F with the bosonic action of Eq. 31 into a total action [18]

$$S = \frac{1}{2} \int_{Q_p} dx \int d\theta \quad D^2 \Phi D \Phi \tag{55}$$

with the superfield

$$\Phi\left(x, \theta\right) = \phi\left(x\right) + \theta\psi\left(x\right) \tag{56a}$$

$$D\Phi = \theta\, \partial\Phi + \frac{\partial}{\partial\theta}\Phi \tag{56b}$$

This respects the following supersymmetry algebra

$$\begin{aligned} \delta\phi &= \psi\xi \\ \delta\psi &= -\left(\partial\phi\right)\xi \\ \left[\delta_1, \delta_2\right]\phi &= -2\xi_1\,\xi_2\partial\phi \end{aligned} \tag{57}$$

Summary

The bosonic open and closed p-adic strings have been calculated consistently at the tree-level. A stable tachyon-free vacuum of the bosonic string is elusive, and the bosonic string seems to be a sick theory like ϕ^3 field theory, perhaps not surprising in view of the cubic-type string field theory. These p-adic calculations represent the first nonperturbative (albeit classical) string calculations.

Future directions of p-adic strings are two: (i) string loops and unitarity and (ii) superstrings where there is, at present, the problem of defining p-adic fermions when p = 1 (module 4).

I have focused on the status of p-adic strings and not discussed other areas of use for p-adic numbers in theoretical physics. For example, p-adic quantum mechanics and p-adic spacetime. The latter seems interesting as a possibility at scales below the Planck length [19]: it is conceivable [17] that the method of arguing for zero cosmological constant $\Lambda = 0$ on the basis of topology fluctuations (wormholes) might be equally presented as fluctuations in geometry (non-Archimedean) or possibly as fluctuations in dimension.

This work was supported in part by the U.S. Department of Energy under Grant No. DE-FG05-85ER-40219.

References

1. M. Gel'fand, M. I. Graev and I. I. Pyatetskii-Shapiro, *Theory and Automorphic Functions*, (Saunders, London, 1969).
J. P. Serre, *A Course in Arithmetic*, (Springer, New York, 1973).

2. I. Volovich, *Class. Q. Grav.* **4** (1987) 183.
3 P. G. O. Freund and E. Witten, *Phys. Lett.* **190B** (1987) 191.
4. P. H. Frampton and Y. Okada, *Phys. Rev. Lett.***60** (1987) 484.
5. L. Brekke, P.G.O. Freund, M. Olson and E. Witten, *Nucl. Phys.* **B302** (1988) 305.
6. P. H. Frampton, Y. Okada and M. R. Ubriaco, *Phys. Lett.* **B213** (1988) 260.
7. E. Marinari and G. Parisi , *Phys. Lett.* **B203** (1988) 52 See also Reference 5.
8. D. Lebedev and A. Morozov, ITEP preprint (1988) 163-88.
9. P. H. Frampton and Y. Okada, *Phys. Rev.* **D37** (1988) 3077.
10. B. L. Spokoiny, *Phys. Lett.* **208B** (1988) 401.
11. H. Nishino and Y. Okada, *Phys. Lett.* **219B** (1989) 258.
12. P. H. Frampton and H Nishino, *Phys. Rev. Lett.* **62** (1989) 1960.
13. P. H. Frampton and H. Nishino, unpublished.
14. A. V. Zabrodin, *Comm. Math. Phys.* **123** (1989) 463.
15. L. O. Chekhov, A. D. Mironov and A. V. Zabrodin, Moscow preprint (1989).
16. P. H. Frampton and H. Nishino, University of North Carolina Report IFP-364-UNC (1989) submitted to Phys. Lett. B.
17. P. H. Frampton, unpublished.
18. M. R. Ubriaco, University of Puerto Rico preprint (October 1989).
19. Some of these topics are discussed in much more detail in another recent talk: P. H. Frampton, *Non-Archimedean Geometry and Applications to Particle Theory,* 18th International Conference on Differential Geometric Methods in Theoretical Physics, Granlibakken, Tahoe City, California, July 1989 (in press).

ARITHMETIC IDEAS IN STRING THEORY

PETER G.O. FREUND

Enrico Fermi Institute and Department of Physics
University of Chicago, Chicago, Illinois 60637

The arithmetic features of string theory have come to be appreciated as of late. These arithmetic features originate in the fact that string world sheets are compact Riemann surfaces and therefore algebraic curves. Therefore, maybe not surprisingly, a new class of string-theories has been constructed[1] in which the world sheets, rather than being algebraic curves over the fields $\mathbf{R}$ or $\mathbf{C}$, of real or complex numbers, are taken as algebraic curves over some field $\mathbf{Q_p}$ of p-adic numbers or over one of its finite extensions.[2] The fields $\mathbf{Q_p}$, with p running over all primes and the field $\mathbf{R} \equiv \mathbf{Q}_\infty$ provide all possibilities of placing a topology on (i.e., of Cauchy completing) the field $\mathbf{Q}$ of rational numbers.[3,4] Therefore theories defined over p-adic fields $\mathbf{Q_p}$ and over the archimedean field $\mathbf{Q}_\infty$ should all be somehow related, reflecting this common origin. A first such relation has been discovered in the form of adelic product formulas[5,6]. Here we wish to comment on the implications of such adelic formulas in higher orders of perturbation theory, or in other words in higher genus. We shall also consider the possibility of non-adelic connections between archimedean and p-adic theories.

To keep things simple, let me consider open strings. At tree level then, integrations over the boundary of the world sheet are involved. In the archimedean case this can be pictured as follows: the world sheet itself can be viewed as the upper half of the complex plane and its boundary as the real axis. A similar picture with the complex plane replaced by a quadratic extension of $\mathbf{Q_p}$ can be implemented in the p-adic case as well. More profitable however is the picture in which one considers the upper half of the complex plane as the homogeneous space $H_\infty = SL(2,\mathbf{R})/SO(2,\mathbf{R})$. Here $SO(2,\mathbf{R})$ is the maximal compact subgroup of $SL(2,\mathbf{R})$. Viewing $\mathbf{R}$ as $\mathbf{Q}_\infty$, H_∞ is found to have a p-adic counterpart $H_p = PGL(2,\mathbf{Q_p})/PGL(2,\mathbf{Z}_p)$. The maximal compact subgroup of $PGL(2,\mathbf{Q_p})$ is $PGL(2,\mathbf{Z}_p)$ (here $\mathbf{Z}_p = \{x \in \mathbf{Q_p} :\mid x \mid_p \leq 1\}$ = set of p-adic integers). Just as the boundary of T_∞ is the real projective line, so the boundary of T_p is the p-adic projective line. The surprise is that though T_p is a p-adic line with the attendant topology, T_p itself is a *discrete* space, it is a Bruhat-Tits-Cayley tree, or Bethe lattice, with $p+1$ edges meeting at each vertex. Starting from a gaussian model on T_p, Zabrodin[7], in an elegant paper, has been able to derive the p-adic string amplitudes of reference 1.

Work supported in part by the NSF: PHY 88-21039

The tree-amplitudes for the open string are factorized, crossing symmetric, meromorphic, as required by relativistic quantum theory, but they involve pairs of complex conjugate poles on the physical sheet in violation of causality. Indeed they correspond to a nonlocal theory. Call $A_0^{(p)}$ the p-adic four-point tree-amplitude and $A_0^{(\infty)}$ the ordinary Veneziano string four-point tree-amplitude (in a normalization such that they are Tate-Gel'fand-Graev beta-functions, so that a proportionality factor equal to the square of a coupling constant has been removed). Then the adelic product formula which connects all $A_0^{(p)}$ and $A_0^{(\infty)}$ is[5]

$$A_0^{adelic} = A_0^{(\infty)} \prod_p A_0^{(p)} = \prod_v A_0^{(v)} = 1. \tag{1}$$

In the product (1) $p \in P =$ set of all prime numbers, $v \in P \cup \{\infty\}$. The product formula can be interpreted as an infinite composite substructure for the ordinary Veneziano string. Alternatively, it can be viewed as defining an extremely simple tree amplitude $A_0^{adelic} = 1$ for an adelic string.

Staying with 4-point functions, one may ask, why this product formula holds at the tree level already. How does it extend to higher genus? To gain an answer[8] to this question let $A_n^{(v)}$ be the amplitudes in genus n, and $A^{(v)}$ the exact (total) amplitudes in the theory based on the field $\mathbf{Q_v}$ ($v \in P \cup \{\infty\}$). Then

$$A^{(v)} = \sum_{n=0}^{\infty} \epsilon^n A_n^{(v)} \tag{2}$$

with ϵ a suitable expansion (coupling) parameter. Similarly at the adelic level $A^{adelic} = \sum \epsilon^n A_n^{adelic}$. Let us now make the theoretically more natural assumption that

$$\prod_v A^{(v)} = A^{adelic}. \tag{3}$$

Inserting the expansion (2) into the exact product formula (3), one obtains an infinity of formulae, one in each genus. At tree level, genus $n = 0$, one recovers Eq. (1), which is thus found compatible with the exact equation (3). At genus $n = 1$ one finds

$$\sum_v \prod_w \frac{A_0^{(w)}}{A_0^{(v)}} A_1^{(v)} = A_0^{adelic} \sum_v \frac{A_1^{(v)}}{A_0^{(v)}} = \sum_v \frac{A_1^{(v)}}{A_0^{(v)}} = A_1^{adelic} \tag{4}$$

which is a ***sum, not a product formula.*** Similar sum formulas are then obtained in higher genus. In fact, they define the, hopefully all simple, adelic genus-n amplitudes A_n^{adelic}. (Similar considerations apply to the adelic formulas[9] for ratios of N-point and $(N+1)$-point amplitudes for $N \geq 5$). It would be interesting to see whether the recent extensions[10] of Zabrodin's approach to higher genus, could shed some light on the nature of A_n^{adelic} for $n \geq 1$.

Beyond the adelic product formulas discussed up to this point, there seems to emerge a second rather unexpected archimedean-nonarchimedean connection to which we now turn. An explicit effective tachyon lagrangian $\mathcal{L}_p$ reproducing the tree amplitudes of the p-adic theory is - unlike for the archimedean case - easy to construct. Such a lagrangian will necessarily be non-local, as mentioned above. Its form is[11,12]

$$\mathcal{L}_p = \frac{p}{g^2}\left\{\frac{1}{2}\phi(\frac{\Box}{2})_q\phi + (\infty)_q\left(\frac{\phi^{p+1}}{p+1} - \frac{\phi^2}{2}\right) + \frac{p}{2(p+1)}\right\} \tag{5a}$$

where $q = 1/p$ and the q-analogue $(A)_q$ of A is defined as

$$(A)_q = \frac{1-q^A}{1-q}. \tag{5b}$$

Here $\phi(x)$ is a scalar tachyon-field in ambient (26-dimensional) space and as is clear from Eq. (5b), the kinetic term $\frac{1}{2}\phi(\frac{\Box}{2})_q\phi$ contains derivatives of arbitrarily high order, it is non-local.

Strictly speaking, Eqs. (5) apply for q the inverse of a prime number. In particular, this places q inside the unit interval. yet, disregarding its origins, we can consider the lagrangian (5) even for other values of p, in particular, as suggested by Spokoiny[13] at $p = 1$. At $p = 1$ the non-locality of the kinetic term disappears and one is faced with a local lagrangian

$$\mathcal{L}_1 = \frac{1}{4g^2}\{\phi\Box\phi + \phi^2(-\frac{1}{2} + \ln\phi) + 1\}. \tag{6}$$

This lagrangian $\mathcal{L}_1$ is *not* the local lagrangian corresponding to the ordinary string theory. Rather, it is the lagrangian of a local field theory out of which the p-adic theory, corresponding to the lagrangian $\mathcal{L}_p$ is obtained by a q-deformation of the kinetic term and an appropriate readjustment of the interaction terms. In the theory of the so-called q-strings[14,15] non-localities of the type (5) also appear, but there the $q = 1$ limit is the ordinary local Veneziano string, not the theory with lagrangian $\mathcal{L}_1$ (equation (6)). There are thus two kinds of deformations that seem to be relevant here: those that yield the q-strings from the Veneziano string, and those that yield the p-adic strings for $q = 1/p$ starting from the $\mathcal{L}_1$ theory at $p = 1$. Correspondingly there are *two* deformation parameters. The question then arises as to the existence of such a two-parameter quantum-group-like object which with one of the parameters yields the p-adic strings. Thus, it now appears that there might exist a connection between p-adics and q-deformations with $q = 1/p$. It may be interesting therefore to point out that precisely such a connection has recently been noticed also by Macdonald[16] in a completely different context. Roughly speaking, given a root system R, he constructs a family of orthogonal (with respect to a suitable weight function) polynomials, which depend rationally on two parameters q and s (our notation differs from Macdonald's to emphasize the connection with Eqs. (5) above; his t is our q and his q is our s)

A) When $s = 0$ and $q = 1/p$ (with p a prime number), Macdonald's polynomials yield the zonal spherical function on a semisimple p-adic Lie group G relative to a maximal compact subgroup K, such that the restricted root system (G, K) is dual to the given root system R.

B) When $q = 1$ and $s = 1$ with $s = q^k$ then, for certain values of k, the Macdonald polynomials give zonal spherical functions on the archimedean real symmetric space G/K.

Macdonald's polynomials provide a"bridge between harmonic analysis on real symmetric spaces and on their p-adic analogues". The connection with q-analogues both for p-adic strings ($q = 1/p$) and for the Macdonald polynomials, then suggests the existence of a quantum group-like object to accommodate all these facts. Macdonald specifically asks for the existence of such an object depending on *two* parameters. Now, the quantum groups usually considered, depend on only one parameter q. Deformations involving two parameters have been encountered,[17,18] though the relevance of these two-parameter objects to Macdonald's and our considerations is not clear.

I wish to thank Professors Yu. I. Manin, B. Srinivasan, E. Witten and D. Zagier for very stimulating conversations, and for calling references 16 and 18 to my attention.

References

1. P. G. O. Freund and M. Olson, Phys. Lett. **B199** (1987) 186.

2. Alternative versions in which ambient space-time is p-adic and even quantum amplitudes are p-adically valued have also been proposed: see I. V. Volovich, Class. Quant. Grav. **4** (1987) L83 and the follow-up paper of B. Grossmann, Phys. Lett. **B197** (1987) 101. We shall not discuss these alternatives here, as no consistent string theory along these lines seems possible.

3. N. Koblitz, *p-adic Numbers, p-adic Analysis and Zeta Functions*, (Springer, Berlin 1986).

4. M. Gel'fand, M. I. Graev and I. I. Piatetskii-Shapiro *Representation Theory and Automorphic Functions* (Saunders, London 1966).

5. P. G. O. Freund and E. Witten, Phys. Lett. **B199** (1987) 191.

6. Yu. I. Manin, Talk at the Poiana-Braşov School on Strings and CFT, 1-14 Sept. 1987.

7. A. V. Zabrodin, Comm. Math. Phys. **123** (1989) 463; Modern Phys. Lett. **A4** (1989) 367.

8. The now following considerations emerged during an interesting conversation with D. Zagier.

9. L. Brekke, P. G. O. Freund, E. Melzer and M. Olson, Phys. Lett. **B216** (1989) 53.

10. L. O. Chekhov, A. D. Mironov and A. V. Zabrodin, Comm. Math. Phys. to appear.

11. L. Brekke, P. G. O. Freund, M. Olson and E. Witten, Nucl. Phys. **B302** (1988) 365.

12. P. H. Frampton and Y. Okada, Phys. Rev. Lett. **60** (1988) 484.

13. B. L. Spokoiny, Phys. Lett. **208B** (1988) 406;

14. D. D. Coon, Phys. Lett. **29B** (1969) 669; Phys. Rev. **186** (1969) 1422; D. D. Coon, S. Yu and M. Baker, Phys. Rev. **D5** (1972) 1429.

15. L. J. Romans, preprint USC-88/HEP014 (1988).

16. I. G. Macdonald, Queen Mary College Preprint 1989; I. G. Macdonald, *Spherical Harmonies on a Group of p-adic Type*, Ramanujan Inst. Lecture Notes No. 2, Madras, 1971;

17. E. K. Sklyanin, Funk. Anal. Appl. **16** (1982) 27.

18. M. Artin, W. F. Schelter and J. Tate, to be published; M. Artin and W. F. Schelter, Adv. Math. **66** (1987) 171.

ON THE GEOMETRICAL SENSE OF P-ADIC STRINGS

BORIS L. SPOKOINY
L. D. Landau Institute for Theoretical Physics
The Academy of Sciences of the USSR
GSP-1, Kosygin Street 2
117940 Moscow V-334, USSR

ABSTRACT

A p-adic analogue of the open string worldsheet is introduced. It is an infinite regular tree with p+1 links in each vertex. The boundary of the tree is in one-to-one correspondence with the set of all p-adic numbers Q_p. Starting from a local quadratic action on this tree, we obtain a non-local action on Q_p and calculate N-point tachyonic amplitudes. They coincide with on-shell tree amplitudes of some non-local scalar field theory. In the limit $p \to 1$, the above theory becomes local and we study its dynamics. We find instantons describing a decay of the metastable state which exists in the theory. Multiloop calculations are also given.

A p-adic number may be represented by a formal series

$$t = p^n(a_0 + a_1 p + a_2 p^2 + \dots) \quad , \tag{1}$$

where p is prime, $0 \leqslant a_i \leqslant p-1$. It may also be represented in the form of a tree as shown in fig. 1 for p = 2.

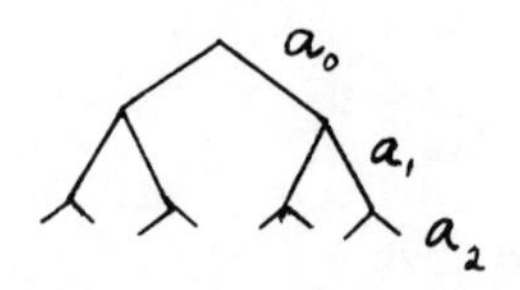

fig. 1

The value of a_0 determines the concrete possibility. Then for each of these p possibilities, we have p possibilities determined by the value of a_1 and so on. Thus, we obtain a branch of the tree. The end points of the tree describe p-adic numbers (1) with fixed n. If the tree is finite, its end points correspond to some finite approximations of p-adic numbers (1) (the cut series). For an infinite tree, the boundary of the tree corresponds to p-adic numbers (1) with fixed n. To take into account all n, we are to take an infinite number of the branches and connect them as shown in fig. 2.

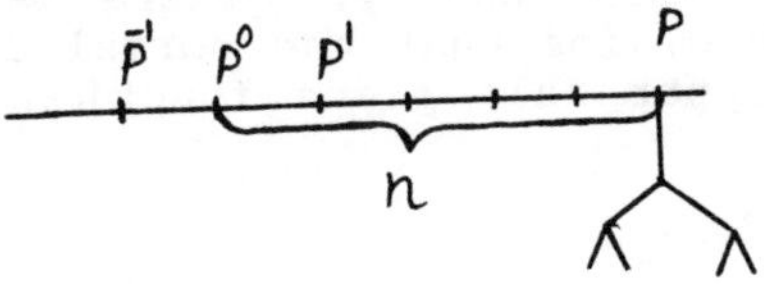

fig. 2

The boundary of the obtained tree corresponds to Q_p or more exactly to p-adic projective line $P^1(Q_p)$. We may draw the tree more symmetrically as in fig. 3.

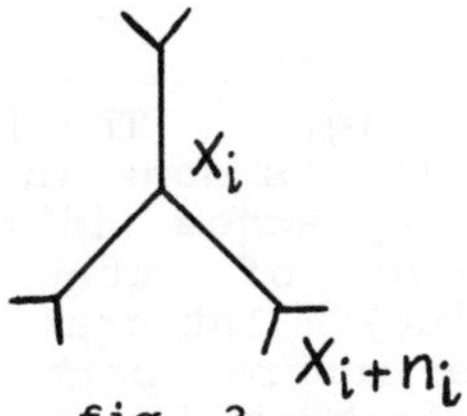

fig. 3

In this case, to construct the correspondence between the boundary of the tree and Q_p, we are to choose some line as in fig. 2. So we have some gauge freedom.

We define a field x_i in the vertices of the tree and choose the simplest local action [1]

$$S = \frac{1}{2T_0} \sum_{i,n_i} (x_i - x_{i+n_i})^2 \quad , \tag{2}$$

where the vertex $i + n_i$ is the nearest neighbor of the vertex i. Representing the tree in the form of fig. 2 and integrating over inner points of the tree, we get

$$\int_{\text{inner } i} e^{-S} DX_i = e^{-S_{eff}} \quad , \tag{3}$$

where

$$S_{eff} = \frac{1}{2T_0} \iint_{Q_p} \frac{[x(t) - x(t')]^2}{|t - t'|_p^2} \, dtdt' \quad . \tag{4}$$

The action (4) was introduced previously in [2-3].

Let us compare the results obtained with those of the archimedean theory. There is the famous Poisson formula which claims that the normal derivative of the harmonic in the upper half plane function calculated on the real axis is

$$\left. \frac{\partial x}{\partial n} \right|_{t'=t} = - \frac{1}{\pi} \int_{-\infty}^{+\infty} \frac{x(t) - x(t')}{|t-t'|^2} \, dt' \quad , \tag{5}$$

see fig. 4. The integral in (5) is done in the Cauchy sense (in the sense of the middle value). Integrating over the inner part of the worldsheet in the theory of open strings, we have

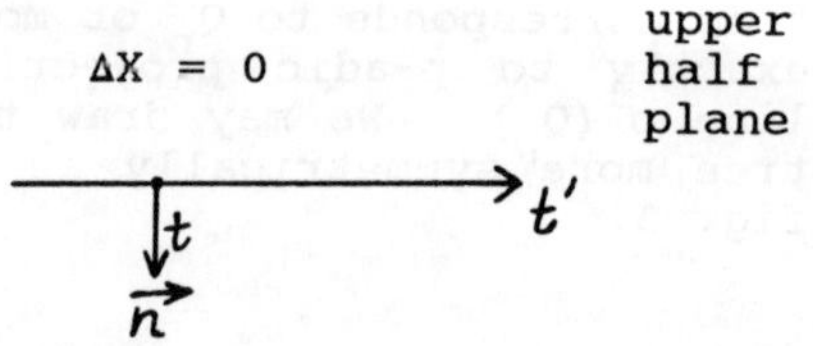

$\Delta X = 0$ upper half plane

t t' $\vec{n}$

fig. 4

$$Z = \int e^{- \frac{1}{T_0} \int (\partial \vec{x})^2 d^2\sigma} D\vec{X} \propto \exp\left(\frac{1}{T_0} \int \vec{x} \frac{\partial \vec{x}}{\partial n} d\ell \right) \quad . \tag{6}$$

If the boundary of the worldsheet is the real axis, then according to (5)

$$Z \propto \exp\left(- \frac{1}{2T_0} \iint dtdt' \frac{[\vec{x}(t) - \vec{x}(t')]^2}{|t - t'|^2} \right) \quad , \tag{7}$$

which is in correspondence with (4). So we see that a tree is a non-archimedean analogue of the worldsheet of the open string.

Now let us calculate the tachyonic amplitudes. They are

$$\tilde{A}(X_i^\mu) = g^{N-2} \int \prod_{i,\mu} \delta(X^\mu(t) - X_i^\mu) e^{-S} DX^\mu(t) \quad , \tag{8}$$

where S is given by (4). After a Fourier transformation, we get

$$A(K_i^\mu) = g^{N-2} \int DX^\mu(t) \prod_j dt_j \exp(- S + ik_j^\mu X^\mu(t_j)) \quad . \tag{9}$$

Then, as usual, we are to find a solution of the classical equations obtained by variation of the expression in the exponent of (9):

$$\frac{2}{T_0} \hat{A}_1 X^\mu(t) + i \sum_{j=1}^{N} K_j^\mu \delta(t-t_j) = 0 \quad , \tag{10}$$

where $\hat{A}_1$ is an integral operator which is a non-archimedean analogue of (5). Eq. (10) is solved by Fourier transformation, and it has a solution only if

$$\sum_{j=1}^{N} K_j^\mu = 0 \quad . \tag{11}$$

Integration over the zero mode $X^\mu(t)$ = const gives the factor

$$\delta^{(D)}\left(\sum_{j=1}^{N} K_j^\mu\right)$$

which ensures (11). We will omit this factor in what follows. The solution of Eq. (10) is of the form

$$X^\mu(t) = -i\alpha \sum_{j=1}^{N} K_j^\mu \, \ell n|t-t_j|_p \quad , \tag{12}$$

where

$$\alpha = \frac{p(p-1)}{2(p+1)\ell np} T_0 \quad . \tag{12a}$$

Substituting (12) into (9), we get

$$N(\vec{k}_i) = g^{N-2} \int \prod_{i<j} |t_i - t_j|^{\alpha \vec{k}_i \vec{k}_j} \, dt_1 \ldots dt_N \, / \, \theta \quad . \tag{13}$$

The Möbius invariance of (13) results in $\vec{k}_i^2 = 2/\alpha$. In Eq. (13), θ is the volume of the Mobius transformation group. Choosing as usual $t_N=0$, $t_{N-1}=1$, $t_{N-2} = \infty$, we obtain

$$A_N(\vec{k}_m) = g^{N-2} \int dt_1 \ldots dt_{N-3} \prod_{i=1}^{N-3} |t_i|_p^{\alpha \vec{k}_i \vec{k}_N} \, |1-t_i|_p^{\alpha \vec{k}_i \vec{k}_{N-1}}$$

$$\prod_{1 \leqslant i<j \leqslant N-3} |t_i - t_j|_p^{\alpha \vec{k}_i \vec{k}_j} \tag{14}$$

as in [4-5]. In these works, the expression (14) was postulated. However, this assumption is not unique. One may take an arbitrary multiplicative character $Q_p \to c$ instead of the norm for p-adic generalization. We obtain a unique amplitude from fundamental action (2).

In the non-archimedean case, one can calculate the integrals (14) explicitly and not only explicitly but also sensibly. The reason why it is possible is the following. In the archimedean case, if we have n points, there are n(n-1)/2 different distances between these points in general. In the non-archimedean case there are at most n-1 different distances. It follows from the main non-archimedean inequality

$$|x + y| \leqslant \max(|x|, |y|) \quad . \tag{15}$$

It is easy to see from (15) that each triangle is isosceles, and then by induction to prove the above results, taking into account that adding a new point we get not more than one new value of the distances between the points as follows from (15).

In (14), we have (N-3) points and unity. So we have (N-2) points, at most (N-3) different distances between them and (N-3) integrations. As a result, the integral (14) is given by a sum of multiple integrals of the type

$$\int_{|t|<1} dt|t|^{\alpha\vec{k}_i\vec{k}_j} = \frac{1-\frac{1}{p}}{p^{\alpha\vec{k}_i\vec{k}_j}-1} = \frac{1-\frac{1}{p}}{p^{\frac{\alpha}{2}(\vec{k}_i+\vec{k}_j)^2-2}-1} = G(\vec{k}_k+\vec{k}_j) \tag{16}$$

where $\vec{k}_i^2 = 2/\alpha$. The integral may be represented as a sum of graphs as shown in fig. 5.

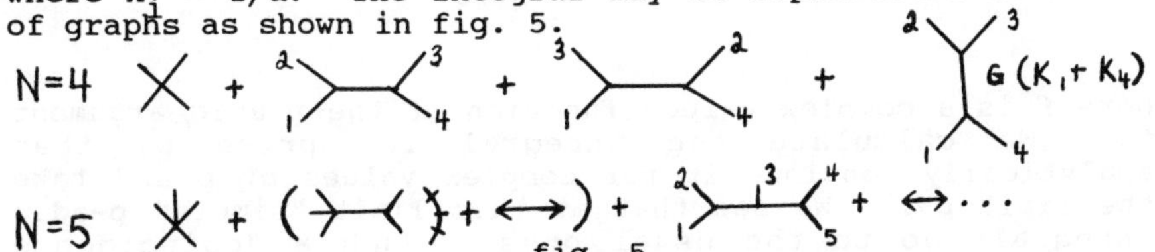

fig. 5

Actually, N non-archimedean points can be represented in the form of some tree as is clear from the representation of Q_p as a boundary of the tree. This is the main reason why the non-archimedean Koba-Nielsen amplitudes give rise to some graphs.

The same amplitudes are also given by tree diagrams of some D-dimensional theory of a scalar field with the lagrangian [5,6,2]

$$L = - \sum_{n=3}^{p+1} \frac{g^{n-2}c_n\Phi^n}{n!} + \frac{1}{2(p-1)} \Phi \left(p^{-\frac{\alpha\Delta}{2}} - p\right)\Phi . \tag{17}$$

The first term in (17) gives us vertices and the second one propagators. After the substitution $\phi = 1 + g\Phi/p$, we obtain

$$S = \int d^Dx\left[\phi p^{-\frac{\alpha\Delta}{2}} \phi - \frac{2}{p+1} \phi^{p+1} - \frac{p-1}{p+1}\right] \frac{p^2}{2g^2(p-1)} . \tag{18}$$

Let us consider some interesting limits. The integrals over the functions which depend only on the norm of the argument will be called "simple". It is easily verified from the definition of a p-adic integral that

$$\int_{|t|\leq 1} f(|t|)dt = \sum_{k=0}^{\infty} \int_{|t|=p^{-k}} f(p^{-k})dt$$

$$= \left(1 - \frac{1}{p}\right) \sum_{k=0}^{\infty} p^{-k} f(p^{-k}) \xrightarrow{p\to 1} \int_0^1 f(x)dx \quad . \tag{19}$$

Here f is a complex valued function of the p-adic argument t. We calculate the integral for prime p, then analytically continue it for complex values of p and take the limit $p\to 1$. We see that at this limit "simple" p-adic integrals go to the usual ones. Such a "coincidence principle" [2].

At p=1, Eq. (18) goes to

$$S = \frac{1}{4g^2} \int d^D x \left[\alpha(\nabla\phi)^2 + \phi^2 \ln \frac{e}{\phi^2} - 1\right] \quad . \tag{20}$$

We obtain a local theory for which the Cauchy problem may be correctly formulated and so we may consider the dynamics. The theory (18) may be considered to be a p-deformation of the local theory (20) [7]. There is an instanton solution

$$\phi(x) = \pm e^{D/2} e^{-x^2/2\alpha} \tag{21}$$

which describes the tunneling from the metastable vacuum $\phi=0$ in the theory (20) (see fig. 6). There is also a "soliton" solution

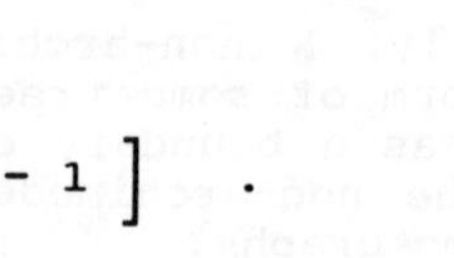

fig. 6

$$\phi(\vec{x}) = \text{const } e^{-\vec{x}^2/2\alpha} \quad , \tag{22}$$

which as shown in [8] is another instanton also describing the decay of the metastable vacuum. There also exists a solution depending only on time

$$\phi(t) = \text{const}\ e^{t^2/2\alpha} \tag{23}$$

which describes rolling down from the top of the potential (fig. 7). In the last case, we continued the solution into Minkowski space-time.

fig. 7

So far, we considered only tree diagrams. In the Schottky parametrization, we can consider p-adic Riemann surfaces in the same way as we do with ordinary Riemann surfaces, but we describe only Mumford curves [9]. (For other references see in the preprint by Volovich [9].) Let us consider a p-adic torus in more detail. As in the usual case, we make an identification $z \sim qz$, where $z, q, \in Q_p$, $|q|_p < 1$. Instead of Eq. (10), we have

$$\frac{2}{T_0} \int_{Q_p} \frac{x^\mu(t) - x^\mu(t')}{|t-t'|_p^2}\, dt' + i \sum_{j=1}^{N} K_j^\mu\, \hat{\delta}(t-t_j) = 0 \quad , \tag{24}$$

where $\hat{\delta}(t-t_j)$ is a "periodic" δ-function (under the transformation $t \to qt$),

$$\hat{\delta}(t-t_i) = \sum_n |q|^{-n}\, \delta(t-q^n t_i) \quad . \tag{25}$$

Then

$$x^\mu(t) = -i\alpha \sum_{j=1}^{N} K_j^\mu\, \ell n \left|\theta\left(q\middle|\ \frac{t}{t_j}\right)\right|_p \quad , \tag{26}$$

where

$$\theta(q|t) = \prod_{n\geq 0} (1-q^n t) \prod_{n>0} \left(1 - \frac{q^n}{t}\right) \qquad (27)$$

is an obvious p-adic generalization of the theta-function in the Weierstrasse form and α is given by (12a).

That factor of the N-point amplitude which depends upon momenta K_j^μ is of the form

$$A(K_j^\mu) = \prod_{i<j} \left|\theta\left(q\left|\ \frac{t_i}{t_j}\right.\right)\right|_p^{\alpha\vec{p}_i\vec{p}_j} \exp\left(-\ \pi\alpha\vec{p}_i\vec{p}_j \ln^2\left|\frac{t_i}{t_j}\right| / \ln\frac{1}{|q|}\right). \qquad (28)$$

The norm of the p-adic θ-function is a rational function. The measure on a moduli space may be obtained algebraically by the Manin-Beilinson-Voronov formulas and directly generalized to the p-adic case. In fact, we are to replace the usual $|\theta(q|t)|$ by the p-adic norm of the p-adic θ-function $|\theta(q|t)|_p$ and the measure is $1/|q|_p^2$, i.e., much simpler than in the archimedean case. For more details, see [10].

Multiloop amplitudes may be also calculated from the tree picture [11]. Let us represent the tree T in the form

$$T = PGL(2,Q_p)/PGL(2,Z_p) \quad , \qquad (29)$$

where Z_p is a set of integer p-adic numbers ($n\geq 0$ in (1)). A surface F is T/Γ, where Γ is a Schottky group. Γ has generators $\gamma_1, \ldots, \gamma_q$, where γ_i are hyperbolic elements. Each γ_i may be represented in the form

$$\gamma = \begin{bmatrix} q & 0 \\ 0 & 1 \end{bmatrix} \quad , \quad |q| = p^m \quad . \qquad (30)$$

Choosing the path shown in fig. 2 to be the invariant axis of γ, we see that the factorization over γ is the identification of points obtained by the shift over m units along the path as shown in fig. 8.

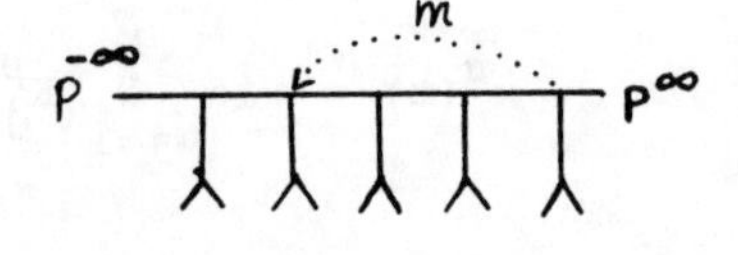

fig. 8

So γ_i gives us a cycle L_i of the length m_i. The number of common links of cycles L_1 and L_2 will be designated $\langle L_1, L_2\rangle$ and is a lattice analogue of $\mathrm{Im}\tau_{12}$ for usual Riemann surfaces. (See fig. 9.)

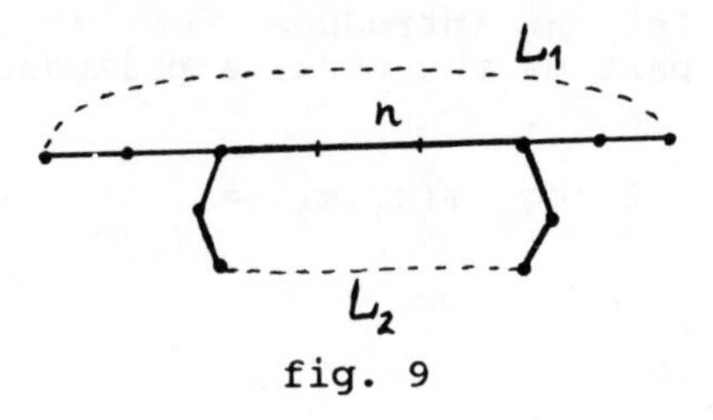

fig. 9

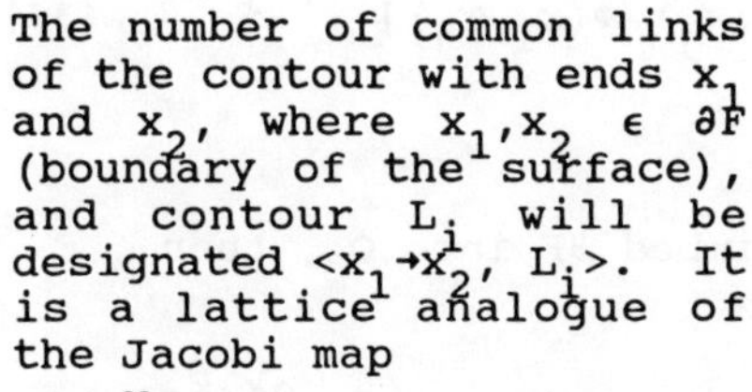

The number of common links of the contour with ends x_1 and x_2, where $x_1, x_2 \in \partial F$ (boundary of the surface), and contour L_i will be designated $\langle x_1 \to x_2, L_i\rangle$. It is a lattice analogue of the Jacobi map $\int_{x_2}^{x_1} \omega_i$. (See fig. 10.)

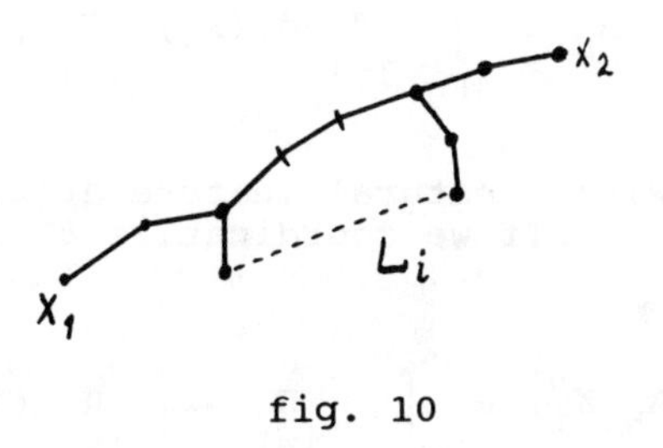

fig. 10

The notion of a reduced graph F^R is also useful. (See fig. 11.)

F → F^R

fig. 11

We cut all "non-essential" links.

An analogue of the logarithm of the prime form is

$$\log_p \psi(x_1,x_2) = \tfrac{1}{2}\langle x_1 \to x_2, x_1 \to x_2\rangle - d(x_1 \to x_2, F^R) \quad , \qquad (31)$$

where

$$d(x_1 \to x_2, F^R) = \inf_{\substack{z_1 \in x_1 \to x_2 \\ z_2 \in F^R}} d(z_1,z_2) \quad . \qquad (32)$$

Let us introduce also an analogue of the non-analytical part of the usual amplitudes:

$$2 \log_p \Phi(x_1, x_2) = \sum_{i,j=1}^{g} \langle x_1 \to x_2, L_i \rangle \langle L_i, L_j \rangle^{-1} \langle L_j, x_1 \to x_2 \rangle . \tag{33}$$

Then

$$A_N \propto \int_{\partial Fg} \prod_{j=1}^{N} d\mu(x_j) \prod_{i<j}^{N} [\psi(x_i, x_j)\ \Phi(x_i, x_j)]^{\alpha \vec{k}_i \vec{k}_j} \tag{34}$$

with a natural measure $d\mu(x_j)$.

If we coordinatize ∂T and imbed ∂F into Q_p, then

$$A_N(K_i^{\mu}) \propto \int_{\partial F \in Q_p} \prod_{j=1}^{N} dx_j \prod_{i<j} \{|E(x_i, x_j)|_p\ \Phi(x_i, x_j)\}^{\alpha \vec{k}_i \vec{k}_j}, \tag{35}$$

where

$$E(x,y) = (x,y) \prod_{\gamma \in \Gamma}{}' \frac{[x-\gamma(y)][y-\gamma(x)]}{[x-\gamma(x)][y-\gamma(y)]} ,$$

$x, y \in Q_p$, which coincides with the standard archimedean formula after the replacement

$$\langle L_i, L_k \rangle \to -2\pi \mathrm{Im}\tau_{ik} ,$$

$$\langle x \to y, L_i \rangle \to \int_x^y \omega_i .$$

Acknowledgements

I am very grateful to Professor L. Clavelli for the hospitality in the University of Alabama where this work was finished.

References

1. A. Zabrodin, Kiev ITP preprint 1989.
2. B.L. Spokoiny, Phys. Lett. B208 (1988) 401.
3. R.B. Zhang, Phys. Lett. B209 (1988) 229.
4. P.G.O. Freund and M. Olson, Phys. Lett. B199 (1987) 186.
5. L. Brekke, P.G.O. Freund, M. Olson and E. Witten, Nucl. Phys. B302 (1988) 365.
6. P.M. Frampton and Y. Okada, Phys. Rev. Lett. 60 (1988) 484.
7. P.G.O. Freund, talk at this workshop.
8. P.M. Frampton, M. Nishino, North Carolina Univ. preprint, September 1989.
9. D. Mumford, Compos. Math. 24 (1972) 129;
   Yu. I. Manin, Sovrem. Problem. Matem., VINITI publ., 3 (1974) (in Russian);
   L. Gerritzen and M. van der Put, Lect. Notes in Math., 817 (1980);
   I.V. Volovich, CERN preprint (1987).
10. D. Lebedev and A. Morozov, preprint ITEP 8-163 (1988).
11. L.O. Chekhov, A.D. Mironov and A.V. Zabrodin, preprint (1989).

TOPOLOGY CHANGE IN QUANTUM GRAVITY

Joseph Polchinski
Department of Physics
University of Texas
Austin, Texas 78712

ABSTRACT

We give a review and status report on the idea that wormholes in spacetime have a profound influence on the constants of nature. We discuss some problems and some directions of recent progress.

In general relativity, the geometry of spacetime is dynamical — gravity results from the curvature of spacetime. Is the *topology* of spacetime fixed, or is it, too, dynamical? For example, can a tiny piece of the universe break off, forming a closed "baby universe", and then reattach itself at some distant spacetime location? The baby universe spacetime looks like a tube, or "wormhole", glued into the larger spacetime at each end. If this is allowed, then it might be happening constantly, as a virtual process, at the Planck scale.

Whether this sort of process occurs, and what effects it has, are among the unanswered questions of quantum gravity. Why are they intruding upon a particle physics meeting at this particular time? Over the past few years, a number of physicists, perhaps most persuasively Sidney Coleman, have argued that the wormhole process would have a profound influence on the constants of nature that we measure. Moreover, it might answer one of the great mysteries about these constants — why the cosmological constant, the energy density of the vacuum, is so close to zero. The bulk of my talk is a review and status report on the wormhole idea, followed by a brief discussion of some other recent ideas.

For the first part of my talk, I will lead you through the following flow chart:

1. Does topology fluctuate?

   YES: Go to question 2.
   NO: Boring!

2. Does topology change imply new *kinds* of low energy physics, or does it just contribute to the usual constants of nature?

   YES: Wow!
   NO: Go to question 3.

3. Are the constants of nature predictable?

   YES: Go to question 4.
   NO: Yuk!

4. Is $\Lambda = 0$ predicted?

   YES: COLEMANIA!
   NO: Return to go.

For the answer to the first question, "does topology fluctuate?", let us consult the authorities. On the ninth floor of the physics building at the University of Texas, one can walk down the hall and talk to two of the pioneers of quantum gravity, John Wheeler and Bryce deWitt. Unfortunately, one walks back to one's office more confused than before, because Wheeler answers with a resounding "Yes!" and DeWitt with an equally forceful "No!" Of course, both physicists offer various thought experiments and other arguments to support their point of view, but in the end it comes down to their respective intuitions about the way physics ought to be. Could it be that both are right, that there are two consistent theories of quantum gravity, one in which topology fluctuates and one in which it is fixed? Or is one of these inconsistent?

One can try to get a handle on these questions by means of thought experiments. It may be possible to set up a situation in which a consistent outcome is possible only if topology change occurs, or, alternatively, only if it is forbidden. At the end of the talk I will return to such a thought experiment.

A second tool is the toy model — a simplified system from which one might gain some insight. One such toy model is $2+1$ dimensional gravity[1]. A second is string theory. The fundamental interaction in string theory is the splitting of one string in two, or two re-joining into one. This is analogous to the splitting off and re-joining of a baby universe in quantum gravity. (Notice that we are talking about string theory as a *model* of quantum gravity, so that each string represents a universe; when we talk about string theory as a *theory* of quantum gravity, a string is a single graviton.) In fact, there is a rather detailed dynamical correspondence between the string model and four dimensional quantum gravity[2]. Experience with topology change in string theory suggests that topology change in quantum gravity is physically consistent and can be understood in a quantitative way. In the case of string theory, there would seem to be two consistent theories, with topology change and without (to be precise, there are more, since in the topology changing case there is the option to include or exclude unoriented topologies), and topologies within nature (speaking here as a string believer), there may be some deep reason why topology change also occurs at the level of universes; see ref. 3 for some interesting speculations in this direction.

There has been some preliminary application of string technology and string intuition to baby universes, but I believe that there is much more to be learned from this model. It is interesting to note that when we say that string theory is consistent, we speak as external observers carrying out experiments on strings, while for the analogy to universes one would be living on the string world-sheet. In string theory there is known to be a close connection between quantum mechanical consistency (unitarity) on the world-sheet and in spacetime. It should also be possible to relate unitarity in a single universe to some sort of multi-universe unitarity. This would be a large step towards bringing some order to the subject.

To conclude, we do not know the answer even to the first, most basic, question. We therefore assume the more interesting answer, that topology change occurs, and proceed to the second question: do wormholes lead to new kind of observable physics? If we are carrying out an experiment near one end of a wormhole, some particles might enter the wormhole and disappear. This would seem to carry away information, leading to incoherence in the quantum theory[4,5]. Since wormholes tend to be quite small (Planck-sized) their ends look like local interactions, and the loss of information is a random average over the corresponding coupling constant. This is new physics of a very radical sort. However, Coleman, and Giddings and Strominger, have argued that coordinate invariance requires the amplitude for the wormhole to be independent of spacetime position[6]. Instead of an independent averaging over the couplings at each spacetime point, as envisioned in ref. 5, one averages only over coupling constants which are constant in spacetime. In our universe this average has long since collapsed onto very precise values of the couplings. The only affect is that the couplings we measure differ from those in the underlying fundamental theory by a random and unpredictable wormhole contribution.

The argument of Coleman, Giddings, and Strominger is quite persuasive if one is considering very small wormholes, but it is less clear if one probes the wormholes with wavelengths short enough to resolve the wormhole structure. This is another kind of thought experiment which may be useful in deciding between different interpretations of topology change. The extreme case is a wormhole large enough for a person to pass through. If the other end is in the past, it would seem that we could travel back in time, with all the attendent paradoxes of science fiction. Morris, Thorne, and Yurtsever[7] have looked at this, and have given tentative arguments which suggest that time evolution in such a spacetime is actually consistent and non-paradoxical — I don't fully understand this, though.

So the answer to the second question is not clear either. Let us assume the more conservative answer, that wormholes affect the low energy theory only through the constants of nature. The third question is, are these constants predictable? According to Coleman, Giddings, and Strominger, the wormhole contribution is random. Our long-standing goal to predict the constants of nature is unreachable, just as we cannot predict when a given radioactive decay will occur, or which arm of a Stern-

Gerlach device an ion will pass through. This is a truly awful conclusion. But what Sidney Coleman takes away with his right hand — predictability — he returns with his left. The path integral for quantum gravity will include spacetimes in which our universe will be connected via wormholes to large virtual universes. Coleman argues that our probability distribution for coupling constants is determined by the path integral over the virtual universes. Further, large virtual universes give a very highly peaked probability distribution, giving in particular a delta-function peak at zero cosmological constant, and possibly delta function peaks for all other coupling constants as well[7]. Predictability is restored!

Coleman's argument is based on the Euclidean path integral formulation of quantum gravity. This theory does not actually exist. The naive Euclidean path integral has a divergence, due to the conformal factor, which is not found in ordinary quantum field theories, and the correct treatment of this divergence is not known. Also, the interpretation of the Euclidean path integral is not clear. Unlike ordinary quantum field theory, it has not been derived from a Hamiltonian description. Coleman cuts through these difficulties with two simple and reasonable assumptions. The first is that in the semiclassical approximation the Euclidean path integral is dominated by the saddle point of least action — in this case, the Euclidean four-sphere. The second is that the path integral is to be interpreted as an expectation value. These assumptions are quite plausible - they certainly hold in ordinary quantum field theories. And they do seem to lead to a delta-function peak in the probability distribution for couplings, restoring predictability.

Unfortunately, closer examination has revealed several problems. These include:

1. The phase problem[9]. If one takes Coleman's assumptions seriously and evaluates the contribution of the Euclidean four-sphere, there is a dimension-dependent phase factor. This eliminates the delta-function peak in four dimensions. More important, the would-be probability distribution for couplings is *complex* in some dimensions, showing Coleman's assumptions to be inconsistent with one another.
2. The large wormhole problem[10]. The probability distribution found by Coleman is so strongly peaked that it really doesn't make sense. One sign of this is the large wormhole problem. One would expect wormholes to be Planck-sized. Large wormholes - a fermi, an Angstrom, a meter, or a light year, are heavily suppressed by their action. This suppression, though, counts for nothing when compared with the enormous factor found in the path integral by Coleman. Fischler and Susskind[10] argue that Coleman's assumptions lead to a spacetime in which wormholes are dense on all scales, looking nothing like the approximately flat spacetime we see.
3. The problem of interpretation. An attempt to derive the Euclidean path integral from a many-inverse Hilbert space theory, by analogy to what is done in ordinary quantum mechanics and field theory, leads to results very different from Coleman's[11]. There is some evidence for a peak at $\Lambda = 0$, but the peak is weaker than Coleman's, and depends upon some metaphysical/anthropic considerations. It seems that the Euclidean path integral, which has been so useful in the past, simply may not play a useful role in quantum gravity with topology change.

Of course, each of these criticisms can be criticized in its turn[12]. Since we are discussing a theory without rules or foundations, all of these arguments involve assumptions about what the finished theory will be. Nevertheless, I believe that each of the criticisms above is quite serious, and that Coleman's idea is unlikely to survive in its present form.

So, we have had no more success answering the third question than the first two. Let us be optimistic, and assume that when all is understood predictability will be restored. Then, is it likely that wormholes are after all the reason for the vanishing of the cosmological constant? The cosmological constant is measured (or bounded) by observing physics at a scale of 10^{10} light years, by the long-distance structure of the universe. On the other hand, it's value is determined, in quantum field theory, by the dynamics of the vacuum at scales somewhere between 10^{-15} and 10^{-32} centimeters. This is the basic problem; how can a measurement at 10^{10} light years feed back to much shorter distance scales? Years of frustrated searching and contrived ideas had produced no such mechanism. Then Coleman pointed out that, with mild and plausible assumptions, precisely this occurs in quantum gravity: the probability distribution for coupling constants is determined by the path integral over cosmological-scale virtual universes, but feeds into the physics we see through Planck-sized wormholes. In a remarkable and elegant fashion quantum gravity accomplishes automatically what many of us had concluded was impossible. For this reason it is very tempting to believe that, in spite of all the uncertainties and objections, Coleman has come close to the truth, and that his mechanism will survive, perhaps in some modified form.

Let me finish by discussing two recent steps forward in wormhole physics. The first concerns the string model for gravity. Recently, three groups have succeeded in carrying out exactly the sum over all topologies in this theory[13]. To be precise, they have taken a discrete sum over surfaces, and by using large-N matrix methods have obtained a continuum limit which appears to be two dimensional gravity, with no matter or with very simple matter (conformally invariant with central charge less than one). Their result is a differential equation for the sum over topologies as a function of the coupling constant in the one universe to two universe splitting amplitude. The power series solution reproduces sphere plus torus plus genus two plus ..., but there are nonperturbative terms as well! The meaning of this is not yet clear. It is not even clear whether we should regard it as a result in statistical mechanics, in quantum gravity, or in string theory. To be precise, in order to model the dynamics of four dimensional strings, one needs a matter central charge greater than 25 — in this case the kinetic term for the conformal part of the metric, induced by quantum effects, has the same sign as in four dimensions. String theory needs a matter central charge of 25 (and the metric provides the 26th unit) or 26 (and the metric decouples, with the same result). The range central charge < 1, where the results[13] hold, is really statistical mechanics. But this is only a first step. Perhaps the methods can be generalized, or the physical meaning of the differential equation and the non-perturbative terms extracted and applied elsewhere.

The second bit of progress is a rather Wheeleresque thought experiment —

the decay of the true vacuum[14,15]. Consider a scalar field whose potential has two local minima — the true vacuum at ϕ_1, $V(\phi_1) = 0$, and the false vacuum at $\phi_2, V(\phi_2) > 0$. The false vacuum is metastable. Its decay, by the nucleation and expansion of a bubble of true vacuum, is a familiar process in quantum field theory[16]. I am now discussing the reverse process: the nucleation and expansion of a region of false vacuum in the midst of the true vacuum. How can this happen? We know that the true vacuum bubble involves competition between volume energy and surface energy. Here both surface energy and volume energy make the bubble want to shrink. But, there is a third effect, gravity, which favors expansion: for a large enough bubble the inflation of the false vacuum outpaces the tendency of the bubble wall to shrink.

There is an extensive literature on the classical motion of the false vacuum bubble — see refs. 14,15 for references. The nucleation process, however, is complicated, because it involves both quantum mechanics[17] and gravity in an essential way. Recently Willy Fischler, Dan Morgan, and I have made a Hamiltonian quantization of the bubble. We show that the nucleation occurs and calculate the barrier penetration factor[14]. Essentially, this involves solving Dirac's constraint equations,

$$\mathcal{H}(\vec{x})|\psi\rangle = \vec{\mathcal{P}}(\vec{x})|\psi\rangle = 0 \ ,$$

where $\mathcal{H}$ and $\vec{\mathcal{P}}$ are the generators of coordinate transformations in the WKB approximation. The result is interesting for at least three reasons:

1. As the possible origin of matter in our universe. Our universe may have once been in a state of empty true vacuum. In particular, many attempts to explain the vanishing of the cosmological constant lead to such a state. The matter we see would then arise from the "free lunch process": a bubble of false vacuum nucleates, expands, and then decays to true vacuum plus matter. This isn't really something for nothing: the matter energy produced is offset by negative gravitational potential energy.

2. As a test of the Euclidean path integral formalism. Farhi, Guth, and Guven[15] have recently studied the same process with the Euclidean bounce formalism. They have shown that a nonsingular bounce solution for this process does not exist. There does exist a singular bounce solution — the metric becomes degenerate (zero eigenvalue) in places. Since the Hamiltonian formalism unambiguously indicates that the tunnelling occurs, this is evidence that such degenerate metrics necessarily appear in the path integral for quantum gravity.

3. As a thought experiment in topology change. The false vacuum bubble has a feature that we have not yet touched upon. In order for the negative gravitation potential energy to offset the positive surface and volume energies, the space around the bubble must be highly curved, so curved, in fact, that when the bubble nucleates it is almost in its own universe, connected to the rest of spacetime by a narrow neck. As the bubble expands, the neck collapses, becoming a (spacelike) Schwarzchild singularity. Essentially, there are two universes, a closed universe containing the false vacuum bubble and a black hole, and the external universe containing a black

hole, with the two black holes being at opposite ends of the Schwarzchild singularity. After the black holes evaporate, the two universes are completely disconnected: the topology of the universe has changed! This is the sort of thought experiment which led Wheeler to give a positive answer to the first question on the flow chart. There are two caveats, however:

First, the evaporation may stop with a Planck-sized black hole: no one knows what happens at this point. The two universes would then remain connected. However, the final stable black hole would have to look more-or-less like an ordinary particle — coordinate invariance would forbid the two universes from exchanging energy or information — and so as a practical matter this would be indistinguishable from topology change. Perhaps more seriously, in this process the topology of *space* has changed, but the topology of *spacetime* is still simply connected. For a wormhole, the disconnected universe would have to be able to reconnect. There is no argument as yet that this would have to be the case.

In conclusion, it is very likely that wormholes have a profound effect on the constants of nature, but we do not know what it is. Coleman's argument does not seem to be complete, or correct, in its present form, but it is tempting to believe that it will somehow survive. It is interesting to note that, just when we believe that we have the solution to the ultraviolet problem of quantum gravity, namely string theory, we are reminded that there are other problems, like topology change, still unsolved. I think that it is a good time to be reconsidering these questions. They are difficult, but they are important; and there are some new ideas to work with.

Acknowledgements

Research supported in part by the Robert A. Welch Foundation and NSF Grant Phy. 8605978.

References

1. E. Witten, Nucl. Phys. **B323** (1989) 113;
   S. Carlip and S.P. deAlwis, "Wormholes in 2+1 Dimensions," IAS preprint (1989).

2. J. Polchinski, Nucl. Phys. **B324** (1989) 123.

3. T. Banks, "Progress Report on Wormhole Physics", Santa Cruz preprint (1989).

4. S.W. Hawking, D.N. Page, and C.N. Pope, Nucl. Phys. **B170** (1980) 283.
   S.W. Hawking, Comm. Math. Phys. **87** (1982) 395;
   A. Strominger, Phys. Rev. Lett. **52** (1984) 1733.

5. G.A. Lavrelashvili, V.A. Rubakov , and P.G. Tinyakov, JETP Lett. **46** (1987) 167;

S.W. Hawking, Phys. Lett. **195B** (1987) 337; Phys. Rev. **D37** (1988) 904;
S. Giddings and A. Strominger, Nucl. Phys. **B306** (1988) 980.

6. S. Coleman, Nucl. Phys. **B307** (1988) 864.
S. Giddings and A. Strominger, Nucl. Phys. **B307** (1988) 854.

7. M.S. Morris, K.S. Thorne, and U. Yurtsever, Phys. Rev. Lett. **61** (1988) 1446.

8. S. Coleman, Nucl. Phys. **B310** (1988) 643.

9. J. Polchinski, Phys. Lett. **B219** (1989) 251.

10. V. Kaplunovsky, unpublished
W. Fischler and L. Susskind, Phys. Lett. **B217** (1989) 48;
J. Polchinski, Nucl. Phys. **B325** (1989) 619.

11. W. Fischler, I. Klebanov, J. Polchinski, and L. Susskind, Nucl. Phys. **B327** (1989) 157.

12. J. Preskill, Nucl. Phys. **B323** (1989) 141;
S. Coleman and K. Lee, Phys. Lett. **B221** (1989) 665;
B. Grinstein, Nucl. Phys. **B321** (1989) 439;
S.-J. Rey, Nucl. Phys. **B319** (1989) 765;
P. Mazur and E. Mottola, "Absence of Phase in the Sum over Spheres", Florida - Los Alamos preprint (1989);
P.A. Griffin and D.A. Kosower, "Curved Spacetime One-Loop Gravity in a Physical Gauge", Fermilab preprint (1989).

13. M. Dougles and S. Shenker, "Strings in Less Than One Dimension." Rutgers preprint (1989);
D. Gross and A. Migdal, "Nonperturbative two Dimensional Quantum Gravity," Princeton preprint (1989);
E. Brezin and V. Kazakov, ENS preprint (1989).

14. W. Fischler, D. Morgan and J. Polchinski, "Quantum Nucleation of False Vacuum Bubbles," Texas preprint (1989).

15. E. Farhi, A.H. Guth, and J. Guven, "Is it Possible to Create a Universe in the Laboratory by Quantum Tunnelling?" MIT preprint (1989).

16. S. Coleman, Phys. Rev. **D15** (1977) 2929.

17. E. Farhi and A.H. Guth, Phys. Lett. **183B** (1987) 149.

STRINGS ON THE ISO(2,1) MANIFOLD

A. STERN
Department of Physics and Astronomy
University of Alabama
Tuscaloosa, AL 35487, U.S.A.

ABSTRACT

Some recent work on the dynamics of particles and strings moving on the 2+1 dimensional Poincare group [ISO(2,1)] manifold is reviewed. An anomaly in the general covariance of 2+1 gravity on a disc is discussed. The anomaly persists after the inclusion of an SO(2,1) Chern-Simons term in the gravity action.

1. Introduction

Recently, there has been some interest in formulating string dynamics on a noncompact group manifold G.[1,2] The case of G=SU(1,1) provides an example of classical string propagation on curved space-time,[1] and has application to the N=2 superconformal algebra.[2] Although no unitary representations were found for the SU(1,1) current algebra, unitary representations could be obtained after moding out the U(1) subgroup.

In this talk, I report, among other things, on a successful attempt (with P. Salomonson and B.-S Skagerstam) at finding unitary representations of the current algebra associated with G=ISO(2,1), i.e. the Poincare group in 2+1 dimensions.[3] This current algebra does not contain the SU(1,1) current algebra because the SU(1,1) central term is absent. Nevertheless, a non zero central term is present in the ISO(2,1) current algebra, so we obtain a central extension of the ISO(2,1) loop group. Unitary representations can be found by the method of induced representations.[4]

The standard action for closed strings on a group manifold is the Wess-Zumino-Novikov-Witten (WZNW) action S_{WZNW}. For the case G=ISO(2,1), we found that S_{WZNW} also describes[3]:

a) A bosonized spinning string propagating on 2+1 Minkowski space. Analogously, a Wess-Zumino particle action for G=ISO(2,1) was shown to describe a bosonized spinning point particle in 2+1 Minkowski space.[5]

b) 2+1 gravity on a disc (x time). This result follows from the correspondence between conformal field theories and topological field theories[6,7], such as 2+1 gravity[8,9].

The Virasoro algebra constructed from the ISO(2,1) current algebra generators was found to have a central charge which is independent of the Kac-Moody central charge. It is simply equal to the dimension of the group G, i.e. c=6. For b), the Virasoro algebra generates diffeomorphisms at the boundary of the disc. The corresponding symmetries are therefore anomalous. This anomaly is cancelled with the addition of conformal matter fields at the boundary.[3] Since ghosts contribute -26 to the central charge and gravity contributes 6, we require that c=20 for the matter fields.

Here we also consider the possibility of adding an SO(2,1) Chern-Simons term to the 2+1 gravity action. This term does not affect the classical field equations for gravity, but it does alter the quantum mechanical description, much like the θ-term in QCD. We show that the above mentioned anomaly persists after the inclusion of this term.

In Section 2, we shall review some basic properties of the ISO(2,1) group, and substitute into the Wess-Zumino particle action in Section 3. In Section 4, we substitute into the WZNW string action. Unitary representations for Kac-Moody algebras which are based on certain semidirect product groups are given in Section 5, and the SO(2,1) Chern-Simons term is discussed in Section 6.

2. The ISO(2,1) Group

Let Λ be a 3x3 Lorentz group matrix, while a is a Lorentz vector parametrizing translations. We can then write ISO(2,1) = $\{(\Lambda,a)\}$, where group elements satisfy the semidirect product rule

$$(\Lambda,a)\cdot(\Lambda',a') = (\Lambda\Lambda',\Lambda a'+a). \tag{2.1}$$

Let t_i and u_i, i=0,1,2, denote the Lorentz and translation generators, respectively. Their Lie brackets are given by

$$[t_i,t_j] = \epsilon_{ijk}t^k$$

$$[t_i,u_j] = \epsilon_{ijk}u^k \tag{2.2}$$

$$[u_i,u_j] = 0 .$$

Indices are raised and lowered with the space-time metric η = diag(-1,1,1).

It is convenient to introduce the 4x4 matrix representation for elements (Λ,a),

$$(\Lambda,a) \rightarrow g = \begin{pmatrix} \Lambda & a \\ 0 & 1 \end{pmatrix} \quad . \tag{2.3}$$

Then the semidirect product rule (2.1) is realized with matrix multiplication, and the Lie brackets (2.2) are realized as commutator brackets. Generators t_i and u_i have nonzero matrix elements

$$[t_i]_{jk} = -\epsilon_{ijk} \qquad [u_i]_{j3} = \eta_{ij} \quad . \tag{2.4}$$

Under the adjoint action of the group, a Lie algebra element v transforms as $v \rightarrow gvg^{-1}$. Two bilinear invariants can be constructed from Lie algebra elements v and w as follows:

$$\text{(i)} \quad \text{Tr } vw \tag{2.5}$$

and

$$\text{(ii)} \quad \text{Tr } \tilde{v}w \quad , \tag{2.6}$$

where $\tilde{v}$ is the dual of v, i.e. $[\tilde{v}]_{AB} = \frac{1}{2} \epsilon_{ABCD} v^{CD}$, $A,B,\ldots = 0,..3$. Invariant (i) is degenerate, while (ii) is nondegenerate. The former follows because $\text{Tr} u_i v = 0$, for all Lie algebra elements v. Thus (i) projects out the SO(2,1) subalgebra. As in ref. [9], the nondegenerate invariant (ii) is off-diagonal in t_i and u_i,

$$\text{Tr } \tilde{u}_i t_j = \eta_{ij} \quad , \text{Tr } \tilde{u}_i u_j = \text{Tr } \tilde{v}_i v_j = 0 \quad . \tag{2.7}$$

Although (i) can be generalized to any number of space-time dimensions, (ii) can only be defined for 4x4 matrices, and hence 2+1 dimensions.

In Sections 3 through 5 we shall apply invariant (i) to particle and string actions, and in Section 6 we consider invariant (ii).

3. Particle dynamics on ISO(2,1)

Let $g=g(\tau)$ coordinatize the particle degrees of freedom on group manifold G. The standard action for a particle moving on G is obtained from the bilinear invariant of the Maurer-Cartan form on G with some constant Lie algebra element K.[3,10] One then integrates over the particle world line. If we take G=ISO(2,1), and the bilinear invariant (ii), this action takes the form

$$S_{part.} = \int \mathrm{Tr}\, \tilde{K}\, g^{-1} dg \quad . \tag{3.1}$$

Upon choosing

$$K = \left(\begin{array}{ccc|c} & & & -s \\ & & -m & \\ & m & & \\ \hline & & & \end{array}\right) \quad , \tag{3.2}$$

the action (3.1) describes a particle of mass m and spin s moving in 2+1 Minkowski space. For this we note the following:

1. $S_{part.}$ is invariant under Poincare transformations.

Under a (left) Poincare transformation

$$g \to g_0\, g \quad , \tag{3.3}$$

[where g_0 denotes a constant element of ISO(2,1)] the Maurer-Cartan form $g^{-1}dg$, and hence the action (3.1), are unchanged.

2. Six constants of the motion follow from (3.1), which can be identified with the particle momenta P_i and angular momenta J^i. Furthermore, J^i contains an intrinsic spin. The equations of motion for the system are obtained upon making infinitesimal variations δg of g. Then

$$\delta S_{part.} = \int Tr\, \tilde{K}\, g^{-1} d(\delta g g^{-1}) g$$

$$= \int Tr\, \widetilde{gKg^{-1}}\, d(\delta g g^{-1}) \quad , \tag{3.4}$$

For arbitrary variations $\delta g g^{-1}$, a minimum of the action implies that

$$gKg^{-1} \equiv P_i t^i - J_i u^i \tag{3.5}$$

are constants of the motion. If we substitute the matrices (2.3) and (3.2) into gKg^{-1}, we have

$$P^i = m\Lambda^i{}_0 \quad , \quad J^i = m\epsilon^{ijk} a_j \Lambda_{k0} + s\Lambda^i{}_0 \quad . \tag{3.6}$$

Here a^i can be interpreted as the space-time coordinate of the particle. Then the first term in J_i is seen to be the orbital angular momentum of the particle, while the second represents an intrinsic spin. The form of P_i and J_i in (3.6) leads to the mass-shell and Pauli-Plebanski conditions for a massive, spinning particle,

$$P^2 = -m^2 \quad , \quad P \cdot J = -ms \quad . \tag{3.7}$$

3. It was shown in ref's [3,10] that the Poisson bracket algebra of the constants of motion corresponds to the 2 + 1 dimensional Poincare algebra,

$$\{ J_i, J_j \} = \epsilon_{ijk} J^k \quad ,$$
$$\{ J_i, P_j \} = \epsilon_{ijk} P^k \quad , \tag{3.8}$$
$$\{ P_i, P_j \} = 0 \quad .$$

In passing to the quantum theory, the relevant Hilbert space carries unitary representations of the ISO(2,1) group. After imposing conditions (3.7) on the states of the Hilbert space, we restrict to the irreducible representations for a massive spinning particle. Other representations of the Poincare group (eg., massless particles and tachyons) can be obtained for different choices of the constant Lie algebra element K.[3,10]

4. String dynamics on ISO(2,1)

A string moving on a group manifold G can be coordinatized by $g = g(\tau,\sigma)$, where τ and σ parametrize the string world sheet. The standard action for a string moving on a group manifold is the Wess-Zumino-Novikov-Witten action, which for G=ISO(2,1) and bilinear invariant (ii), takes the form

$$S_{string} = \frac{1}{2\lambda^2}\int_{\partial M} d\sigma\, d\tau\, \widetilde{\mathrm{Tr}\,(g^{-1}\partial_\mu g) g^{-1}\partial^\mu g} + \frac{n}{12\pi}\int_M \widetilde{\mathrm{Tr}\,(g^{-1}dg)(g^{-1}dg)^2} \quad , \tag{4.1}$$

where $M = D^2 x R^1$, whose boundary ∂M is the string world sheet. (We will only consider closed strings.) After substituting the matrix form for g in (2.3), and assuming that the couplings satisfy $-1/\lambda^2 = n/4\pi \equiv \kappa/2$, the action can be written entirely on the boundary ∂M

$$S_{string} = -\frac{\kappa}{4}\int_{\partial M} d\sigma\, d\tau\, \epsilon^{ijk}(\partial_+\Lambda\Lambda^{-1})_{ij}\, \partial_- a_k \quad , \tag{4.2}$$

where $\partial_\pm \equiv \partial/\partial\tau \pm \partial/\partial\sigma$. Unlike the case with compact G, the coupling κ is not quantized, as will be evident later.

Analogous to the point particle action (3.1), eq.(4.2) describes a (bosonized) spinning string moving in 2+1 Minkowski space. For this we note that:

1. Like $S_{part.}$, S_{string} is invariant under global Poincare transformations (either from the left or the right). More generally, (4.1) is invariant under

$$g \to h(\sigma^+)\; g\; \bar{h}(\sigma^-) \qquad , \qquad (4.3)$$

where $h,\bar{h}$ take values in ISO(2,1) and $\sigma^{\pm} = \tau \pm \sigma$. It is also invariant under diff x $\overline{\text{diff}}$.

2. There are twelve conserved currents which follow from (4.1). They can be identified with left- and right-moving momentum and angular momentum currents. Upon performing variations of a and Λ in (4.2), we find

$$\delta S_{string} = \frac{\kappa}{4}\int d\sigma\, d\tau\; \epsilon^{ijk}\Big\{\partial_+(\Lambda^{-1}\partial_-\Lambda)_{ij}(\Lambda^{-1}\delta a)_k + (\Lambda^{-1}\delta\Lambda)_{ij}\partial_+(\Lambda^{-1}\partial_- a)_k\Big\} \quad . \qquad (4.4)$$

Extrema of the action occur for $\partial_+ J^i_- = \partial_+ P^i_- = 0$, where

$$J^i_- = \kappa(\Lambda^{-1}\partial_- a)^i \quad , \quad P^i_- = \frac{\kappa}{2}\epsilon^{ijk}(\Lambda^{-1}\partial_-\Lambda)_{jk} \quad , \qquad (4.5)$$

or equivalently $\partial_- \bar{J}^i_+ = \partial_- \bar{P}^i_+ = 0$, where

$$\bar{J}^i_+ = \kappa[\Lambda\partial_+(\Lambda^{-1} a)]^i \quad , \quad \bar{P}^i_+ = \frac{\kappa}{2}\epsilon^{ijk}(\partial_+\Lambda\Lambda)_{jk} \quad . \qquad (4.6)$$

We identify P^i_μ and $\bar{P}^i_\mu$ with momentum currents, and J^i_μ and $\bar{J}^i_\mu$ with angular momentum currents. Like the particle case, the latter two cannot be written solely in terms of orbital angular momentum currents; So we conclude that a (bosonized) spin current is present.

3. The Poisson bracket algebra of the current densities yields two central extensions of the Poincare loop group. From ref. [3],

$$\begin{aligned}
\{P^i(\sigma)\ ,\ P^j(\sigma')\} &= 0 \quad , \\
\{J^i(\sigma)\ ,\ P^j(\sigma')\} &= -\epsilon^{ijk}P_k(\sigma)\delta(\sigma-\sigma') + \kappa\eta^{ij}\partial_\sigma\delta(\sigma-\sigma') \quad , \\
\{J^i(\sigma)\ ,\ J^j(\sigma')\} &= -\epsilon^{ijk}J_k(\sigma)\delta(\sigma-\sigma') \quad ,
\end{aligned} \tag{4.7}$$

$$\begin{aligned}
\{\bar{P}^i(\sigma)\ ,\ \bar{P}^j(\sigma')\} &= 0 \quad , \\
\{\bar{J}^i(\sigma)\ ,\ \bar{P}^j(\sigma')\} &= -\epsilon^{ijk}\bar{P}_k(\sigma)\delta(\sigma-\sigma') - \kappa\eta^{ij}\partial_\sigma\delta(\sigma-\sigma') \quad , \\
\{\bar{J}^i(\sigma)\ ,\ \bar{J}^j(\sigma')\} &= -\epsilon^{ijk}\bar{J}_k(\sigma)\delta(\sigma-\sigma') \quad ,
\end{aligned} \tag{4.8}$$

and all other Poisson brackets are zero. (For convenience, we have dropped the world sheet indices on the currents.) The central term occurs in the P_i, J_i bracket, due to the off diagonal structure of the invariant bilinear (2.7).

Before discussing representations of the corresponding quantum algebra, we briefly mention the connection to gravity on $R^1 \times D^2$. Witten showed the correspondence between conformal field theory defined with a WZNW action, and topological field theory defined by a Chern-Simons action[6]. He further showed the equivalence of the Chern-Simon action for ISO(2,1) and the Einstein-Hilbert action for gravity in 2+1 dimensions[9]. Using bilinear invariant (ii), the former becomes

$$S_{CS} = \kappa \int_{R^1 x D^2} \mathrm{Tr}\ \tilde{A}\left(dA + \frac{2}{3} A^2\right) \quad , \tag{4.9}$$

where A is an ISO(2,1) connection one-form, and κ can be identified with the gravitational constant. Upon substituting $A = \omega^i\, t_i + e^i u_i$ into (4.9), it takes the form as given in ref. [9].

Following ref's [7], if we substitute the solutions of the field equations resulting from (4.9) back into the action, we get an effective field theory defined on the boundary of D^2xR^1. The effective action is the "chiral" WZNW action, i.e.

$$S_{cs} = \frac{\kappa}{2} \int_{R^1xS^1} d\sigma\, d\tau\, \epsilon^{ijk}(\partial_\tau \Lambda\Lambda^{-1})_{ij}\partial_\sigma a_k \quad . \tag{4.10}$$

Eq. (4.10) differs from (4.2) in that light-cone variables σ^+ and σ^- are replaced by σ and τ. Since (4.10) is first order in τ derivatives, constraints exist in the Hamiltonian formalism which essentially cut the number of degrees of freedom in half. These constraints state that $\bar{J}^i_+ \approx \bar{P}^i_+ \approx 0$, while the variables J^i_- and P^i_-, defined in (4.5), can be taken to be the remaining degrees of freedom.

5. Unitary Representations

We now exhibit the unitary representations for the Kac-Moody group associated with ISO(2,1), or more generally the semidirect product group IH. By IH, we mean the group generated by t_i and u_i, where t_i generate the subgroup H, and now i = 1,2, ... D = dim(H). Furthermore, in eq. (2.2) and (4.7), we replace ϵ_{ijk} and η_{ij}, respectively, by the structure constants C_{ijk} and invariant metric g_{ij} for H. u^i transforms as an adjoint vector under H. It follows that IH is the Poincare group only for the case H = SO(2,1).

Unitary representations for the IH Kac-Moody group are easily found using the method of induced representations[4]. The latter can be applied since the quantum operators $P_i(\sigma)$ are diagonalizable. Let $\hat{P}_i(\sigma)$ be a standard point in the space of all momentum densities. Then we define the state $|\hat{P}, \alpha\rangle$, such that

$$\mathbf{P}_i(\sigma)\,|\hat{P},\alpha\rangle = \hat{P}_i(\sigma)\,|\hat{P},\alpha\rangle \quad . \tag{5.1}$$

Remaining states in the Hilbert space $\mathcal{K}$ are obtained by application of the loop group elements $\Lambda(\sigma)$, mapping S^1 to H. For this use

$$U(\Lambda)^{-1}\mathbf{P}_i U(\Lambda) = (\Lambda\mathbf{P})_i - \frac{\kappa}{2} C_{ijk}(\partial_\sigma \Lambda\Lambda^{-1})^{jk} \quad , \tag{5.2}$$

where $U(\Lambda)$ denotes a unitary representation for $\Lambda(\sigma)$. Eq. (5.2) follows from the quantum version of brackets (4.7). Faithful representations of the loop group of H exist since no central term appears in the subalgebra spanned by the $J(\sigma)$'s.

α in the state $|\hat{P},\alpha\rangle$ is a degeneracy index. It serves to label matrix representations $\{D_{\alpha\beta}\}$ of the little group $H_{\hat{P}} = \{\tilde{\Lambda}\}$ of $\hat{P}(\sigma)$, i.e.

$$U(\tilde{\Lambda})\,|\hat{P},\alpha\rangle = D_{\beta\alpha}(\tilde{\Lambda})\,|\hat{P},\beta\rangle \quad , \tag{5.3}$$

$$(\tilde{\Lambda}\hat{P})_i - \frac{\kappa}{2} C_{ijk}(\partial_\sigma \tilde{\Lambda}\tilde{\Lambda}^{-1})^{jk} = 0 \quad . \tag{5.4}$$

For the special case where $\hat{P} = 0$, (5.4) shows that $\tilde{\Lambda}$ is independent of σ. Hence, $H_{\hat{P}=0}$ is isomorphic to H. As in ref. [3], $H_{\hat{P}}$, in general, is a finite group; either H or one of its subgroups.

The representations of the Kac-Moody group associated with IH can be classified using the invariants of $H_{\hat{P}}$. In addition, an invariant can be constructed from momentum operators $\mathbf{P}(\sigma)$. Namely,

$$\mathcal{P}\ \mathrm{Tr}\ e^{\frac{1}{\kappa}\int d\sigma\ \tilde{\mathbf{P}}} \quad , \qquad \tilde{\mathbf{P}}_{ij} = C_{ijk}\mathbf{P}^k(\sigma) \quad , \tag{5.5}$$

where $\mathcal{P}$ denotes path ordering. Eq. (5.5) is unchanged under transformations (5.2), and can be used to classify representations.

A Virasoro algebra can be defined from the Kac-Moody generators utilizing the standard Sugawara construction. Let $\mathbf{P}_n^i$ and $\mathbf{J}_n^i$ be Fourier components of the currents $\mathbf{P}^i(\sigma)$ and $\mathbf{J}^i(\sigma)$, respectively. Then set

$$\mathbf{L}_m = \frac{1}{2\pi\kappa} \sum_r :\mathbf{P}_r \cdot \mathbf{J}_{m-r}: - \frac{C_2}{8\pi^2\kappa^2} \sum_r \mathbf{P}_r \cdot \mathbf{P}_{m-r} \quad , \tag{5.6}$$

where the colons denote normal ordering and $C_2 \delta^i{}_\ell = c^{ijk} c_{jk\ell}$. From (5.6), we have

$$[\mathbf{J}_n^i, \mathbf{L}_m] = n\mathbf{J}_{n+m}^i \qquad [\mathbf{P}_n^i, \mathbf{L}_m] = n\mathbf{P}_{n+m}^i \quad . \tag{5.7}$$

The Virasoro algebra obtained from (5.6) has a central charge c which is independent of κ as well as C_2. We find c = 2D, corresponding to the dimension of IH. Then for H= ISO(2,1), c = 6. Analogous Virasoro generators $\bar{\mathbf{L}}_m$ can be constructed from Fourier components of the operators associated with the current densities $\bar{J}^i(\sigma)$ and $\bar{P}^i(\sigma)$.

In the above discussion, we have ignored effects due to ghosts. Ghosts arise when we start with the reparametrization invariant version of (4.1), and fix a gauge using the Faddeev-Popov procedure. As usual, the ghosts contribute -26 to the Virasoro central charge. The fact that the total central charge for H = SO(2,1) is not zero signals an anomaly in the general covariance of 2+1 gravity at the boundary of $M = R^1 x D^2$. This is since $\bar{\mathbf{L}}_m \approx 0$ are the first class constraints generating reparametrizations of τ at the boundary ∂M of M. The anomaly is cancelled with the inclusion of matter fields at ∂M. The latter are required to define a conformal field theory which contributes 20 to the total Virasoro central charge.

6. SO(2,1) Topological Terms

In the preceding, we have only utilized the bilinear invariant (ii). We now consider adding terms constructed from invariant (i). More specifically, we include

$$S'_{string} = \frac{\kappa'}{2}\int_{\partial M} d\sigma d\tau \, \mathrm{Tr}\, g^{-1}\partial_\mu g \, g^{-1}\partial^\mu g$$

$$+ \frac{\kappa'}{6}\int_M \mathrm{Tr}(g^{-1}dg)^3 \quad , \tag{6.1}$$

in the string action. The total action is then (4.1) plus (6.1). (6.1) corresponds to an SO(2,1) WZNW action. Upon substituting (2.3) into (6.1),

$$S'_{string} = \frac{\kappa'}{2}\int_{\partial M} d\sigma d\tau \, \mathrm{Tr}\, \Lambda^{-1}\partial_\mu \Lambda \, \Lambda^{-1}\partial^\mu \Lambda$$

$$+ \frac{\kappa'}{6}\int_M \mathrm{Tr}(\Lambda^{-1}d\Lambda)^3 \quad . \tag{6.2}$$

Analogously, we can add an SO(2,1) Chern-Simons term

$$S'_{CS} = \frac{\kappa'}{2}\int_{R^1 x D^2} \mathrm{Tr}\, A\left(dA + \frac{2}{3}A^2\right) \tag{6.3}$$

to the 2+1 gravity action (4.9). Upon following the procedure in ref's [7], this term can be shown to be equivalent to the chiral version of (6.1).

The terms (6.1) and (6.3) do not affect the classical equations of motion of the system, but do change the resulting quantum mechanics. This change manifests itself in the appearance of new central terms in the current algebra. Replace the current densities J^i_- and $\bar{J}^i_+$ in eq's (4.5) and (4.6) by

$$J_-^i = \kappa(\Lambda^{-1}\partial_- a)^i + \kappa'/2\ \epsilon^{ijk}(\Lambda^{-1}\partial_-\Lambda)_{jk} \quad , \quad \text{and}$$

$$\bar{J}_+^i = \kappa[\Lambda\ \partial_+(\Lambda^{-1}a)]^i + \kappa'/2\ \epsilon^{ijk}(\partial_+\Lambda\Lambda^{-1})_{jk} \quad . \tag{6.4}$$

Then the last Poisson brackets in (4.7) and (4.8) are replaced by

$$\{J^i(\sigma),J^j(\sigma')\} = -\epsilon^{ijk}\ J_k(\sigma)\delta(\sigma-\sigma') + \kappa'\eta^{ij}(\partial_\sigma-\partial_{\sigma'})\delta(\sigma-\sigma'),$$

$$\{\bar{J}^i(\sigma),\bar{J}^j(\sigma')\} = -\epsilon^{ijk}\ \bar{J}_k(\sigma)\delta(\sigma-\sigma') - \kappa'\eta^{ij}(\partial_\sigma-\partial_{\sigma'})\delta(\sigma-\sigma'), \tag{6.5}$$

respectively. (Again, we have dropped the world sheet indices on the currents.) Thus, $J^i(\sigma)$ and $\bar{J}^i(\sigma)$ define two SU(1,1) Kac-Moody subalgebras[1,2]. On the other hand, P^i and J^i generate the most general central extension of the ISO(2,1) loop group.

Possible unitary representations for the Kac-Moody group generated by J^i and P^i are obtained by taking tensor products of representations discussed in Section 5, with those of ref. [2].

The SO(2,1) topological term does alter the definition of the reparametrization generators. The Virasoro generators in (5.6) are replaced by

$$\mathbf{L}_m = \frac{1}{2\pi\kappa}\sum_r :\mathbf{P}_r\cdot\mathbf{J}_{m-r}: + \frac{4\pi\kappa' - C_2}{8\pi^2\kappa^2}\sum_r \mathbf{P}_r\cdot\mathbf{P}_{m-r} \quad , \tag{6.6}$$

where $C_2 = -2$ for $H=SO(2,1)$. Nevertheless, the central charge associated with the Virasoro algebra is unaltered. That is, we again find c=6 for H=SO(2,1). Consequently, the anomaly discussed in Section 5 persists when the SO(2,1) Chern-Simons term is added to the 2+1 gravity action.

Acknowledgement

The author wishes to thank L. Clavelli, P. Salomonson and B.-S. Skagerstam for helpful discussions.

References

1. J. Balog, L. O'Raifeartaigh, P. Forgacs and A. Wipf, Nucl. Phys. B325 (1989) 225.

2. L. Dixon, M. Peskin and J. Lykken, Nucl. Phys. B325 (1989) 329.

3. P. Salomonson, B.-S. Skagerstam and A. Stern, University of Alabama preprint UAHEP 897.

4. E. Wigner, Ann. Math. 40 (1939) 149.

5. B.-S. Skagerstam and A. Stern, University of Alabama preprint UAHEP 895 (to appear in Int. J. Mod. Phys.).

6. E. Witten, Comm. Math. Phys. 117 (1988) 353.

7. G. Moore and N. Seiberg, Phys. Lett. 220B (1989) 422; S. Elitzur, G. Moore, A. Schwimmer and N. Seiberg; Nucl. Phys. B326 (1989) 108.

8. A. Achucarro and P. K. Townsend, Phys. Lett. 180B (1986) 89.

9. E. Witten, Nucl. Phys. B311 (1988) 46; B323 (1989) 113.

10. A. P. Balachandran, G. Marmo, B.-S. Skagerstam and A. Stern, *Gauge Theory and Fibre Bundles*, Lecture Notes in Physics, Vol. 188 (Springer-Verlag, Berlin, 1983).

Axionic String Instantons and Their Low-energy Implications

Soo-Jong Rey
Department of Physics
University of California, Santa Barbara CA 93106 USA

ABSTRACT

In four-dimensional compactified heterotic superstring theory with $N=1$ spacetime supersymmetry, we construct a class of axionic instanton solutions. The instantons are shown to be self-dual and characterized by integer-valued Kalb-Ramond 'magnetic' charges and by Euler characteristics. They saturate a Bogomolnyi bound, that follows naturally from the spacetime supersymmetry. The instantons induce local operators in the low-energy effective Lagrangian, and provide a new mechanism to violate the superpotential nonrenormalization theorem. We also find a (super) conformal field theory of Euclidean spacetime that may be identified as an exact solution of the axionic instanton. We finally speculate on possible interpretations of the instantons as solutions of a two-dimensional quantum gravity (noncritical strings) coupled to matter conformal field theory of central charge $c<25$.

1. Introduction

String theory is the first consistent theory of quantum gravity, that may also unify other fundamental interactions. Starting from ten-dimensional superstring theories, one may compactify internal six-dimensions on a Calabi-Yau manifold or an equivalent conformal field theory [1]. The uncompactified four-dimensions are usually assumed to be flat Minkowski spacetime. On the other hand, it is of interest to investigate nontrivial spacetime backgrounds other than Minkowski spacetime. Such examples would include full-fledged string analog of the Schwarzschild solution, expanding universes [2], instantons, monopoles, cosmic strings [3] and domain walls [4]. They will shed light, for example, on the nature of spacetime singularities in string theory [5], string production in a strong background field [6] and many other related questions. I expect that nonperturbative physics is the arena where stringy effects play a significant role, and our intuition based on field theory limit approximation fails.

In this talk, I report some progress I made along the above direction [7]. To be specific, I consider heterotic superstrings compactified on $N = 2$, $c = 9$ superconformal field theory [1] that leads to four-dimensional $N = 1$ spacetime supersymmetry. A class of axionic string instantons are found from the low-energy field theory point of view. These instantons are solutions of a generalized self-duality equation. In the low-energy effective Lagrangian, the instantons induce local operators that renormalize superpotentials. Thus, the string nonrenormalization theorem [8] is violated by nonperturbative string effects. Furthermore, once the instanton fluctuations become significant, the heterotic superstrings, being axionic strings [9], are confined to become boundaries of axionic domain walls. I also report an exact conformal field theory whose proper long-wavelength limit reduces to the axionic string instantons. This is based on a collaboration with G.T. Horowitz [10]. Finally, I comment on a possible interpretation of the axionic instantons from the viewpoint of the two-dimensional quantum gravity (noncritical strings).

2. Axionic Instantons from the Low-energy Point of View

Heterotic strings compactified on $c = 9$ (2,2) superconformal field theories give rise to many moduli fields. Among them there are two dilaton-axion moduli fields $Z_a = e^{\beta_a D} + i\beta_a A$, $a = S, T$, whose low-energy dynamics are described by the Kähler potential

$$\mathcal{K} = -\sum_{a=S,T} {\beta_a}^{-2} \ln(Z_a + \bar{Z}_a). \tag{1}$$

The dilaton 'coupling constants' are denoted by β_a and take values $\beta_S{=}1$ and $\beta_T{=}\frac{1}{\sqrt{3}}$ respectively. Their numerical values are completely determined by the scale invariance at the string tree level and spacetime supersymmetry [11].

The low-energy Lagrangian reads

$$e^{-1}L = -\mathcal{R} + \sum_{a=S,T} {\beta_a}^{-2} \frac{\nabla Z_a \nabla \bar{Z}_a}{(Z_a + \bar{Z}_a)^2}. \tag{2}$$

Let us first concentrate on the dilaton-axion fields by turning off the gravity. After introducing the 'dual' Kalb-Ramond field to the axion field A, $dA = e^{2\beta D} H^*$, the action reads

$$\begin{aligned} S_m &= \int d^4x \, \frac{1}{2}[(\nabla D)^2 + \frac{1}{3!} e^{2\beta D} H^2] \\ &= \int d^4x \frac{1}{2}[\nabla D \pm e^D H^*]^2 \mp \int d^4x e^{\beta D} \nabla D \wedge H. \end{aligned} \tag{3}$$

Thus, the minimum action configuration is achieved if and only if

$$\nabla_\mu D = \pm e^{\beta D} H^*_\mu, \qquad dH = 0. \tag{4}$$

The classical equations of motion are automatically satisfied once Eq.(4) is solved. The Eq.(4) is a generalized (anti) self-duality equation between the dilaton and the Kalb-Ramond gauge fields.

There are several ways to understand the generalized self-duality condition Eq.(4) by drawing analogies to the familiar self-duality condition of four-dimensional Yang-Mills gauge theory. First, the dilaton field dD is a one-form while the Kalb-Ramond gauge field H is a three-form. In four dimensions, they can be dual-transformed into each other through the generalized self-duality Eq.(4). Second, the dilaton field is a scalar field while the Kalb-Ramond field in four-dimensions is dual to the pseudoscalar axion field A. The interchange of a scalar field and a pseudoscalar field thus defines the generalized self-duality Eq.(4).

It is easy to find the following $SO(3)$ symmetric solutions of Eq.(4)

$$ds^2 = d\tau^2 + \tau^2 d\Omega_3^2,$$

$$\exp(-\beta D)(\tau) = A + \frac{|Q|}{2\pi\tau^2}, \qquad \text{and} \qquad H_{abc} = \epsilon_{abc}\frac{Q}{\pi\tau^3} \tag{5}$$

in which τ denotes the radius of three-sphere Ω_3. There are two classes of instantons: type I in which $A = 1$ and type II in which $A = 0$, depending on the boundary conditions of the dilaton field at spacetime infinity. Geometrically, the solution describes a Kalb-Ramond 'magnetic' charge Q sitting at the origin and the coupling constant e^{-D} increasing near the center of the instanton. Consistent string propagation on a nontrivial Kalb-Ramond field background restricts the charge Q to be interger-valued (stringy Dirac quantization condition) [12]. If the instanton comes from the moduli field Z_T, the six-dimensional internal space decompactifies as one moves outward from the center of the instanton.

Furthermore, the energy momentum tensor vanishes identically when the above self-duality condition is met. Thus, the above solution Eq.(5) remains exact even after gravity is turned on. One also finds multi-instanton solutions by linear superpositions of Eq.(5)

$$\exp(-\beta D) = A + \sum_{a=1}^{N} \frac{Q_a/2\pi}{(x - x_a)^2}, \qquad \text{and} \qquad H_{abc} = \sum_{a=1}^{N} \epsilon_{\mu\nu\lambda\sigma} \frac{Q_a}{\pi} \frac{(x - x_a)^\sigma}{(x - x_a)^4}. \tag{6}$$

The underlying physics is analogous to the Prasad-Sommerfield limit of non-abelian magnetic monopoles: the dilaton mediated attractive force is balanced by the Kalb-Ramond field mediated repulsive force. In fact, this can be seen as a saturation of the Bogomolnyi bound [13] that comes out naturally in the present context due to the underlying supersymmetry of the Lagrangian Eq. (2).

Locally, the metric is completely flat. However, the global topology of the instanton is $R \times S_3$, since the singular origin $\tau = 0$ is removed. Therefore, the Euler characteristic of the instanton is $\chi_E = 0$. (An analogous example is Euclideanized Rindler spacetime interpreted as a gravitational instanton [14].) Indeed, the axionic instantons are characterized by two topological data: the Kalb-Ramond 'magnetic' charge Q and the Euler characteristic χ_E. The vacuum configuration corresponds to $|Q=0, \chi_E=1>$, while the axionic instantons to $|Q \in \mathbf{Z}, \chi_E \neq 1>$. In fact, axionic spacetime instantons are different from most of the known gravitational instantons for two reasons. First, our axionic instantons are globally Euclidean, while the gravitational instantons are only locally Euclidean at asymptotic infinity. Second, since axionic instantons carry Kalb-Ramond 'magnetic' charges that are additive, the cluster decomposition argument requires one should include the instanton anti-instanton pairs too. Thus, isolated instanton backgrounds must be taken into account also. This is not the case for the gravitational instantons, which leaves their physical interpretations unclear.

For both types of instantons, the action of a single instanton is calculated to be

$$S_I = \frac{2\pi}{\beta} Q[e^{\beta D(\infty)} + i\beta A(\infty)] \, . \tag{7}$$

This is the action required to paste an instanton to otherwise vacuum spacetime with the dilaton and the axion expectation values $D(\infty)$ and $A(\infty)$ respectively. One also finds that there is no contribution to the action from the core region of the instanton. This suggests that many physical features of the axionic instanton are insensitive to its small scale details. In other words, the singularity of the solution Eq.(5) is 'soft' enough.

One can find out fermionic zero modes by applying supersymmetry transformations to the instanton solution Eq.(5). The dilatino supersymmetry transformation in the axionic instanton background reads

$$\delta\lambda = [\gamma^\mu \nabla_\mu D + e^{\beta D} H_{\mu\nu\lambda}\gamma^{\mu\nu\lambda}]\epsilon(x) = (\gamma^\mu \nabla_\mu D)[1 \pm \gamma_5]\epsilon(x). \tag{8}$$

In the second equality, we have used the self-duality condition Eq(4). Thus, in the axionic instanton background, we find the supersymmetry transformation annihilates half of the states and gives rise to only two, chiral dilatino zero modes. Again, this is in complete analogy with supersymmetric Yang-Mills theory in which the supersymmetry transformation of gaugino χ reads

$$\delta\chi^a = F^a_{\mu\nu}\sigma^{\mu\nu}\epsilon = \frac{1}{2}F^a_{\mu\nu}\sigma^{\mu\nu}[1 \pm \gamma_5]\epsilon. \tag{9}$$

Due to the self-duality condition $F = \pm F^*$, there are two chiral gaugino zero modes.

Similarly, from the gravitino supersymmetry transformation,

$$\delta\psi_\mu = [\nabla_\mu - \frac{1}{2\beta}e^{\beta D} H_{\mu\nu\lambda}\gamma^{\nu\lambda}]\epsilon(x), \tag{10}$$

one finds that the nontrivial Kalb-Ramond background field provides a spacetime torsion, so that the new covariant derivative is defined with respect to a connection $\omega - \frac{1}{\beta}H$. In the axionic instanton background, one finds that there exist two, covariantly constant, chiral spinors that approach a constant at spacetime infinity. The supersymmetry transformation annihilates half of the gravitino states once the self-duality condition is met. Again, there are only two chiral, gravitino zero modes.

With the instanton action Eq.(7), the local operators that are induced in the low-energy effective Lagrangian read

$$\sum_{a=S,T} \int d^4x\, J_a \prod_i \mathcal{F}_a^{(i)} \exp(-\frac{2\pi}{\beta_a} Q_a Z_a) + \text{h.c.} \in S_{\text{eff}}. \tag{11}$$

Here, the integration over the position of the instanton and the Jacobians J_a are provided by the bosonic zero modes through a change of variables to the instanton collective coordinates. Local fermionic operators $\mathcal{F}_a^{(i)}$ are needed to soak up the fermionic zero modes. This is the term that should come from an instanton induced superpotential. If one approximates the low-energy effective Lagrangian to be globally supersymmetric, the superpotential induced by the axionic instantons is approximately

$$\mathcal{W} \approx \sum_{a=S,T} C_a \exp(-\frac{2\pi}{\beta_a} Q_a Z_a). \tag{12}$$

The calculable, numerical pre-exponential coefficients are denoted by C_a.

3. (Super)-Conformal Field Theory of The Axionic String Instanton

In collaboration with G.T. Horowitz [10], we have also found a (super) conformal field theory, whose leading gradient approximation reduces to the type II axionic instanton.

Let us assume that Euclidean string theory is described by a Wick rotation of the string time coordinate. Exact classical solutions of Euclidean string theory should correspond to some conformal field theory.

We found a relevant conformal field theory to the axionic instanton is a direct product of Feigen-Fuks Coulomb gas with an imaginary background charge q and $SU(2)$ Wess-Zumino-Witten (WZW) model at level $Q \in \mathbf{Z}$. Their central charges are $C_1 = 1 + 12q^2$ and $C_3 = \frac{3Q}{Q+2}$ respectively. (The Minkowski version of this theory has been investigated by Antoniadis *et al.*, and interpreted as an expanding Universe [2].) This conformal field theory is assumed to be coupled to some internal, $c = 9$, (2,2) superconformal field theories. Cancellation of the total central charge is achieved if and only if

$$12q^2 - \frac{6}{Q+2} = 0. \tag{13}$$

Since the worldsheet fermions are coupled free, the same condition Eq.(13) applies for both left and right moving sectors. In fact, Eq.(13) holds for bosonic, heterotic and type II superstring theories.

One can interpret these conformal field theories geometrically in terms of a nonlinear sigma model of string propagation defined on a Euclidean spacetime and other nontrivial massless background fields. The imaginary charge q of the Feign-Fuks Coulomb gas describes the dilaton background growing linearly in imaginary time, while the WZW model at level Q provides a constant background Kalb-Ramond field on a three-sphere of radius $2Q$. Thus,

$$g_{\mu\nu} = [dX_o^2 + 2Qd\Omega_3^2],$$

$$D(X_o) = -2qX_o \qquad \text{and} \qquad H_{abc} = \epsilon_{abc}Q. \tag{14}$$

The spacetime geometry is cylindrical, along which the dilaton field increases linearly. Furthermore, this geometrical interpretation suggests that Eq.(13) be viewed as a self-duality condition in the context of conformal field theory: the charge q of the Feign-Fuks Coulomb gas is related to the level Q (Kalb-Ramond 'magnetic' monopole charge) of the WZW model.

To relate the conformal field theory to the axionic instanton background Eq.(5) found from the low-energy field theory, we transform Eq.(14) into those of canonical backgrounds with a correctly normalized Einstein-Hilbert action. The results are

$$G_{\mu\nu} \equiv e^D g_{\mu\nu} = [d\tau^2 + C\tau^2 d\Omega_3^2],$$

$$D(\tau) = -2\ln Q\tau \qquad \text{and} \qquad H_{abc} = \epsilon_{abc}Q. \tag{15}$$

The constant C in the metric turns out to be

$$C = 2q^2 Q. \tag{16}$$

Thus, for any finite Q, the canonical spacetime has a deficit angle around the center of the instanton, defined by $\tau = 0$. In other words, the canonical metric has a conical singularity with deficit angle $\delta\Omega \approx \frac{4\pi^2}{Q+2}$. Even though such a deficit angle is harmful to point particle propagation, it is perfectly acceptable to string theory, since the above conformal field theory describes a consistent, exact background on which a string propagates.

In the limit $Q \to \infty$, we find that this conformal field theory approaches the instanton solution Eq.(5) with $A = 0$. Thus, we regard the above conformal field theory as an exact solution of Euclidean string theory to which our type II axionic instanton provides a leading approximation.

Finally, with the above example of nontrivial conformal field theory, we may be able to understand several important issues such as how to define zero modes, collective coordinate quantization, and topological charge in terms of full-fledged string theory / conformal field theory.

4. Low-Energy Implications of the Axionic Instantons

What would be the effects of the axionic string instantons on the low-energy physics?

First, since the instantons carry nonzero axion charges, the axion feels a nonperturbative potential generated by these instantons. In the dual description, the Kalb-Ramond gauge field acquires a nonperturbative mass gap. The heterotic string is known to be an axionic string [14]. Thus the instantons will force the four-dimensional strings to be confined by forming axionic domain walls, but with nontrivial dilaton field configuration [4]. The wall thickness is around the compactification scale, which is again around the Planck scale.

More interestingly, the instantons will induce nonperturbative superpotentials in the effective Lagrangian. We have derived them in Eq.(12). Since the potential depends upon the dilaton and axion fields, the nonrenormalization theorem of the superpotential [8] is violated. Unfortunately, the potential vanishes exponentially as the dilaton field runs off to infinity and thus appears unrealistic.

However, there may be a way to generate a nontrivial minimum of the dilaton potential in the weak coupling regime, $<D>\gg 1$. One possibility is through combined dilatino and gluino condensations induced by axionic instantons and Yang-Mills instantons from the hidden E_8' sector of compactified heterotic string theory. In that case, the superpotential in the low-energy effective Lagrangian induced by these nonperturbative effects reads

$$\mathcal{W} \approx C_d \exp(-2\pi Z_s) + C_g S \exp(-\frac{3}{2\beta_o} Z_s). \tag{17}$$

The complex-valued coefficients C_d and C_g denote pre-exponential factors, that are calculable. The β_o is the one-loop beta function of a subgroup $G \in E_8'$. Their relative phase generically takes arbitrary value. A nontrivial minimum of the dilaton potential will arise if and only if

$$\frac{D\mathcal{W}}{DZ_s} \equiv [\frac{\partial}{\partial Z_s} + \frac{\partial K}{\partial Z_s}]\mathcal{W} = 0 \tag{18}$$

has a nontrivial solution in the weak coupling regime. After a short calculation [7], one can see such a solution exists once suitable conditions on the coefficients

$\frac{C_d}{C_g}$ are met: $C_d/C_g \gg 1$ and $\arg(C_d/C_g) \approx \pi$. The dilaton settles down to the minima of the potential and the dynamical supersymmetry breaking is triggered. A complete investigation of the proposed mechanism, however, is yet to be made.

5. Noncritical String Interpretations

The dilaton background with linear Euclidean time dependence in the axionic instanton and the conformal field theory appears to be rather ad hoc. However, we would like to show that this particular background is quite natural from the point of view of noncritical string theory (two-dimensional quantum gravity coupled to matter) in dimension $D < 25$.

The two-dimensional quantum gravity coupled to a conformally invariant matter of central charge D is described by a Liouville theory [15]

$$S_L = \frac{1}{8\pi}\int d^2\sigma\sqrt{\hat{g}}[\hat{g}^{ab}\partial_a\rho\partial_b\rho + \mathcal{R}^{(2)}2q\rho + \lambda e^{\alpha\rho}]. \tag{19}$$

A freedom in splitting the worldsheet metric into a background metric and a fluctuating conformal factor $g_{ab} = e^{\sigma}\hat{g}_{ab}$ fixes uniquely

$$q^2 = (25 - D)/12. \tag{20}$$

If we imagine that the coupled matter consists of a direct product of a $C_{int} = 22$ internal conformal field theory and the $SU(2)$ WZW model at level Q, we find that Eq.(20) becomes identical to the self-duality condition Eq.(13). Therefore, the two-dimensional curvature coupling to the linear ρ field in Eq.(19) is quite a natural consequence from the two-dimensional gravity point of view.

Futhermore, it has been observed [16, 17] that the ρ field plays a role of dynamically generated 'Minkowski time' in the critical string theory as long as $D > 25$. From this point of view, Eq.(20) follows by demanding cancellation of total central charge of the ghosts, the conformal factor ρ, and the matter fields. Our discussions in the previous paragraph suggest that the 'Euclidean time' of the critical string theory is a consequence of $D < 25$ matter coupling to the two-dimensional quantum gravity. In particular, if part of the matter is an $SU(2)$ WZW model at level Q, the two-dimensional quantum gravity formally conicides with the axionic string instantons. The transition from the 'Euclidean' string theory to the 'Minkowskian' one occurs across the critical point, $q = 0$.

This new interpretation has its own problem, however. Since the matter central charge $1 \ll D = 25 - 6/(Q+2) < 25$, the exponent α becomes complex-valued,

$$\alpha = \alpha_r + i\alpha_i = \frac{1}{2\sqrt{3}}[\sqrt{25 - D} + i\sqrt{D - 1}]. \tag{21}$$

One may try to modify [17] the Lagrangian Eq.(19) so that $e^{\alpha\rho}$ is replaced by $e^{\alpha_r}\cos\alpha_i$. This gives a well-defined, unitary theory of two-dimensional quantum gravity. It remains, however, to be seen whether this modification leads to a universality class relevant to the Euclideanized critical string theory.

6. Acknowledgement

For numerous discussions on the topics, I thank Michael Dine, Jeffrey A. Harvey and, especially, Gary T. Horowitz. For enjoyable atmosphere, I am also grateful to the organizers of the Workshop. This work is supported in part by the National Science Foundation under the Grant PHY-85-06686.

7. References

1. P. Candelas, G.T. Horowitz, A. Strominger and E. Witten, Nucl. Phys. **B258** (1984) 46; D. Gepner, Phys. Lett. **B199** (1987) 370, Nucl. Phys. **B296** (1988) 757.
2. I. Antoniadis, C. Bachas, J. Ellis and D.V. Nanopoulos, CERN-TH 89/5231, to appear in Nucl. Phys. **B** (1989): see also S.P. de Alwis, J. Polchinski and R. Schimmrigk, Phys. Lett. **B218** (1989) 449.
3. A. Dabholkar and J.A. Harvey, Phys. Rev. Lett. **63** (1989) 478; B. Greene, A. Shapere, C. Vafa and S.T. Yau, *Stringy Cosmic Strings and Noncompact Calabi-Yau Manifolds*, HUTP 89/A047 preprint (1989).
4. S.-J. Rey and C. Pryor, *Dilaton-Axion Electrodynamics in Superstring Theory*, UCSB-TH 89/45 preprint (1989).
5. G.T. Horowitz and A.Steif, *Spacetime Singularity in String Theory*, UCSB-TH 89/43 preprint (1989).
6. S.-J. Rey, *String Pair Production in Axion Background*, UCSB-TH 89/52 preprint (1989), to appear.
7. S.-J. Rey, *Confining Phase of Superstrings and Axionic Strings*,UCSB-TH-89/23 preprint (1989).
8. M. Dine and N. Seiberg, Phys. Rev. Lett. **B57** (1986) 2625.
9. E. Witten, Phys. Lett. **B153** (1984) 243.
10. G.T. Horowitz and S.J. Rey, *String Theory Instantons with Nontrivial Spacetime Topology*, UCSB-TH 89/53 preprint (1989), to appear.
11. E. Witten, Phys. Lett. **B155** (1984) 151.
12. R. Rohm and E. Witten, Ann. Phys. **170** (1986) 454.

13. E.B. Bogomolnyi, Sov. J. Nucl. Phys. **24** (1976) 449; S. Coleman *et al.*, Phys. Rev. **D15** (1977) 544.

14. S.M. Christensen and M.J. Duff, Nucl. Phys. **B146** (1978) 11.

15. A.M. Polyakov, Mod. Phys. Lett. **A2** (1987) 893; V.G. Knizhnik, A.M. Polyakov and A.A. Zamolodchikov, Mod. Phys. Lett. **A3** (1988) 819; F. David, Mod. Phys. Lett. **A3** (1988) 1651; J. Distler and H. Kawai, Nucl. Phys. **B321** (1989) 509.

16. F. David and E. Guitter, Europhys. Lett. **3** (1987) 1169, Nucl. Phys. **B293** (1988) 332; S.R. Das, S. Naik and S.R. Wadia, Mod. Phys. Lett. **A4** (1989) 1033; T. Banks and J. Lykken, *String Theory and Two-Dimensional Quantum Gravity*, UCSC-SCIPP 89/19 preprint (1989).

17. J. Polchinski, *A Two-Dimensional Model of Quantum Gravity*, Univ. of Texas UTTG-02-89 preprint (1989).

W–ALGEBRAS, W–STRINGS, and QUANTUM GROUPS

Jean-Loup GERVAIS

Laboratoire de Physique Théorique de l'Ecole Normale Supérieure
24, rue Lhomond - 75231 PARIS Cédex 05 - FRANCE

ABSTRACT

Recent developments related to non-linear extensions of the Virasoro algebras (W–algebras) are reviewed. The critical properties of the W–strings are first displayed, together with their cohomology properties. The intimate relationship between W–algebras and Toda field theories is next discussed, and this allows us to establish the connection between the former and quantum groups.

1. INTRODUCTION

New types of infinite-dimensional algebras have been discovered [1 – 8] in the current developments of two-dimensional critical models. These, so called W–algebras, are non-linear extensions of the Virasoro algebra in the sense that, although the latter algebra is part of them, the commutation relations do not close in the usual way: Commutators may be reexpressed only in terms of symmetric non-linear polynomials of the generators. W–strings, discussed by Bilal and myself [9], are new types of string theories which have W–algebras as world sheet symmetries. Although one is unable to construct them explicitly at the present time, section 2 shows, following ref. 9, that they exist and enjoy strikingly novel properties since their intercept is larger than one (two) for open (closed) strings. Next, in section 3, I explain how the W–algebras can be sytematically constructed by quantizing the Toda theories associated with ordinary (ungraded) simple Lie algebras. Thus the W–strings and W–algebras are classified in parallel with these simple Lie algebras, the Veneziano model being asssociated with $A_1 = SU(2)$. Finally section 4 displays the connection between the chiral algebra of the quantum Toda theories just mentioned and a quantum deformation of the associated simple algebra. This establishes a link between W–strings and quantum group which, I believe, will be intrumental in the new string scattering amplitudes.

2. BASIC FEATURES OF THE W STRINGS

2.1 Basic features of the purely bosonic strings

As a preparation for the discussion of W-strings, it is useful to first recapitulate basic features of string theories, related to the Virasoro algebra, on the example of the purely bosonic strings.

Properties A : The critical central charge and intercept

In the conformal gauge, the physical states are specified by the conditions

$$(L_0 - \epsilon_0)|\text{phys} >= 0; \qquad L_n|\text{phys} >= 0, \quad n > 0. \tag{2.1}$$

These so-called ghost-killing conditions remove two degrees of freedom iff the Virasoro central charge C and the intercept ϵ_0 are chosen as

$$C = C_{crit} \equiv 26, \qquad \epsilon_0 = 1. \tag{2.2}$$

It is well known that the choice (2.2) is the only one leading to consistent bosonic string theories. The special properties of the strings at these values can be seen in several ways:

A.1 The appearance of spurious states (see e.g. ref. 10): One may discuss them, in a model-independent way, by only using the representation theory of the Virasoro algebra. Denote by Δ the eigenvalues of L_0. At level $R = 1$ in the string Hilbert space there is a state:

$$|\Delta = 1\,,\, R = 1 >= L_{-1}|\Delta = 0\,,\, R = 0 >, \tag{2.3}$$

which satisfies condition (2.1) with $\epsilon_0 = 1$. It is thus physical and of vanishing norm. At the next level, $R = 2$, the critical values (2.2) are such that there is another similar physical state

$$|\Delta = 1\,,\, R = 2 >= (L_{-2} + \frac{3}{2}L_{-1}^2)|\Delta = -1\,,\, R = 0 > . \tag{2.4}$$

This goes on at higher levels. Thus condition (2.2) is such that states appear which are physical, that is satisfy condition (2.1), and are generated by powers of the L_{-n}'s. Because of this last property, they have a vanishing norm and decouple from the physical S-matrix, even though they are physical. They were called spurious states in the early days of string theory. If (2.2) holds, the decoupling from the S-matrix becomes precisely such that th number of degrees of freedom is reduced by two at every level. In particular, for the Veneziano model, that is, for the trivial Minkowski background metric, this is the correct counting since the value $\epsilon_0 = 1$ leads to massless spin-one and spin-two particles.

Since this argument was first given long ago, it has become possible to discuss the existence of spurious states abstractly by means of Kac's formula [11]. Consider, in general, the Hilbert space generated by the Virasoro generators L_{-n}, $n > 0$ applied to a highest-weight vector with weight Δ such that $(L_0 - \Delta)|\ \Delta\ ,\ 0 >= 0$, $L_n|\ \Delta\ ,\ 0 >= 0$, $n > 0$. Kac's theorem concerns the vanishing of the determinant of the corresponding metric (inner product matrix). Define

$$\Delta(r,s;C) = \frac{C - C_0}{24} - \frac{1}{96}\left((r+s)\sqrt{C - C_0} + (r-s)\sqrt{C - C_1}\right)^2$$

$$\text{with} \qquad C_0 = 1, \quad C_1 = 25. \tag{2.5}$$

If $\Delta = \Delta(r,s;C)$ for some positive r and s, Kac's determinant vanishes at levels larger than or equal to $r \times s$. The above particular form of Kac's formula was given in ref. 12. One may discuss the spurious states of the string directly from Kac's formula. Indeed a Kac's zero corresponds to a state, which is generated by the $L_{(-)}$'s and is annihilated by the $L_{(+)}$'s. Consider now the corresponding states obtained from the Virasoro generators of the string theory. Since the representation is reducible, these states do not vanish. Those which are eigenstates of L_0 with eigenvalue one, are the spurious states we just discussed. For instance, the values of Δ on the right hand side of equations (2.3,4) for the first two spurious states are given by $\Delta(1,1;26)$ and $\Delta(1,2;26)$, respectively. This procedure will straightforwardly extend to W-strings.

A.2 The Brink-Nielsen argument[13]: For the Veneziano model, the intercept ϵ_0 can be interpreted as the regularized sum of the zero-point (Casimir) energies of the string harmonic oscillators. At $D = D_c = C_{crit}$ space-time dimensions, there are $D_c - 2$ degrees of freedom and the Casimir energy is

$$(D_c - 2)\frac{1}{2}\sum_1^\infty n = -\frac{D_c - 2}{24} = -\epsilon_0 \tag{2.6}$$

in agreement with eq. (2.2).

A.3 Modular invariance: For the Veneziano model, the partition function of the closed string sector, with $D-2$ physical free bosonic degrees of freedom, is modular invariant iff equation (2.2) holds. This modular invariance of course works as well in the more complicated cases, for instance, after compactification of certain directions, or when the Liouville mode is included [14].

A.4 Mirror properties of Kac's formula and fractal gravity: It is useful to remark that

$$C_{crit} = C_0 + C_1 . \tag{2.7}$$

Then one verifies the following remarkable "mirror" property:

$$\Delta(r, s; C) + \Delta(r, -s; C_{crit} - C) = \frac{C_1 - C_0}{24} = 1 = \epsilon_0 . \tag{2.8}$$

As is well known, the Kac formula (2.5) plays a central role in certain classes of conformally-invariant field theories in two dimensions since for suitably restricted values of r, s, and C, it gives the weights of the primary fields. For $C < 1$ (see ref. 15), and $C > 25$ (see refs. 16, 17), the rational conformal theories have central charges

$$C_{p,p'} = 1 - 6(p-p')^2/pp' , \tag{2.9}$$

for integer p and p'. This formula also enjoys a mirror property:

$$C_{p,p'} + C_{p,-p'} = C_{crit} . \tag{2.10}$$

For $C < 1$, (resp $C > 1$) the conformal weights are given by (2.5) with $r > 0$, $s > 0$ (resp $r > 0$, $s < 0$). $C > 1$ is the natural region for the Liouville theory [16] (where its Planck constant is positive), which describes the quantum behavior of the Weyl component of the world-sheet metric. The mirror properties (2.8,10) thus show that (2.1,2) are natural conditions for the coupling of rational theories with $C < 1$ to the world-sheet metric, the Virasoro generators L_n being given by the direct sum. In this way, critical systems on a random lattice (fractal gravity [18]) appear as generalizations of str theories in the conformal gauge.

Properties B : String field cohomology

B.1 The nilpotency of the BRST operator: Kato and Ogawa [19] have shown that there exists a nilpotent BRST operator, $Q^2 = 0$, iff equation (2.2) holds. This is the modern powerful way to recover property A.1 and to show that eq. (2.2) is necessary for the consistency of the string quantization.

B.2 The ghost system: The Virasoro generators being fields of conformal spin 2 (up to the central term) they are associated with a ghost pair of spin 2 and -1. In general, such a ghost pair of spin j and $1-j$ has a central charge [20]

$$C_{gh}(j) = -2\left(1 - 6j(1-j)\right), \tag{2.11}$$

and, in particular,

$$C_{gh}(2) = -26. \tag{2.12}$$

Thus the total Virasoro generators, $L_n + l_n^{gh}$, have vanishing central charge if the L_n's have the critical central charge, $C = C_{crit} = -C_{gh}$. Moreover, when the ghosts are included, there must be no violation of conformal invariance. This relates the intercept ϵ_0 to the ghost number of the physical states. Bosonizing the ghost system, the ghost Virasoro generators read

$$l_n^{gh} = \frac{1}{2}\sum_m : a_m a_{n-m} : + \sqrt{2}\alpha_0 n a_n - \alpha_0^2 \delta_{n,0}, \qquad \text{with} \quad \sqrt{2}\alpha_0 = \frac{3}{2}, \tag{2.13}$$

where the a_n are harmonic oscillators ($[a_n, a_m] = n\delta_{n,-m}$). The Virasoro generators (2.13) are those of the Coulomb gas formalism [21], or equivalently of the quantum Liouville theory [16] corresponding to a Planck constant of the Liouville system equal to $\hbar = -1/8\alpha_0^2$. Using eq. (2.13), the mass-shell condition (2.1) can be rewritten as

$$(L_n + l_n^{gh})\big|\text{phys} \otimes \text{ghost}\big\rangle = 0 \quad \forall\, n \geq 0, \tag{2.14}$$

since the ghost component of the physical states is such that

$$a_0\big|\text{phys} \otimes \text{ghost}\big\rangle = -\frac{1}{2}\big|\text{phys} \otimes \text{ghost}\big\rangle. \tag{2.15}$$

Hence the intercept ϵ_0 is automatically contained in the l_n^{gh}'s:

$$l_0^{gh}\big|\text{phys} \otimes \text{ghost}\big\rangle = -\epsilon_0\big|\text{phys} \otimes \text{ghost}\big\rangle. \tag{2.16}$$

B.3 The Chern-Simon-Witten action: The ghost-number of the ghost vacuum, eq. (2.15) is related to the grading $-1/2$ associated with the string field A in Witten's string field theory [22] with action

$$I = \int \left(A \star QA + \frac{2}{3} A \star A \star A\right); \tag{2.17}$$

and the value of $-\sqrt{2}\alpha_0$, eq. (2.13), is connected with the grading $-3/2$ of the integration procedure. This may be seen [23] as follows. The overlap δ-functional describing the integration procedure for the ordinary string is given by

$$I[X(.)] \equiv \prod_{0\leq\sigma\leq\pi/2} \delta\big(X(\sigma) - X(\pi-\sigma)\big). \tag{2.18}$$

Writing this functional in the form of a ket in the string Fock space, the authors of ref. 23 show that, in D dimensions,

$$K_{2n}|I\rangle = \frac{n}{2}(-1)^n(-D)|I\rangle, \qquad K_{2n} \equiv L_{2n} - L_{-2n}. \tag{2.19}$$

The ghost part of the integration is defined as

$$I^{gh} \equiv \exp\big(-i\frac{3}{2}\varphi(\frac{\pi}{2})\big)\prod_\sigma \delta(\cdots),$$

where the last term is a δ-functional overlap similar to (2.18), and φ is the ghost field. Thanks to the mid-point insertion, the ghost integration functional satisfies

$$K^{gh}_{2n}|I^{gh}\rangle = \frac{n}{2}(-1)^n 26|I^{gh}\rangle, \qquad K^{gh}_{2n} \equiv l^{gh}_{2n} - l^{gh}_{-2n}. \tag{2.20}$$

If we combine (2.19) and (2.20), we see that the total K operator gives zero at the critical dimension $D = 26$. The coefficient $-\frac{3}{2}$ of the mid-point insertion gives the grading $-\frac{3}{2}$ back for the integration procedure of (2.17). We have generalized this argument to the Liouville string [24], showing that it is not limited to the Veneziano model considered in ref. 23. This cancellation of anomalies only depends upon the criticality of the central charge.

2.2 The W-strings

The W-algebras: The W-algebras are non-linear extensions of the Virasoro algebra [1 – 8]. In general, every W-algebra is associated with a simple Lie algebra g, and for each independent Casimir invariant of g of order k there is a conformal spin k generator $(W_{(k)})_n$ in the W-algebra. These higher-spin generators can be very systematically obtained either by the GKO coset construction [2] based on the Kac-Moody algebra $\hat{g}$, for the generalized minimal series, or from the Toda theories [3,4] associated with g. This last method will be reviewed in section 3. If $g = A_N (= SU_{N+1})$ e.g. there are generators of spin 2 (Virasoro), $3, 4, \ldots N+1$.

The simplest example, corresponding to SU_3, is Zamolodchikov's spin 3 algebra for the operators $W_{(2)} \equiv L W_{(3)} \equiv W$:

$$[L_n, L_m] = (n-m)L_{n+m} + \frac{C}{12}n(n^2-1)\delta_{n,-m} \tag{2.21}$$

$$[L_n, W_m] = (2n-m)W_{n+m} \tag{2.22}$$

$$\begin{aligned}[W_n, W_m] = {} & \beta(n-m)Z_{n+m} + \frac{C}{360}n(n^2-1)(n^2-4)\delta_{n,-m} \\ & + (n-m)\Big(\frac{1}{15}(n+m+2)(n+m+3) \\ & - \frac{1}{6}(n+2)(m+2)\Big)L_{n+m}\end{aligned} \tag{2.23}$$

$$\text{where} \quad Z_n = \sum_m : L_m L_{n-m} : -\frac{1}{20}(n^2-4-5\rho_n)L_n \tag{2.24}$$

$$\beta = \frac{16}{22+5C}, \quad \rho_n = n \bmod 2. \tag{2.25}$$

More complicated algebras have been studied corresponding to arbitrary simply-laced [1, 2] g and simply- or non-simply-laced [3, 4] g. In the following we will restrict ourselves to the case of simply-laced g.

The ghost system: In a W-string theory where the W-algebra (e.g. eqs. (2.21-5) for $g = SU_3$) replaces the Virasoro algebra, a ghost pair of conformal spin k and $1-k$ should be associated with each spin k generator. The contribution of each pair to the central charge C is $C_{gh}(k)$, see eq. (2.11), and the total ghost central charge is

$$C_{gh} = -2\sum_k (1 - 6k(1-k)), \tag{2.26}$$

where k runs over the orders of the Casimir invariants of g ($k = e+1$ where e runs over the exponents of g). In ref. 2 it was remarked that, by eq. (2.26), $-C_{gh} \equiv C_{crit}$ equals

$$C_{crit} \equiv -C_{gh} = N + N(1+2h_c)^2. \tag{2.27}$$

Here h_c is the Coxeter number of g : $N+1$ for $A_N = SU_{N+1}$, $2N-2$ for $D_N = SO_{2N}$, 12, 18 and 30 for E_6, E_7 and E_8. (N always denotes the rank of g.)

The generalized Kac formula: The value (2.27) of the critical central charge is also confirmed by the generalization of the Kac formula (2.5). In the present general scheme, it corresponds to $g = SU_2$ For arbitrary simply-laced g, it is replaced by:

$$\Delta(\tilde{r}, \tilde{s}; C) = \frac{C - C_0}{24} - \frac{1}{8Nh_c(h_c+1)}$$

$$\times \sum_{p,q} \mathcal{G}_{pq} \left((r_p + s_p)\sqrt{C - C_0} + (r_p - s_p)\sqrt{C - C_1} \right)$$

$$\times \left((r_q + s_q)\sqrt{C - C_0} + (r_q - s_q)\sqrt{C - C_1} \right). \tag{2.28}$$

$\mathcal{G}$ denotes the inverse of the Cartan matrix of g, $\widetilde{r} = (r_1, \ldots r_N)$. The r_p's and the s_p's are integers. The above formula was derived in refs. 1,2 for the rational cases. We arrived at the present general form in refs. 3,4. The latter articles also deal with non-simply-laced g. The values of C_0 and C_1 are

$$C_0 = N, \quad C_1 = N(1 + 2h_c)^2. \tag{2.29}$$

By analogy with eq. (2.7) we expect that C_{crit} should be given by

$$C_{crit} = C_0 + C_1, \tag{2.30}$$

which indeed gives the value (2.27) back.

In section 2.1 we remarked that the intercept ϵ_0 of the ordinary bosonic string may be derived from the mirror property of Kac's formula, (property A.4). In the present case, the generalized Kac formula (2.28) does satisfy a generalization of (2.8)

$$\Delta(\widetilde{r}, \widetilde{s}; C) + \Delta(\widetilde{r}, -\widetilde{s}; C_{crit} - C) = \frac{C_1 - C_0}{24}. \tag{2.31}$$

This indicates that the intercept ϵ_0 should be given by

$$\epsilon_0 = \frac{C_1 - C_0}{24} = \frac{N h_c(h_c + 1)}{6}. \tag{2.32}$$

Note that in all cases, this is an integer.

As already remarked, the W-algebra generators of a combined system (e.g. physical + ghosts) are not a priori given by the sum of the individual generators. This is due to the non-linearity of the algebra, see, e.g. eqs. (2.23,4). However, the mirror formula (2.31) indicates that at least L_0 should be simply given by the sum of the L_0's of the individual systems. Moreover, since the central charges of (2.31) add up to C_{crit}, the additivity of the generators seems to hold for all L_n's as well. On the other hand, the fact that $C_{crit} + C_{gh} = 0$ indicates that this is also true when one puts a "physical" system and the ghosts together. This additivity of the L_n's is to be expected since the Virasoro *sub*algebra is *linear*.

Let us mention that the series of central charges for the rational conformal theories (W-generalized BPZ series or Z_N series),

$$C_{p,p'} = N \left(1 - \frac{h_c(h_c + 1)(p - p')^2}{pp'} \right), \quad p, p' \epsilon Z, \tag{2.33}$$

still satisfy the mirror property (2.10) :

$$C_{p,p'} + C_{p,-p'} = C_{crit}\,, \tag{2.34}$$

supporting again the additivity of the L_n's. The two mirror properties (2.31) and (2.34) should be interpreted as allowing a consistent coupling of a (W-) minimal model theory (2.33) to the quantum fluctuations of the $2d$ world-sheet or to a random lattice ("fractal W-gravity"). This shows that the Toda theories should play the role of a generalized world-sheet metric for the W-strings.

The nilpotency of the BRST operator: The values (2.30) and (2.32) for the critical central charge and the intercept have been confirmed for $g = SU_3$: Thierry-Mieg [8] has shown that there exists a nilpotent BRST operator for the algebra (2.21-25) if and only if $C = C_{crit} = 100$ and $\epsilon_0 = 4$.

The appearance of spurious states: By analogy with the usual situation, we expect that a physical W-string state obeys the ghost-killing conditions

$$(L_0 - \epsilon_0)|\text{phys}>= 0, \qquad L_n|\text{phys}>= 0, \quad n > 0$$

$$(W_{(k)})_n|\text{phys}>= 0, \quad n > 0, \tag{2.35}$$

where $W_{(k)}$ are the higher spin generators ($W_{(3)} \sim W$ for $g = SU_3$). Conditions (2.35) just mean that a physical state is a highest-weight state of the W-algebra with L_0 eigenvalue ϵ_0. The W-string Fock space will provide a representation of this algebra which remains to be constructed. Again, the criticality of the central charge should manifest itself by the appearance of spurious states. Now they should obey (2.35) and be created from a highest weight state $|\Delta, 0>$ by application of the L_{-n}'s or $(W_{(k)})_{-n}$'s. The same reasoning as in A.1 shows that we may foresee what will happen directly from the generalized Kac formula (2.28). For the standard case, the signal of criticality was that $\Delta(1, 2; 26) = -1$ differs from $\epsilon_0 = 1$ by minus two so that a spurious state appears at the second level. For W-algebras there is a similar phenomenon, namely, formula (2.28) leads to

$$\Delta(\vec{r}^{(j)}, \vec{s}^{(j)}; C_{crit}) = -2\sum_q \mathcal{G}_{jq}, \qquad \text{for} \quad r_q^{(j)} = 1, \quad s_q^{(j)} = 1 + \delta_{q,j}\,.$$

Since $2\sum_q \mathcal{G}_{jq}$ and ϵ_0 are both integers, there is again the possibility that spurious states appear at the levels $\epsilon_0 + 2\sum_q \mathcal{G}_{jq}$, confirming that (2.30,32) are indeed the critical values for the W-strings.

The Brink-Nielsen argument and modular invariance: Trying to generalize the Brink-Nielsen argument [13], one immediatly sees that the critical centra charge C_{crit} (2.30) and the intercept ϵ_0 (2.32) are related by

$$\frac{C_{crit}-2N}{24}=\frac{C_1-C_0}{24}=\epsilon_0. \tag{2.36}$$

The l.h.s. can again be interpreted as the zero-point (Casimir) energy of $C_{crit}-2N$ sets of harmonic oscillators. This indicates the decoupling of $2N$ degrees of freedom due to the appearance of spurious states we just noted. This decoupling is certainly necessary for the correct counting o physical degrees of freedom. Concerning modular invariance, equation (2.36) ensures that the partition function

$$\begin{aligned}Z(\tau)\equiv & Tr\exp\left(2\pi i\tau(L_0-\epsilon_0)-2\pi i\bar{\tau}(\overline{L_0}-\epsilon_0)\right)\\ =&\exp\left(4\pi(\epsilon_0-\frac{C-2N}{24})Im\tau\right)\left((Im\tau)^{1/2}|\eta(\tau)|^2\right)^{-(D-2N)}\end{aligned} \tag{2.37}$$

is modular invariant at $C=C_{crit}$. Of course, modular invariance can be achieved in many other ways. For example, part or all of the free fields could be replaced by the Toda fields [3,4] generalizing the situation encountered for the Liouville strings [14,24]. Thus the present argument is certainly more general than the free-field case considered which may be too naive for the W-strings.

The W-string field theories: Let us finally speculate how the W-string field action might look like. In order to do so, we remark that the complete system of the $2N$ ghost fields can be bosonized to yield N bosonic ghost fields. They are equivalent to the N Toda fields [3,4] at a coupling constant which is such that $C_{Toda}=C_{gh}=-C_{crit}$. Each ghost pair of conformal spin k and $1-k$ has its bosonized Virasoro generator given by eq. (2.15):

$$l_n^{gh,k}=\frac{1}{2}\sum_m :a_m^k a_{n-m}^k:+\sqrt{2}\alpha_0(k)na_n^k-(\alpha_0(k))^2\delta_{n,0} \tag{2.38}$$

with α_0 replaced by

$$\sqrt{2}\alpha_0(k)=k-\frac{1}{2}. \tag{2.39}$$

Their contribution to the central charge is $1-24(\alpha_0(k))^2$ which coincides with (2.11). As already discussed in B.2, conformal invariance must be restored when the ghosts are included, and the total L's have to vanish:

$$(L_n+\sum_k l_n^{gh,k})|\text{phys}\otimes\text{ghost}>=0\quad\forall n\geq 0. \tag{2.40}$$

For $n = 0$ this yields

$$\sum_k \left(\frac{1}{2}(a_0^k)^2 - (\alpha_0(k))^2\right) = -\epsilon_0 \quad \Rightarrow \quad \sum_k (a_0^k)^2 = \frac{N}{4} \tag{2.41}$$

which is satisfied if $a_0^k = \pm\frac{1}{2}\ \forall k$.

One thus sees that the intercept is such that the physical part of the string field A has k^{th} ghost number $\pm\frac{1}{2}$. What is the corresponding grading of the integration procedure? This may be seen by generalizing Gross and Jevicki's [23] argument recalled in B.3. It is reasonable to assum the W-string theory will be such that equation (2.19) is replaced by

$$K_{2n}|I\rangle = \frac{n}{2}(-1)^n(-C_{crit})|I\rangle, \qquad K_{2n} \equiv L_{2n} - L_{-2n}. \tag{2.42}$$

It follows from our general calculation of the K-anomaly in Liouville theories [24] that, if we define, for the kth ghost field,

$$I^{(k)} \equiv \exp\big(i\frac{(1-2k)}{2}\varphi^{(k)}(\frac{\pi}{2})\big)\prod_\sigma \delta(\cdots); \tag{2.43}$$

the contribution to the K-anomaly (the coefficient of the $\frac{n}{2}(-1)^n$ term) is equal to $2(1-6k(1-k))$. It thus follows from (2.6,7) that the integration measure $I\prod_k I^{(k)}$ has no K-anomaly. This leads to a grading $\frac{1}{2} - k$ of the integration with respect to the kth ghost field.

The BRST operator (see, e.g., ref. 8) has a total ghost number 1:

$$Q = c_0^{(2)}(L_0 - \epsilon_0) + \sum_{k\geq 3} c_0^{(k)}((W_{(k)})_0 - \epsilon_0^k) + \cdots, \tag{2.44}$$

(where the $c^{(k)}$'s are the fermionic ghosts). Its grading is well-defined only with respect to the total ghost number, and we expect that only the overall grading (total ghost number) should add up to zero in the W-string-field action. What is the grading of A ? Most likely, physical states should correspond to:

$$A \sim b_0^{(2)} A_{phys} \tag{2.45}$$

where A_{phys} satisfies $c_0^k A_{phys} = 0$, and the b's are the antighost fields. The reason is as follows: We want the cohomology condition $QA = 0$ to give $(L_0 - \epsilon_0)A_{phys} = 0$ as the only condition involving zero components of the W-generators. Indeed the other possible conditions $(W_{(k)} - \epsilon_0^k)A_{phys} = 0$, $k > 2$ would involve the W-string momenta raised to powers higher than two, and cannot appear in the equations of motion. With the choice (2.45) the zero-ghost-mode parts of Q shown in (2.44) automatically vanish except for $k = 2$. The overall grading of A is thus given by

$$gr(A) = -\frac{1}{2} + \frac{1}{2} + \frac{1}{2} + \cdots + \frac{1}{2} = \frac{N-2}{2}. \tag{2.46}$$

For the integration procedure we have (restricting ourselves in the following to A_N, for convenience of notation) $gr(\int) = \sum_k(\frac{1}{2} - k) = -N(N+2)/2$. Finally, the field strength $F = QA + A \star A$ must have a well-defined grading. One must have $gr(Q) = gr(A) + gr(\star)$, thus $gr(\star) = 2 - \frac{N}{2}$.

We see that the Chern-Simon action (cf ref. 25)

$$\mathcal{S} = (n+1) \int \int_0^1 dt \big(A(\star tQ\, A + \star t^2 A \star A)^n\big) \tag{2.47}$$

has grading 0, $gr(\mathcal{S}) = 0$, provided

$$n = \frac{N(N+1)}{4} + \frac{1}{2}. \tag{2.48}$$

Of course this is only possible if n comes out as an integer, which is the case for $N = 1$ or 2 mod 4. Then, the action (2.47) yields

$$F \equiv Q\, A + A \star A = 0 \tag{2.49}$$

as a consequence of the W-string field equations of motion.

If eq. (2.48) does not yield an integer value for n, i.e. for $N = 3$ or 4 mod 4, one can instead construct the topological action

$$\mathcal{S} = \int F(\star F)^m \qquad m = \frac{N(N+1)}{4}. \tag{2.50}$$

This type of topological action can be gauge fixed by imposing $\widetilde{F} \sim (\star F)^m$, leading to an action of the type $\int F\widetilde{F}$. In this game, however, one must introduce a metric, contrary to the odd case. Altogether, the cohomologies which appear are similar to the ones of ν-dimensional manifolds with $\nu = 2 + N(N+1)/2$.

2.3 Outlook

Although we are unable at present to carry out the construction of the W-strings, the hints presented here for their existence are compelling. The striking novel feature is that the intercepts are larger than one and would of course probably double if we considered the left-moving and the right-moving modes together. The spectrum cannot be predicted at the present time, but unitarity tells us that Regge trajectories are linear at the tree level. Thus many more tachyons seem to be present. By analogy with the usual strings one hopes that all these unwanted particles will disappear from the appropriate sectors of the W-superstring theories. Moreover, one is led to massless particles with spins larger than one (for the open strings) or two (for the closed ones). Unless they all decouple, this gives a picture of the asymptotic physics where interactions of a novel nature would take place. They would have to turn on as one goes up in energy. This may lead to new interesting physics for the coming very-high-energy accelerators.

3. TODA THEORIES, W–ALGEBRAS, QUANTUM GROUPS

3.1 Conformally invariant Toda theories and W–algebras

Classically, the dynamics is governed by the 2D action

$$S \propto \int d_2\xi \Big(\frac{1}{2}(\vec{\nabla}\phi_a)\mathcal{K}_{ab}(\vec{\nabla}\phi_b) - \sum_a (\prod_b \exp(\mathcal{K}_{ab}\phi_b))\Big). \tag{3.1}$$

For A_N which we choose for simplicity, there are N fields ϕ_a, $a = 1, \cdots, N$. $\mathcal{K}$ is the Cartan matrix. The general classical solution is of the form:

$$exp(-\Phi_1) = \sum_{i=1}^{N+1} \psi_i(x_+)\chi_i(x_-), \; \ldots$$

$$exp(-\Phi_r) = \sum_{i_1, \cdots i_r} \psi_{i_1, \cdots, i_r}(x_+)\, \chi_{i_1, \cdots, i_r}(x_-), \; \ldots$$

where $\psi_{i_1, \cdots, i_r}$ is the Wronskian of $\psi_{i_1}, \psi_{i_2}, \cdots, \psi_{i_r}$. $x_\pm$ are the 2D light cone variables and ψ_i and χ_i are the basic chiral fields. We only discuss the former explicitly. It follows from the Toda field equations that they satisfy the one-variable differential equation

$$\Big(-\frac{d^{N+1}}{(d\sigma)^{N+1}} + \sum_{k=2}^{N+1} W_{(k)} \frac{d^{N-k+1}}{(d\sigma)^{N-k+1}}\Big)\psi_a(\sigma) = 0. \tag{3.2}$$

The fields $W_{(k)}$ have conformal weights k up to central terms. They are the generators of the W–algebra. This comes out as follows. There exist fields $p_k(\sigma)$,

$k = 1, \ldots, N+1$. It follows from the canonical Poisson brackets associated with (3.1) that they obey the free field Poisson bracket structure:

$$\Big\{p_k(\sigma_1), p_l(\sigma_2)\Big\} = 4\pi \mathcal{K}_{kl}\, \delta'(\sigma_1 - \sigma_2). \tag{3.3}$$

$\vec{\alpha}_k$, $k = 1, \ldots, N$ are a set of simple roots (arrows denote vectors in the weight space of A_N). The Cartan matrix is $\mathcal{K}_{kl} = \vec{\alpha_k} \cdot \vec{\alpha_l}$. A standard construction of the simple roots of A_N is to take N+1 orthonormal vectors $\vec{e}_r$ and to let $\vec{\alpha}_r = \vec{e}_r - \vec{e}_{r+1}$, $r = 1, \ldots, N$. $W_{(2)}(\sigma)$ is the stress-energy tensor. In terms of the p–fields it is given by:

$$W_{(2)}(\sigma) \equiv \gamma \sum_{k,l=1}^{N+1} \left(\mathcal{K}^{-1}\right)_{kl} p_k(\sigma)\, p_l(\sigma) + \sqrt{\gamma}\, \frac{N+1}{2} \sum_{l=1}^{N+1} \left(\mathcal{K}^{-1}\right)_{ll} p_l'(\sigma). \tag{3.4}$$

In view of (3.3) the Virasoro generators

$$L_n \equiv \frac{1}{8\pi\gamma} \int_0^{2\pi} d\sigma T(\sigma) e^{in\sigma} + \frac{N(N+1)(N+2)}{48\gamma} \delta_{n,0} \tag{3.5a}$$

satisfy

$$i\Big\{L_m, L_n\Big\} = (m-n) L_{m+n} + \frac{C}{12}\left(m^3 - m\right) \delta_{m,-n}, \tag{3.5b}$$

with $C = N(N+1)(N+2)/2\gamma$. γ is a free parameter that plays the role of a coupling constant. It is convenient to introduce

$$\vec{\Lambda}(\sigma) \equiv \sum_{l=1}^{N} \vec{\lambda}_l\, p_l(\sigma), \tag{3.6}$$

with $\vec{\lambda}_j \cdot \vec{\alpha}_k = \delta_{j,k}$. The $\vec{\lambda}_j$ are a set of fundamental weights of A_N. Equation (3.4) may be rewritten as

$$W_{(2)}(\sigma) \equiv \gamma \vec{\Lambda} \cdot \vec{\Lambda} + +\sqrt{\gamma} \sum_{l=1}^{N+1} \vec{\lambda}_l \cdot \vec{\Lambda}'. \tag{3.7}$$

The expression of the other $W_{(k)}$ fields is derived by reexpressing (3.2) in the form:

$$\left(\frac{d}{d\sigma} - \vec{e}_{N+1} \cdot \vec{\Lambda}\right)\left(\frac{d}{d\sigma} - \vec{e}_N \cdot \vec{\Lambda}\right) \cdots \left(\frac{d}{d\sigma} - \vec{e}_1 \cdot \vec{\Lambda}\right) \psi_a = 0, \tag{3.8}$$

so that

$$W_{(k)} \equiv (-1)^{k+1} \sum_{m_1>\cdots>m_k=1}^{N+1} (\vec{e}_{m_1}\cdot\vec{\Lambda})\,(\vec{e}_{m_2}\cdot\vec{\Lambda})\cdots(\vec{e}_{m_k}\cdot\vec{\Lambda}) + \cdots . \tag{3.9}$$

One may verify from these expressions, and making use of (3.3), that the $W_{(k)}$'s satisfy a Poisson bracket realization of a W–algebra which is such that the Poisson bracket $\left\{W_{(k)}, W_{(l)}\right\}$ is a symmetric polynomial of the $W_{(m)}$'s.

After proper quantization, the corresponding quantum algebra closes in a non linear way. This gives a systematic realisation of the W–algebras such as (2.21-25).

3.2 Exchange algebras and quantum groups

The p–fields are periodic:

$$p_k(\sigma) = \sum_n p_n^{(k)} e^{-in\sigma} \tag{3.10}$$

and it is useful to introduce $\vec{\Lambda}_0 \equiv \sum_{l=1}^N \vec{\lambda}_l\, p_0^{(l)}$. A convenient set of independent solutions of (3.2) is of the form

$$\psi_1(\sigma) = exp\Big(\sqrt{\gamma}\sum_j (\mathcal{K}^{-1})_{1j}\, [q_0^{(j)} + p_0^{(j)}\sigma + i\sum_{n\neq 0} \frac{p_n^{(j)}}{n} e^{-in\sigma}]\Big), \tag{3.11}$$

and $\psi_r(\sigma) = \mathcal{S}_{r+1}(\sigma)\,\psi_1(\sigma)$, where $\mathcal{S}_{r+1}(\sigma)$ are classical screening operators such that $V_r(\sigma+2\pi) = exp\big(2\pi\vec{\Lambda}_0\cdot\vec{\lambda}^{(r)}\big)\, V_r(\sigma)$. The vectors $\vec{\lambda}^{(r)}$ are the weights of the defining representation. $q_0^{(j)}$ and $p_0^{(k)}$ are conjugate dynamical variables.

In the quantum case, eqs. (3.3) and (3.11) become

$$\Big[p_k(\sigma_1), p_l(\sigma_2)\Big] = 4\pi i \mathcal{K}_{kl}\, \delta'(\sigma_1-\sigma_2), \tag{3.12}$$

$$\psi_1(\sigma) =: \exp\Big(\sqrt{\frac{h}{\pi}}\sum_j (\mathcal{K}^{-1})_{1j}\, [q_0^{(j)} + p_0^{(j)}\sigma + i\sum_{n\neq 0} \frac{p_n^{(j)}}{n} e^{-in\sigma}]\Big) :, \tag{3.13}$$

where there are two choices for the number h:

$$h_\pm = \frac{\pi}{4\gamma}\Big(1 - 4\gamma \pm \sqrt{1-8\gamma}\Big) = \frac{2\pi}{C_1 - N}\Big(C - \frac{N+C_1}{2} \pm \sqrt{(C-C_1)(C-N)}\Big) \tag{3.14}$$

and $C_1 = N(2N+3)^2$. The second expression of h follows from the fact that the central charge is given by $C = N + N(N+1)(N+2)/(2\gamma)$. Equations (3.13,14) are such that an operator equation similar to (3.2) holds. The study of its monodromy properties leads[4] to the exchange algebra ($\vec{\varpi} \equiv -i\sqrt{2\pi/h}\,\vec{\Lambda}_0/2$):

$$\psi_j(\sigma)\psi_k(\sigma') = \sum_{\substack{l=1,\dots,N+1\\ m=1,\dots,N+1}} S_{jk}^{lm}(\vec{\varpi})\,\psi_l(\sigma')\,\psi_m(\sigma).$$

$$S_{jj}^{jj} = e^{-ih\epsilon N/(N+1)}, \quad S_{jk}^{jk}(\vec{\varpi}) = \frac{\sin h}{\sin(h\varpi_{jk})} e^{ih\epsilon(1/(N+1)-\varpi_{jk})},$$

$$S_{jk}^{kj}(\vec{\varpi}) = \frac{\sin(h(\varpi_{jk}+1))}{\sin(h\varpi_{jk})} e^{ih\epsilon/(N+1)}, \tag{3.15}$$

where ϵ is the sign of $\sigma - \sigma'$, and the other matrix elements vanish.

Finally, we display the connection with quantum groups by the example of the A_1 Toda (Liouville) theory. Introduce

$$\xi_\alpha = \sum_{i=1,2} u_\alpha^i\,\psi_i \qquad \text{for } \alpha = 1,\, 2, \quad \text{with}$$

$$u_1^1 = u_2^2 = e^{ih\omega/2}; \qquad u_2^1 = e^{-ih(\omega+1)/2}; \qquad u_1^2 = e^{-ih(\omega-1)/2}. \tag{3.16}$$

After some computation, one deduces [7] from (3.15) that the ξ fields so defined satisfy the exchange algebra

$$\xi_\alpha(\sigma)\,\xi_\beta(\sigma') = \sum_{\gamma,\,\delta} \rho_{\alpha\beta}^{\gamma\delta}\,\xi_\gamma(\sigma')\,\xi_\delta(\sigma)$$

where, for $\epsilon = 1$, the only non vanishing matrix elements are

$$\rho_{11}^{11} = \rho_{22}^{22} = e^{-ih/2}; \quad \rho_{12}^{21} = \rho_{21}^{12} = e^{ih/2}; \quad \rho_{21}^{21} = e^{-ih/2} - e^{3ih/2}. \tag{3.17}$$

This last expression coincides with the simplest R matrix of $SL(2)_q$. (spin $J = 1/2$). The general formulae are [26]:

$$\xi_M^{(J)}(\sigma) \equiv \sum_{-J \le m \le J} |J,\omega)_M^m\,\psi_m^{(J)}(\sigma); \quad \text{where} \tag{3.18}$$

$$|J,\omega)_M^m = \sqrt{\binom{2J}{J+M}}\, e^{ihm/2}$$

$$\times \sum_{\left(\frac{J-M+m-s}{2}\right)\ \text{integer}} e^{ihs(\omega+m)} \binom{J-M}{(J-M+m-s)/2} \binom{J+M}{(J+M+m+s)/2}. \tag{3.19}$$

The q-deformed binomial coefficients have been introduced by the formula:

$$\binom{P}{Q} \equiv \frac{\lfloor P \rfloor !}{\lfloor Q \rfloor ! \lfloor P-Q \rfloor !}; \quad \lfloor n \rfloor ! \equiv \prod_{r=1}^{n} \lfloor r \rfloor; \quad \lfloor r \rfloor \equiv \frac{\sin hr}{\sin h}. \tag{3.20}$$

The operators so defined satisfy the general exchange algebra [26]

$$\xi_M^{(J)}(\sigma)\, \xi_{M'}^{(J')}(\sigma') = \sum_{-J \le N \le J;\, -J' \le N' \le J'} (J, J')_{M\, M'}^{N'\, N}\, \xi_{N'}^{(J')}(\sigma')\, \xi_N^{(J)}(\sigma). \tag{3.21}$$

The coefficients $(J, J')_{M\, M'}^{N'\, N}$ of the R–matrix only depend upon the sign of $\sigma - \sigma'$. This matrix is conveniently written in terms of quantum group generators as follows: Introduce a set of Hilbert spaces $\mathcal{H}_J$ with states noted $|J, M>$, $-J \le M \le J$; together with operators $J_\pm$, J_3 such that:

$$J_\pm |J, M> = \sqrt{\lfloor J \mp M \rfloor \lfloor J \pm M + 1 \rfloor}\, |J, M \pm 1> \quad J_3 |J, M> = M\, |J, M> . \tag{3.22}$$

These operators satisfy the $SL(2)_q$ commutation relation

$$[J_+, J_-] = \lfloor 2J_3 \rfloor. \tag{3.23}$$

For $\sigma > \sigma'$, one has, in particular,

$$(J, J')_{M\, M'}^{N'\, N} = \Big(< J, M| \otimes < J', M'|\Big)\, \mathbf{R}\, \Big(|J, N> \otimes |J', N'>\Big), \tag{3.24}$$

$$\mathbf{R} = e^{(-2ihJ_3 \otimes J_3)} \big(1 + \sum_{n=1}^{\infty} \frac{(1 - e^{2ih})^n\, e^{ihn(n-1)/2}}{\lfloor n \rfloor !} e^{-ihnJ_3} (J_+)^n \otimes e^{ihnJ_3} (J_-)^n \big). \tag{3.25}$$

The last formula exactly reproduces the universal R matrix of $SL(2)_q$.

A similar discussion may be carried out for the A_N case[27]. However the quantum deformation is not the standard one.

5. References

1 V. Fateev and S. Lykyanov, Int. J. Mod. Phys. A3 (1988) 507.

2 F. Bais, P. Bouwknegt, M. Surridge and K. Schoutens, Nucl. Phys. B304 (1988) 348, 371.

3 A. Bilal and J.-L. Gervais, Phys. Lett. B206 (1988) 412.
A. Bilal and J.-L. Gervais, Nucl. Phys. B314 (1989) 646.

4 A. Bilal and J.-L. Gervais, Nucl. Phys. B318 (1989) 579.

A. Bilal and J.-L. Gervais, LPTENS Proceedings of the Conference "Infinite dimensional Lie algebras and Lie groups", Marseille 1988 (World-Scientific, Singapore).

5 V. Fateev and A.B. Zamolodchikov, Nucl. Phys. B280 [FS18] (1987) 644.

6 P. Mathieu, Phys. Lett. B208 (1988) 101; I. Bakas, Phys. Lett. B213 (1988) 313; K. Yamagishi, Phys. Lett. B205 (1988) 466.

7 O. Babelon, Phys. Lett. B215 (1988) 523.

8 J. Thierry-Mieg, Phys. Lett. B197 (1987) 368.

9 A. Bilal and J.-L. Gervais, Nucl. Phys. B326 (1989) 222.

10 S. Mandelstam, Phys. Reports 13C (1974) 259.

11 V.G. Kac, Proc. Int. Congr. Math., Helsinki 1978; Lecture Notes in Physics, vol 94, p 441 (Springer, Heidelberg, 1979).

12 J.-L. Gervais and A. Neveu, Nucl. Phys. B257 [FS14] (1985) 59, Comm. Math. Phys. 100 (1985) 15.

13 L. Brink and H.B. Nielsen, Phys. Lett. B45 (1973) 332.

14 A. Bilal and J.-L. Gervais, Phys. Lett. B187 (1987) 39.

15 A. Belavin, A.M. Polyakov and A.B. Zamolodchikov, Nucl. Phys. B241 (1984) 333.

16 J.-L. Gervais and A. Neveu, Nucl. Phys. B224 (1983) 329, B238 (1984) 125, 396.

17 J.-L. Gervais and A. Neveu, Nucl. Phys. B199 (1982) 59, B209 (1982) 125.

18 A.M. Polyakov, Mod. Phys. Lett. A2 (1987) 893.
V.G. Knizhnik, A.M. Polyakov and A.B. Zamolodchikov, Mod. Phys. Lett. A3 (1988) 819.

19 M. Kato and K. Ogawa, Nucl. Phys. B212 (1983) 443.

20 A.M. Polyakov, Phys. Lett. B103 (1981) 211; O. Alvarez, Nucl. Phys. B216 (1983) 125.

21 V. Dotsenko and V. Fateev, Nucl. Phys. B240 [FS12] (1984) 312.

22 E. Witten, Nucl. Phys. B268 (1986) 253.

23 D.J. Gross and A. Jevicki, Nucl. Phys. B283 (1987) 1.

24 A. Bilal and J.-L. Gervais, Nucl. Phys. B284 (1987) 397.

25 B. Zumino, Y.-S. Wu and A. Zee, Nucl. Phys. 239 (1984) 477.

26 J.-L. Gervais, "The quantum group structure of 2D gravity and minimal models" LPTENS 89/14 preprint, to be published in Com. in Math. Phys.

27 E. Cremmer and J.-L. Gervais "The quantum group structure associated with non-linearly extended Virasoro algebras" LPTENS preprint 89/19.

Integrable Quantum Chains and Quantum Groups

Rafael I. Nepomechie
Department of Physics
University of Miami, Coral Gables, FL 33124

Abstract

It has been observed that an open spin 1/2 XXZ quantum Heisenberg chain has the quantum symmetry $U_q[SU(2)]$. For q a root of unity, the central charge coincides with that of the $c < 1$ conformal minimal models. We review the solution of this model by the quantum inverse scattering method, and discuss the generalization to a spin 1 chain.

Quantum groups have begun to interest both mathematicians and physicists. This interest has been stimulated by the appearance of quantum groups in q analysis (see, e.g., Ref. 1), knot theory (see, e.g., Ref. 2), rational conformal field theory (see, e.g., Ref. 3), and as symmetries of dynamical systems. In an effort to better understand quantum groups, Mezincescu, Rittenberg, and the author have pursued [4] the latter application. We have focused on quantum spin chains, since these seem to provide the simplest examples of systems which have quantum algebras as symmetries.

After a brief introduction to quantum groups, we describe an open spin 1/2 quantum chain which, as shown by Pasquier and Saleur [5], has the quantum symmetry $U_q[SU(2)]$. This model is known [6] to be completely integrable. We review the solution of this model by the quantum inverse scattering method, as generalized by Sklyanin [7] for Hamiltonians with surface terms. An interesting feature of this model is that, for values of q equal to roots of unity, the central charge coincides with that of the $c < 1$ conformal minimal models. Finally, we discuss the generalization of these results to a spin 1 chain, for which one expects the central charge to be that of the $c < 3/2$ superconformal minimal models. This work is described in more detail in Ref. 4.

Quantum Groups

Detailed expositions of quantum groups have been presented in Refs. 8, 9, and 10. Here we focus on the simplest example, namely, the quantum universal enveloping algebra $U_q[SU(2)]$, where q is a complex number. This algebra is generated by J^z, J^+, J^-, which obey the relations

$$[J^z\,, J^{\pm}] = \pm J^{\pm}\,,$$
$$[J^+\,, J^-] = \frac{q^{2J^z} - q^{-2J^z}}{q - q^{-1}} \equiv [2J^z]_q\,. \tag{1}$$

In the limit $q \to 1$, these relations become the usual $SU(2)$ algebra. Thus, Eq. (1) defines a deformation of $SU(2)$.

One can readily verify that

$$C_2 = J^- J^+ + \left([J^z + \frac{1}{2}]_q\right)^2 \tag{2}$$

is the Casimir operator.

The representation theory of $U_q[SU(2)]$ has been studied in detail [11,5]. For q not a root of unity, the representations are equivalent to those of $U[SU(2)]$. However, for q a root of unity, the representation theory is more complicated. In particular, the Casimir operator becomes periodic, which signals the fact that there are only a finite number of distinct finite-dimensional irreducible representations.

For $U_q[SU(2)]$ there is an analogue of addition of angular momentum: given two representations $\vec{J}_1$, $\vec{J}_2$ of the algebra (1), the generators

$$J^z = J^z_1 \otimes 1 + 1 \otimes J^z_2\,,$$
$$J^{\pm} = J^{\pm}_1 \otimes q^{J^z_2} + q^{-J^z_1} \otimes J^{\pm}_2\,, \tag{3}$$

also obey this algebra.

Quantum Spin Chains

Consider a system of N spins arranged in a line, which interact according to the Hamiltonian

$$H = \sum_{n=1}^{N-1} \left\{ \sigma^x_n \sigma^x_{n+1} + \sigma^y_n \sigma^y_{n+1} + \frac{1}{2}(q + q^{-1}) \sigma^z_n \sigma^z_{n+1} \right\}$$
$$- \frac{1}{2}(q - q^{-1}) \left(\sigma^z_1 - \sigma^z_N \right), \tag{4}$$

where $\vec{\sigma}_n$, $n = 1, 2, \ldots, N$, are Pauli matrices. This model is known as an open spin 1/2 XXZ Heisenberg chain. In addition to the bulk terms consisting of nearest-neighbor interactions, the Hamiltonian also has surface terms which involve the spins at the two ends of the chain. For $q = 1$, it becomes the Hamiltonian of the XXX chain

$$H = \sum_{n=1}^{N-1} \vec{\sigma}_n \cdot \vec{\sigma}_{n+1} \,, \tag{5}$$

which has manifest $SU(2)$ symmetry.

Pasquier and Saleur [5] have recently made the interesting observation that the model (4) has $U_q[SU(2)]$ symmetry. To see this, one uses the rules (3) to "add" the spins on the N sites, obtaining *

$$\begin{aligned} J^z &= \sum_{n=1}^{N} \frac{\sigma_n^z}{2} \,, \\ J^{\pm} &= \sum_{n=1}^{N} q^{-\frac{\sigma^z}{2}} \otimes \cdots \otimes q^{-\frac{\sigma^z}{2}} \otimes \frac{\sigma_n^{\pm}}{2} \otimes q^{\frac{\sigma^z}{2}} \otimes \cdots q^{\frac{\sigma^z}{2}} \,. \end{aligned} \tag{6}$$

Computing the Casimir (2) for these generators, one finds that it is proportional to the Hamiltonian (4), up to an additive constant. It follows that this Hamiltonian has $U_q[SU(2)]$ symmetry:

$$[H \,, J^z] = 0 \,, \qquad [H \,, J^{\pm}] = 0 \,. \tag{7}$$

Before proceeding to the solution of this model, it is useful to review the solution of the *closed* XXZ chain, which is considerably easier.

Closed spin 1/2 XXZ chain

Instead of having N spins arranged in a line, let us suppose that they are arranged in a circle. Moreover, we take the Hamiltonian to be

$$H = \sum_{n=1}^{N} \left\{ \sigma_n^x \sigma_{n+1}^x + \sigma_n^y \sigma_{n+1}^y + \frac{1}{2}(q + q^{-1}) \sigma_n^z \sigma_{n+1}^z \right\} , \tag{8}$$

with the periodic boundary condition $\vec{\sigma}_{N+1} = \vec{\sigma}_1$. This defines the closed spin 1/2 XXZ chain.

---

* The generators $J_n^z = \sigma_n^z/2$, $J_n^{\pm} = \sigma_n^{\pm}/2 = (\sigma_n^x \pm i\sigma_n^y)/2$ satisfy the algebra (1). In general, the fundamental representation of the deformed algebra is the same as for the undeformed algebra.

The problem now is to find the eigenvectors and eigenvalues of this Hamiltonian. There are two known approaches to this problem. One approach is the "coordinate Bethe Ansatz", which was invented by Bethe [12] to solve the XXX chain, and which was generalized to the XXZ chain by Orbach [13] and Yang and Yang [14]. (The XXZ model is equivalent to the 6-vertex model, which was solved by Lieb [15].)

Here we describe an alternative approach to this problem, which is known as the Quantum Inverse Scattering Method (QISM). This approach has its roots in the work of Baxter (see, e.g., Ref. 16), and has been extensively developed by the Leningrad group [17,18,19]. In the QISM, a central role is played by the so-called R matrix. For the XXZ model, the R matrix is the 4 by 4 matrix

$$R(u) = \frac{1}{\operatorname{sh}\eta} \left(\begin{array}{cc|cc} \operatorname{sh}(u+\eta) & & & \\ & \operatorname{sh} u & \operatorname{sh}\eta & \\ \hline & \operatorname{sh}\eta & \operatorname{sh} u & \\ & & & \operatorname{sh}(u+\eta) \end{array} \right)$$

$$= \sum_{a=0}^{3} w_a(u) \sigma^a \otimes \sigma^a \,, \tag{9}$$

where $q = e^\eta$, and the functions $w_a(u)$ are given by

$$w_0 = \operatorname{sh}(u+\frac{\eta}{2})/2\operatorname{sh}\frac{\eta}{2}\,, \qquad w_1 = w_2 = 1/2\,, \qquad w_3 = \operatorname{ch}(u+\frac{\eta}{2})/2\operatorname{ch}\frac{\eta}{2}\,. \tag{10}$$

The variable u is called the spectral parameter. From the second representation of R in Eq. (9), we see that R acts on the tensor product of vector spaces $V_1 \otimes V_2$, where $V_1 = V_2 = V$ is a two-dimensional vector space which is called the auxiliary space. The R matrix is constructed so as to satisfy the Yang-Baxter equation

$$R_{12}(u-v)\, R_{13}(u)\, R_{23}(v) = R_{23}(v)\, R_{13}(u)\, R_{12}(u-v)\,. \tag{11}$$

Here $R_{12} \equiv R \otimes \begin{pmatrix} 1 & \\ & 1 \end{pmatrix}$, $R_{23} \equiv \begin{pmatrix} 1 & \\ & 1 \end{pmatrix} \otimes R$, etc. are 8 by 8 matrices which act on the tensor product of three vector spaces $V_1 \otimes V_2 \otimes V_3$, with $V_i = V$.

The QISM is based on two observations. The first observation is that to every R matrix there corresponds a completely integrable one-dimensional quantum spin system. To see this, let W_n, $n = 1, 2, \ldots, N$, denote the two-dimensional Hilbert space at site n; and for convenience, let W_0 denote the auxiliary space V. We now introduce the quantity $L_n(u)$, which coincides with the R matrix $R(u)$ in the space $W_0 \otimes W_n$:

$$L_n(u) \equiv R_{0n}(u) = \sum_{a=0}^{3} w_a(u) \sigma^a \otimes \sigma_n^a$$

$$= \begin{pmatrix} w_0 + w_3\sigma_n^3 & w_1\sigma_n^- \\ w_1\sigma_n^+ & w_0 - w_3\sigma_n^3 \end{pmatrix}, \tag{12}$$

where the functions $w_a(u)$ are given by Eq. (10). Thus, L_n is a 2 by 2 matrix of operators at the site n. Next, let us replace the third vector space V_3 on which the Yang-Baxter equation (11) acts by the quantum space W_n. Correspondingly, we make the replacements

$$R_{13} \to \overset{1}{L}_n \,, \qquad R_{23} \to \overset{2}{L}_n \,, \tag{13}$$

where $\overset{1}{L} \equiv L \otimes \begin{pmatrix} 1 & \\ & 1 \end{pmatrix}$, $\overset{2}{L} \equiv \begin{pmatrix} 1 & \\ & 1 \end{pmatrix} \otimes L$. It follows from the Yang-Baxter equation that L_n obeys the relation

$$R_{12}(u-v)\, \overset{1}{L}_n(u)\, \overset{2}{L}_n(v) = \overset{2}{L}_n(v)\, \overset{1}{L}_n(u)\, R_{12}(u-v)\,. \tag{14}$$

We next define the monodromy matrix $T(u)$ as the product of all the L matrices:

$$T(u) \equiv L_N(u) L_{N-1}(u) \cdots L_1(u)\,. \tag{15}$$

It follows from the previous equation that the monodromy matrix obeys

$$R_{12}(u-v)\, \overset{1}{T}(u)\, \overset{2}{T}(v) = \overset{2}{T}(v)\, \overset{1}{T}(u)\, R_{12}(u-v)\,. \tag{16}$$

This important equation is sometimes referred to as the "fundamental relation." Finally, we define the transfer matrix $t(u)$ as the trace (over the auxiliary space) of the monodromy matrix:

$$t(u) \equiv \operatorname{tr} T(u)\,. \tag{17}$$

It follows from the fundamental relation that the transfer matrices commute

$$[t(u)\,, t(v)] = 0\,. \tag{18}$$

(To see this, simply multiply both sides of the fundamental relation by the inverse ${R_{12}}^{-1}$ and take the trace, using the cyclic property of the trace.) One can verify the important fact that the XXZ Hamiltonian (8) is proportional to the logarithmic derivative of the transfer matrix

$$H \sim \frac{d}{du} \ln t(u) \Big|_{u=0} + const\,. \tag{19}$$

(See, e.g., Ref. 18.) From the commutativity of the transfer matrices, we see that

$$[H\,, t(u)] = 0\,. \tag{20}$$

We conclude that we have a one-parameter family of commuting, conserved operators. This signals that the model is completely integrable.

We have seen how to construct a completely integrable Hamiltonian from a solution of the Yang-Baxter equation. The second observation of the QISM is that the Hamiltonian can be diagonalized by the so-called "algebraic Bethe Ansatz." Since the monodromy matrix $T(u)$ is a 2 by 2 matrix of operators, it is of the form

$$T(u) = \begin{pmatrix} A(u) & B(u) \\ C(u) & D(u) \end{pmatrix} . \tag{21}$$

The fundamental relation (16) implies a set of commutation relations among these entries, including

$$\begin{aligned} [B(u)\,, B(v)] &= 0\,, \\ A(u)B(v) &= f(v,u)B(v)A(u) - g(v,u)B(u)A(v)\,, \\ D(u)B(v) &= f(u,v)B(v)D(u) + g(v,u)B(u)D(v)\,, \end{aligned} \tag{22}$$

where the functions f and g are given by

$$f(u,v) = \frac{\mathrm{sh}(u-v+\eta)}{\mathrm{sh}(u-v)}\,, \qquad g(u,v) = \frac{\mathrm{sh}\,\eta}{\mathrm{sh}(u-v)}\,. \tag{23}$$

One can construct a "vacuum" state $|0>$ which is annihilated by $C(u)$

$$C(u)|0>= 0\,, \tag{24}$$

and which is an eigenstate of $A(u)$ and $D(u)$

$$A(u)|0>= \alpha(u)|0>\,, \qquad D(u)|0>= \delta(u)|0>\,, \tag{25}$$

for all values of u. Using the B operators as creation operators, one can further construct the state

$$|v_1,\ldots,v_M>\equiv B(v_1)\ldots B(v_M)|0>\,, \tag{26}$$

where $\{v_1,\ldots,v_M\}$ are a set of parameters which are to be determined. Using the commutation relations (22), one can now prove the important result that this state is an eigenstate of the transfer matrix $t(u) = \mathrm{tr}\,T(u) = A(u) + D(u)$

$$t(u)|v_1,\ldots,v_M>= \tau(u)|v_1,\ldots,v_M>\,, \tag{27}$$

with eigenvalue $\tau(u)$ given by

$$\tau(u) = \alpha(u)\prod_{k=1}^{M} f(v_k,u) + \delta(u)\prod_{k=1}^{M} f(u,v_k)\,, \tag{28}$$

provided that the parameters $\{v_1, \ldots, v_M\}$ satisfy the Bethe equations

$$\frac{\alpha(v_m)}{\delta(v_m)} = \prod_{\substack{k=1 \\ k \neq m}}^{M} \frac{f(v_m, v_k)}{f(v_k, v_m)}, \quad m = 1, \ldots, M. \tag{29}$$

For a given value of M, the Bethe equations can in principle be solved for $\{v_1, \ldots, v_M\}$. Once these parameters are determined, the eigenvalues of the Hamiltonian (8) follow from Eqs. (28) and (19).

It is known that the XXZ model is critical for values of q given by $q = e^{i\gamma}$, $0 < \gamma < \pi$; and for the ground state $M = N/2$. (See, e.g., Ref. 16.) Let us consider the "thermodynamic limit" $N \to \infty$. In this limit, the Bethe equations can be solved analytically, and one can calculate [20] the ground state energy E_0 as a function of N. Since the model is conformally invariant, it is characterized by a value of the central charge c. Conformal invariance implies [21] that the ground state energy scales with N according to

$$E_0(N) = N e_\infty - \frac{\pi \zeta c}{6N} + O(N^{-1}). \tag{30}$$

Here ζ is a factor, which for the XXZ model, is given by [22]

$$\zeta = \frac{\pi \sin \gamma}{\gamma}. \tag{31}$$

From the calculated expression for $E_0(N)$, it follows that the central charge of the XXZ model is $c = 1$.

Chains with spin $s > 1/2$

In order to solve quantum spin chain models with spin greater than 1/2, the only known approach is the QISM. The closed spin s XXZ model was solved by this method in Ref. 24; the corresponding XXX model was solved earlier in Ref. 25.

Let us consider the closed spin 1 XXZ model. The R matrix is given by [26,27]

$$ {}_{ss}R(u) = \left(\begin{array}{ccc|ccc|ccc}
1 & & & & & & & & \\
 & a & & c & & & & & \\
 & & b & & d & & e & & \\ \hline
 & c & & a & & & & & \\
 & & d & & f & & d & & \\
 & & & & & a & & c & \\ \hline
 & & e & & d & & b & & \\
 & & & & & c & & a & \\
 & & & & & & & & 1
\end{array} \right), \tag{32}$$

where

$$a = \frac{\operatorname{sh} u}{\operatorname{sh}(u+2\eta)}\,, \qquad b = \frac{\operatorname{sh} u \operatorname{sh}(u-\eta)}{\operatorname{sh}(u+\eta)\operatorname{sh}(u+2\eta)}\,, \qquad c = \frac{\operatorname{sh} 2\eta}{\operatorname{sh}(u+2\eta)}\,,$$

$$d = \frac{\operatorname{sh} u \operatorname{sh} 2\eta}{\operatorname{sh}(u+\eta)\operatorname{sh}(u+2\eta)}\,, \qquad e = \frac{\operatorname{sh} \eta \operatorname{sh} 2\eta}{\operatorname{sh}(u+\eta)\operatorname{sh}(u+2\eta)}\,, \qquad f = a + e\,. \tag{33}$$

This matrix acts on the tensor product of vector spaces $V_s \otimes V_s$, where the auxiliary space V_s is three-dimensional.

The corresponding monodromy matrix ${}_sT(u)$ and transfer matrix ${}_st(u)$ can be constructed following the procedure described above. Moreover, one finds that the corresponding Hamiltonian is [26]

$$H = \sum_{n=1}^{N} H_{n,n+1}\,,$$

$$H_{n,n+1} = \Big\{ \vec{J}_n \cdot \vec{J}_{n+1} - (\vec{J}_n \cdot \vec{J}_{n+1})^2$$

$$- (q + q^{-1} - 2)\big[(J_n^x J_{n+1}^x + J_n^y J_{n+1}^y)(J_n^z J_{n+1}^z) + (J_n^z J_{n+1}^z)(J_n^x J_{n+1}^x + J_n^y J_{n+1}^y)\big]$$

$$+ \frac{1}{2}(q - q^{-1})^2 \big[J_n^z J_{n+1}^z + (J_n^z)^2 + (J_{n+1}^z)^2 - (J_n^z J_{n+1}^z)^2\big]\Big\}\,, \tag{34}$$

where $\vec{J}_n$ are spin 1 generators of $SU(2)$, and $\vec{J}_{N+1} = \vec{J}_1$. Notice that the spin 1 Hamiltonian contains quartic terms $(\vec{J}_n \cdot \vec{J}_{n+1})^2$, which are not present in the spin 1/2 model.

By construction, this model is completely integrable. However, the Hamiltonian cannot be diagonalized by the algebraic Bethe Ansatz approach which was described above. The difficulty is that the monodromy matrix ${}_sT(u)$ is a 3 by 3 matrix instead of 2 by 2. This difficulty is surmounted by constructing an associated "spin 1/2" monodromy matrix ${}_\sigma T(u)$, whose auxiliary space is two-dimensional. The two monodromy matrices intertwine according to the relation

$${}_{\sigma s}R_{12}(u-v)\; {}_\sigma\overset{1}{T}(u)\; {}_s\overset{2}{T}(v) = {}_s\overset{2}{T}(v)\; {}_\sigma\overset{1}{T}(u)\; {}_{\sigma s}R_{12}(u-v)\,, \tag{35}$$

where ${}_{\sigma s}R$ is given by

$${}_{\sigma s}R(u) = \left(\begin{array}{ccc|ccc} a & & & & & \\ & b & & d & & \\ & & c & & d & \\ \hline & d & & c & & \\ & & d & & b & \\ & & & & & a \end{array}\right), \tag{36}$$

where

$$a = \mathrm{sh}(u+3\eta/2)\,, \qquad b = \mathrm{sh}(u+\eta/2)\,, \qquad c = \mathrm{sh}(u-\eta/2)\,, \qquad d = (\mathrm{sh}\,\eta\,\mathrm{sh}\,2\eta)^{1/2}\,.$$

This matrix, which obeys a generalized Yang-Baxter equation, acts on the tensor product of vector spaces $V_\sigma \otimes V_s$, where the auxiliary space V_σ is two-dimensional.

It follows from the relation (35) that the corresponding "spin 1/2" transfer matrix

$$_\sigma t(u) \equiv \mathrm{tr}\,_\sigma T(u) \tag{37}$$

commutes with the spin 1 transfer matrix

$$[_\sigma t(u)\,,\, _s t(v)] = 0\,. \tag{38}$$

This implies that two transfer matrices have common eigenvectors. The transfer matrix $_\sigma t(u)$ can be diagonalized by the algebraic Bethe Ansatz described above. Finally, the eigenvalues of the spin 1 transfer matrix $_s t(u)$ can be determined using the fact that[19]

$$_s t(u) = {_\sigma t}(u-\eta/2)\ {_\sigma t}(u+\eta/2) - \Delta\{_\sigma T(u)\}\,, \tag{39}$$

where $\Delta\{T(u)\}$ denotes the quantum determinant [19] of the matrix $T(u)$.

Open spin 1/2 XXZ chain

Having reviewed the QISM for closed chains, we now return to our original problem of the open spin 1/2 XXZ chain, with the Hamiltonian (4). A generalization of this model, involving more general surface terms, was first solved by Alcaraz, et al. [6] using the coordinate Bethe Ansatz approach, and later by Sklyanin [7] using QISM. Here we follow the latter approach, since we are also interested in the higher-spin generalization.

Sklyanin found that surface terms require a nontrivial generalization of the QISM. It is necessary to introduce 2 by 2 c-number matrices $K_-(u)$, $K_+(u)$, which obey the relations

$$R_{12}(u-v)\overset{1}{K}_-(u)R_{12}(u+v)\overset{2}{K}_-(v) = \overset{2}{K}_-(v)R_{12}(u+v)\overset{1}{K}_-(u)R_{12}(u-v)\,, \tag{40}$$

and

$$\begin{aligned} R_{12}(-u+v)\overset{1}{K}{}_+^{t_1}(u)R_{12}(-u-v-2\eta)\overset{2}{K}{}_+^{t_2}(v) \\ = \overset{2}{K}{}_+^{t_2}(v)R_{12}(-u-v-2\eta)\overset{1}{K}{}_+^{t_1}(u)R_{12}(-u+v)\,, \end{aligned} \tag{41}$$

respectively. For the XXZ model, these matrices are given by

$$K_-(u) \sim \begin{pmatrix} \mathrm{sh}(u+\xi_-) & \\ & -\mathrm{sh}(u-\xi_-) \end{pmatrix}, \quad K_+(u) \sim \begin{pmatrix} \mathrm{sh}(u+\eta+\xi_+) & \\ & -\mathrm{sh}(u+\eta-\xi_+) \end{pmatrix}, \tag{42}$$

where ξ_-,ξ_+ are arbitrary parameters. These matrices enter into the definition of the transfer matrix $t(u)$, which is given by

$$t(u) = \mathrm{tr}\, K_+(u) T(u) K_-(u) T^{-1}(-u) \,. \tag{43}$$

It can be shown that these transfer matrices commute

$$[t(u)\,, t(v)] = 0 \,. \tag{44}$$

Identifying the Hamiltonian

$$H \sim \frac{d}{du} t(u)\Big|_{u=0} + const \,, \tag{45}$$

it follows that the transfer matrices commute with the Hamiltonian

$$[H\,, t(u)] = 0 \,, \tag{46}$$

and that the model is completely integrable.

Explicit evaluation of the expression (45) for the Hamiltonian yields

$$H = \sum_{n=1}^{N-1} \Big\{ \sigma_n^x \sigma_{n+1}^x + \sigma_n^y \sigma_{n+1}^y + \frac{1}{2}(q+q^{-1}) \sigma_n^z \sigma_{n+1}^z \Big\} + \frac{1}{2}(q-q^{-1}) \Big(\sigma_1^z \,\mathrm{cth}\, \xi_- + \sigma_N^z \,\mathrm{cth}\, \xi_+ \Big) \,. \tag{47}$$

This is indeed the Hamiltonian of an open XXZ chain, but with more general surface terms than those in Eq. (4). In particular, we see that $U_q[SU(2)]$ symmetry is realized when the parameters take the values $\xi_- = -\infty$, $\xi_+ = \infty$.

The Hamiltonian can be diagonalized by the algebraic Bethe Ansatz (with some modifications which are due to the presence of the K matrices). Set $q = e^{i\pi/(\mu+1)}$, and $M = N/2$. The central charge can again be determined by a finite-size scaling analysis. The ground state energy E_0 as a function of N, in the thermodynamic limit, was calculated numerically by Alcaraz, et al. [6], and later analytically by Hamer, et al. [23]. On the other hand, for an open chain, conformal invariance implies that [21]

$$E_0(N) = N e_\infty + f_\infty - \frac{\pi \zeta c}{24 N} + O(N^{-1}) \tag{48}$$

(cf. Eq. (30)). In this way the central charge is found to be

$$c = 1 - \frac{6}{\mu(\mu+1)} \,. \tag{49}$$

For $\mu = 2, 3, \ldots$, these are precisely the values of c of the unitary conformal minimal models.

Open spin 1 XXZ chain

Finally we come to the open spin 1 XXZ chain, which we have recently investigated [4]. We have solved the Sklyanin relations (40), (41) for the K matrices, and have found

$$K_-(u) \sim \begin{pmatrix} \mathrm{sh}(u+\xi_-)\,\mathrm{sh}(u-\eta+\xi_-) & & \\ & -\,\mathrm{sh}(u-\xi_-)\,\mathrm{sh}(u-\eta+\xi_-) & \\ & & \mathrm{sh}(u-\xi_-)\,\mathrm{sh}(u+\eta-\xi_-) \end{pmatrix},$$

$$K_+(u) \sim \begin{pmatrix} \mathrm{sh}(u+\eta-\xi_+)\,\mathrm{sh}(u+2\eta-\xi_+) & & \\ & -\,\mathrm{sh}(u+\eta+\xi_+)\,\mathrm{sh}(u+2\eta-\xi_+) & \\ & & \mathrm{sh}(u+\xi_+)\,\mathrm{sh}(u+\eta+\xi_+) \end{pmatrix}. \tag{50}$$

As for the spin 1/2 case, the K matrices are diagonal, and depend on parameters $\xi_\pm$. In fact, these K matrices can be expressed in terms of the spin 1/2 K matrices, in a manner similar to that by which the R matrices (32), (36) can be expressed in terms of the spin 1/2 R matrix (9) [19].

Moreover, we have constructed the spin 1 transfer matrix ${}_st(u)$. The corresponding Hamiltonian

$$\begin{aligned} H = \sum_{n-1}^{N-1} H_{n,n+1} + \frac{1}{4}(q^2 - q^{-2})\Big\{ \\ \big[\mathrm{cth}\,\xi_- + \mathrm{cth}(\xi_- - \eta)\big] J_1^z + \big[\mathrm{cth}\,\xi_- - \mathrm{cth}(\xi_- - \eta)\big] (J_1^z)^2 \\ - \big[\mathrm{cth}\,\xi_+ + \mathrm{cth}(\xi_+ - \eta)\big] J_N^z + \big[\mathrm{cth}\,\xi_+ - \mathrm{cth}(\xi_+ - \eta)\big] (J_N^z)^2 \Big\} \,, \end{aligned} \tag{51}$$

has the same bulk terms as the Zamolodchikov-Fateev Hamiltonian (34), but has surface terms as well. Notice that the surface terms include quadratic terms $(J^z)^2$, which are not present in the spin 1/2 model. This Hamiltonian has $U_q[SU(2)]$ symmetry for $\xi_- = -\infty$, $\xi_+ = -\infty$ [5,28].

In order to diagonalize the Hamiltonian, we have followed the strategy used to solve the higher-spin closed chains; namely, we have constructed an associated spin 1/2 transfer matrix ${}_\sigma t(u)$ which commutes with the spin 1 transfer matrix ${}_s t(u)$. We have diagonalized ${}_\sigma t(u)$ by the algebraic Bethe Ansatz, and in this way, we have been led to the following Bethe equations

$$\frac{\mathrm{sh}(v_m - \xi_+)\,\mathrm{sh}(v_m + \xi_- - \eta)}{\mathrm{sh}(v_m + \xi_+)\,\mathrm{sh}(v_m - \xi_- + \eta)} \left[\frac{\mathrm{sh}^2(v_m + \eta)}{\mathrm{sh}^2(v_m - \eta)}\right]^N \tag{52}$$
$$= \prod_{\substack{k=1 \\ k\neq m}}^{M} \frac{\mathrm{sh}(v_m - v_k + \eta)\,\mathrm{sh}(v_m + v_k + \eta)}{\mathrm{sh}(v_m - v_k - \eta)\,\mathrm{sh}(v_m + v_k - \eta)}\,, \quad m = 1, \ldots, M\,.$$

Moreover, we have determined the eigenvalues of ${}_s t(u)$, using a generalization of the relation (39). We are now performing a finite-size scaling analysis, in order to determine the central charge.

Further Questions

It appears that for q a root of unity, the spectrum of the open (spin 1/2) XXZ chain becomes truncated to that of a $c < 1$ model, corresponding to the truncation of the representations of $U_q[SU(2)]$. This phenomena has already been studied in some detail (see Refs. 5 and 29), and merits further investigation.

The fact that quantum spin chains are integrable presumably can be explained by an equivalence to free fields (current algebra) in 1+1 dimensions. Indeed, the fact that surface terms lower the central charge of the quantum spin chain is reminiscent of the Coulomb gas formalism of Dotsenko and Fateev [30]. It would be interesting to understand the relationship more precisely, since this might point to a role of $U_q[SU(2)]$ in the Coulomb gas.

Finally, we remark that it should be possible to investigate quantum chains with larger quantum symmetries, such as $U_q[SU(N)]$, along these lines.

Acknowledgments

The work described here was done in collaboration with L. Mezincescu and V. Rittenberg. This work was supported in part by the National Science Foundation under Grant No. PHY-87 03390.

References

1. T. Masuda, K. Mimachi, Y. Nakagami, M. Noumi and K. Ueno, C.R. Acad. Sci. Paris, *307*, Série I (1988) 559; L.L. Vaksman and Ya. S. Soibel'man, Funct. Anal. and Appl. *22* (1989) 170.

2. M. Wadati and Y. Akutsu, Prog. Theor. Phys. Supp *94* (1988); V.G. Turaev, Invent. Math. *92* (1988) 527; N. Yu. Reshetikhin, LOMI preprints E-4-87, E-17-87; A. N. Kirillov and N. Yu. Reshetikhin, LOMI preprint E-9-88.
3. A. Tsuchiya and Y. Kanie, Adv. Stud. in Pure Math. *16* (1988) 297; G. Moore and N. Seiberg, Phys. Lett. *B212* (1988) 451; L. Alvarez-Gaumé, C. Gomez and G. Sierra, Phys. Lett. *B220* (1989) 142; E. Witten, IASSNS-HEP-89/32; A. LeClair, Phys. Lett. *B230* (1989) 103.
4. L. Mezincescu, R.I. Nepomechie and V. Rittenberg, Miami preprint UMTG-150.
5. V. Pasquier and H. Saleur, Saclay preprint SPhT/89-031.
6. F.C. Alcaraz, M.N. Barber, M.T. Batchelor, R.J. Baxter and G.R.W. Quispel, J. Phys. *A20* (1987) 6397.
7. E.K. Sklyanin, J. Phys. *A21* (1988) 2375.
8. L.D. Faddeev, N. Yu. Reshetikhin and L.A. Takhtajan, Algebraic Analysis, *1* (1988) 129.
9. M. Jimbo, Int'l J. Mod. Phys. *A4* (1989) 3759.
10. V.G. Drinfel'd, J. Sov. Math. *41* (1988) 898.
11. P. Roche and D. Arnaudon, Lett. Math. Phys. *17* (1989) 295.
12. H. Bethe, Z. Phys. *71* (1931) 205.
13. R. Orbach, Phys. Rev. *112* (1958) 309.
14. C.N. Yang and C.P. Yang, Phys. Rev. *150* (1966) 321, 327.
15. E.H. Lieb, Phys. Rev. *162* (1967) 162.
16. R.J. Baxter, *Exactly Solved Models in Statistical Mechanics* (Academic Press, 1982).
17. P.P. Kulish and E.K. Sklyanin, Phys. Lett. *70A* (1979) 461.
18. L.D. Faddeev and L.A. Takhtajan, Russ. Math Surv. *34* (1979) 11.
19. P.P. Kulish and E.K. Sklyanin, *Lecture Notes in Physics*, Vol. 151 (Springer, 1982) 61.
20. H.J. de Vega and F. Woynarovich, Nucl. Phys. *B251*[FS13] (1985) 439; F. Woynarovich and H-P Eckle, J. Phys. *A20* (1987) L97.
21. H.W.J. Blöte, J.L. Cardy and M.P. Nightingale, Phys. Rev. Lett. *56* (1986) 742; I. Affleck, Phys. Rev. Lett. *56* (1986) 746.
22. C.J. Hamer, J. Phys. *A18* (1985) L1133.
23. C.J. Hamer, G.R.W. Quispel and M.T. Batchelor, J. Phys. *A20* (1987) 5677.
24. K. Sogo, Phys. Lett. *104A* (1984) 51; H.M. Babujian and A.M. Tsvelick, Nucl. Phys. *B265* [FS15] (1986) 24; A.N. Kirillov and N. Yu Reshetikhin, J. Phys. *A20* (1987) 1565; J. Sov. Math. *35* (1986) 2627.

25. H.M. Babujian, Nucl. Phys. *B215* [FS7] (1983) 317; L.A. Takhtajan, Phys. Lett. *87A* (1982) 479.
26. A.B. Zamolodchikov and V.A. Fateev, Sov. J. Nucl. Phys. *32* (1980) 298.
27. P.P. Kulish and N. Yu. Reshetikhin, J. Sov. Math. *23* (1983) 2435.
28. M.T. Batchelor, L. Mezincescu, R.I. Nepomechie and V. Rittenberg, J. Phys. *A*, in press.
29. F.C. Alcaraz, U. Grimm and V. Rittenberg, Nucl. Phys. *B316* (1989) 735.
30. Vl.S. Dotsenko and V.A. Fateev, Nucl. Phys. *B240* (1984) 312; *B251*[FS13] (1985) 691.

PARTICLE PHYSICS AND CALABI-YAU COMPACTIFICATION

R. ARNOWITT
Center for Theoretical Physics, Department of Physics
Texas A&M University, College Station, TX 77843-4242, USA

and

PRAN NATH
Department of Physics, Northeastern University
Boston, MA 02115, USA

ABSTRACT

The relation between the heterotic string compactified on a three generation Calabi-Yau manifold and low energy particle physics is described. Possible origins of supersymmetry breaking, intermediate scale breaking and the determination of the complex structure are discussed. Models possessing flux breaking of E_6 to $[SU(3)]^3$, matter parity (to preserve adequate proton stability) and intermediate scale breaking to the Standard Model, are shown to possess four additional $SU(3) \times SU(2) \times U(1)$ neutral states of mass $\lesssim$ O (1 TeV) [in addition to the three light generations of the Standard Model]. Two of these states mix with the Standard Model neutrinos leading to neutrinos with Majorana masses which are qualitatively estimated to be O (10^{-6} eV), O (1 eV), and O (1 eV).

1. Introduction

Over the past two years, there has been increased activity in efforts to extract particle physics consequences from string theory. One reason for this is that string theory is the major current contender for a grand unified theory, and thus it is important to see if it is in accord with existing low energy phenomena, and if it implies any new physics. Thus while supergravity GUTs[1] is in agreement with all existing data (and even requires that there be only three light generations to be experimentally consistent[2]) it does not resolve most of the unanswered questions about the Standard Model (e.g. the origin of Yukawa couplings) as well as still possessing virulent gravitational infinities.

Several superstring models have been examined in some detail. These differ mainly in assumptions on how the heterotic string[3] compactifies:

(i) Calabi-Yau/Conformal field theory models (e.g. the $CP^3 \times CP^3/Z_3$ Tian-Yau model[4] and the $CP^3 \times CP^2/(Z_3 \times Z_3')$ Schimmrigk/Gepner model[5])

(ii) Orbifold compactification models[6]

(iii) Fermionic 4-D string models (e.g.[7] flipped $SU(5) \times U(1) \times [U(1)]^3$).

We discuss in this report the Calabi-Yau/Conformal field theory models.

2. Overview of Calabi-Yau Models

We assume here that the 10 dimensional heterotic string with residual $E_6 \times E_8$ internal symmetry compactifies at scale $M_c \approx M_{P\ell} = 2.4 \times 10^{18}$ GeV to $M_4 \times K$, where M_4 is Minkowski space and K is a Calabi-Yau manifold. Physically interesting cases are the three generation manifolds. Two examples are the $CP^3 \times CP^3/Z_3$ Tian-Yau[4] and the Schimmrigk[5] $CP^3 \times CP^2/(Z_3 \times Z_3')$ manifolds. These can be described by the intersection of polynomials and are given respectively by

$$\begin{aligned} P_1 &\equiv \Sigma\, x_\alpha^3 + a_1(x_0 x_1 x_2) + a_2(x_0 x_1 x_3) = 0 \\ P_2 &= x_0 y_0 + \Sigma\, c_i x_i y_i + c_4 x_2 y_3 + c_5 x_3 y_2 = 0 \\ P_3 &= \Sigma\, y_\alpha^3 + b_1(y_0 y_1 y_2) + b_2(y_0 y_1 y_3) = 0 \end{aligned} \tag{1.1}$$

for Tian-Yau and

$$\begin{aligned} P_1 &\equiv \Sigma\, z_\alpha^3 + a_0(z_1 z_2 z_3) + a_1 z_0 \Sigma\, z_i z_{i+1} = 0 \\ P_2 &= \Sigma\, z_i x_i^3 + b_0 z_0 x_1 x_2 x_3 + b_1 z_0 \Sigma\, x_i^3 + b_2 \Sigma\, z_i x_{i+1}^3 \\ &\quad + b_3 \Sigma\, z_i x_{i+2}^3 + b_4(\Sigma\, z_1) x_1 x_2 x_3 = 0 \end{aligned} \tag{1.2}$$

for the Schimmrigk case. In Eqs. (1.1) and (1.2) $\alpha = 0,1,2,3$ while $i = 1,2,3$. The complex moduli a_i, b_i, c_i etc. are VEVs of the moduli fields e.g. $a_i = < A_i >$, $b_i = < B_i >$ etc. Phenomenological properties of *symmetric examples* of these cases have been studied. Here the Yukawa couplings can at least in part be calculated. Thus the symmetric Tian-Yau manifold, $\Sigma\, x_\alpha^3 = 0 = \Sigma\, y_\alpha^3$; $x_0 y_0 + x_1 y_1 + c(x_2 y_2 + x_3 y_3) = 0$, has been examined in some detail[8], and more recently Ross[9] has examined the Gepner 1×16^3 conformal field theory model† $\Sigma\, z_\alpha^3 = 0 = \Sigma\, y_\alpha^3$.

† It is generally argued[10] that conformal field theory models correspond to Calabi-Yau

However, there is reason to believe that the symmetric models may not be physically realistic, and thus it is of interest to examine the general manifolds. However, it is not possible at present to calculate all the Yukawa couplings for the general manifolds (either by conformal field theory methods or topological methods). One can ask, instead, the following question:

> What constraints must the complex structure and Yukawa couplings obey if the model is to reduce to the Standard Model at low energy, and what *model independent predictions* exist for any model that does reduce to the Standard model.

We find below for the three generation models the following model independent predictions:

(i) The existence of four new $SU(3) \times SU(2) \times U(1)$ neutral "exotic" particles of electroweak size mass which interact with Higgs and lepton particles.

(ii) Existence of Majorana neutrino masses whose size can be qualitatively estimated.

Since any acceptable string model *must* reduce to the Standard Model at low energy, one may say that either this class of string models are incorrect or the above predictions hold.

3. Conditions for Achieving the Standard Model

For string theory to reduce to the Standard Model at low energy it must obey several relatively specific constraints. We summarize here what these conditions are, and discuss in more detail in Secs. 4 and 5 how they may be achieved.

At compactification, the 4D symmetry of the theory is $E_6 \times (N = 1$ supergravity) with the zero mass particles lying in the 27 and $\overline{27}$ representations of E_6[13]. The non-simply connected nature of the manifolds allows for "flux breaking" to occur at M_c. The only possibilities that leave the Standard Model unbroken are $E_6 \to [SU(3)]^3$, $SU(6) \times U(1)$, E_6. The latter two possibilities are not promising (the absence of an adjoint representation makes it difficult to see how one may break them beyond $SU(5)$) and so we assume that the flux

---

compactification at a specific point in moduli space. Conformal field theory allows a rigorous calculation of Yukawa couplings for genus zero, while the topological approach[11] omits non-perturbative effects (e.g. instanton contributions) and are unnormalized. Note, however, that the non-perturbative effects vanish for $(27)^3$ couplings[12].

breaking yields

$$E_6 \to SU(3)_C \times SU(3)_L \times SU(3)_R \equiv [SU(3)]^3 \tag{3.1}$$

where C = color and L, R = left, right. Under $[SU(3)]^3$ the 27 representation decomposes into lepton L, quark Q and conjugate quark Q^c nonets:

$$27 = L_r^\ell(1,3,\bar{3}) \oplus Q_\ell^a(3,\bar{3},1) \oplus (Q^c)_a^r(\bar{3},1,3) \tag{3.2}$$

where $a, \ell, r = 1,2,3$ are $SU(3)_{C,L,R}$ triplet indices. One may express the L, Q, Q^c fields in terms of the conventional particles of low energy physics. One has

$$\begin{aligned} L_i &= [\ell_i = (\nu_e, e)_i;\ e_i^c;\ H_i \equiv (\phi_i^+, \phi_i^0);\ H_i' \equiv (\phi_i^{0'}, \phi_i^{-'});\ \nu_i^c, N_i] \\ Q_i &= [q_i^a = (u^a, d^a)_i;\ H_3 \equiv D_i^a];\ Q^c = [u_{ai}^c;\ d_{ai}^c;\ H_3' \equiv D_{ai}^c]\ . \end{aligned} \tag{3.3}$$

Here i is a generation index H, H' are the usual SUSY Higgs doublets and D^a, D_a^c are the Higgs color triplets of $SU(5)$ GUT theory. ν^c and N are $SU(5)$ singlets and N is an O (10) singlet. A similar decomposition holds for the $\overline{27}$.

After flux breaking one is left with the following massless states[8,9]:

$$9 \text{ generations of } L\ ; \quad 6 \text{ mirror generations of } \overline{L}\ . \tag{3.4}$$

In addition, for the Tian-Yau manifold there are 7 generations of Q and Q^c and 4 mirror $\bar{Q}$ and $\bar{Q}^c$, and for the Schimmrigk manifold there are 3 generations of Q and Q^c and no remaining light mirror quarks.

To obtain the Standard Model, one needs a further symmetry breaking at an intermediate scale M_I, reducing $[SU(3)]^3$ to $SU(3) \times SU(2) \times U(1)$. This can be achieved dynamically, if the two $SU(3) \times SU(2) \times U(1)$ singlets of the 27 and $\overline{27}$ i.e. N, ν^c and $\bar{N}$, $\bar{\nu}^c$ grow VEVs. The dynamical interactions available to bring this about are the following:

(i) Supersymmetry Breaking

This is assumed to lead to soft breaking masses m = O (10^3 GeV) and to occur at or near M_c.

(ii) Renormalizable Interactions

These contribute W_r to the superpotential of the form $(27)^3$ and $(\overline{27})^3$.

(iii) Non-Renormalizable Interactions

These contribute W_{nr} to the superpotential of the form $(27\ \overline{27})^n/M_c^{2n-3}$, $n \geq 2$ and arise from integrating out the heavy fields.

The intermediate scale breaking can then be expected to arise in the following way[8]. The Renormalization Group equations of the renormalizable interactions W_r are assumed to turn the soft breaking (mass)^2 negative in one or more generations at the scale M_I. This allows the non-renormalizable interactions W_{nr} to cause N and ν^c VEV growth.

4. Some Difficult Questions

The above picture of intermediate scale breaking raises a number of important questions, e.g. What is the origin of supersymmetry breaking? Do the R.G. equations turn the soft breaking (mass)^2 negative sufficiently rapidly so that $M_I \gtrsim 10^{16}$ GeV (which is phenomenologically required as will be discussed below)? Within a given three generation manifold, what is the mechanism for fixing the complex structure? These are difficult problems whose solutions are only partially understood, and we discuss each briefly.

4.1 *Supersymmetry Breaking*

From both a fundamental and phenomenological viewpoint, supersymmetry breaking is a crucial question for string theory. Thus while supersymmetry will preserve a gauge symmetry once it has been set up, it is the supersymmetry breaking scale $m \ll M_P$ [where M_P is the Planck mass and m = O (1 TeV)] that is the origin of the gauge hierarchy.

We consider here the suggestion that supersymmetry breaking arises from a gaugino condensate in the hidden (E_8) sector (an idea that predates string theory). In this case Renormalization Group arguments suggest that[14]

$$< S\lambda\lambda > \approx h^3 \exp[-3S/(2b_0)]; \quad h \simeq M_P \tag{4.1}$$

where λ is the gaugino field, S and T are the dilaton and scaling chiral fields (S_R, the real part of S is related to the string constant g by $g^2 = 1/S_R$) and b_0 is the β function coefficient of the hidden sector R.G. equations. As a simple model, chose a superpotential of the form[15] $W = P(\Phi, T) + < S\lambda\lambda >$ where

$$P(\Phi, T) = \lambda(T)\Phi^3 + M(T)\Phi^2 \tag{4.2}$$

and Φ is an E_6 singlet field [i.e. from H_1 ($End\ T$)]. The mass M and coupling constant λ arise from non-perturbative world-sheet instanton effects[16]. They may be parameterized by

$$M = M_0(T)\exp[-\alpha T/M_P]\ ; \quad \lambda = \lambda_0(T)\exp[-\beta T/M_P] \tag{4.3}$$

with $\alpha, \beta = O(1)$. As is well-known, the tree effective potential, $V^{(0)}$, does not have a stable minimum. The one loop contribution, $V^{(1)} \cong (M_c^2/32\pi^2) Str M^2$, however, can stabilize the vacuum[15] since $Str M^2 \simeq -m_{3/2}^2 + n m_\varphi^2$, where $m_{3/2}$ is the gravitino mass, n is the number chiral scalar fields ϕ receiving soft breaking mass contributions, and m_ϕ is their mass. Minimizing $V = V^{(0)} + V^{(1)}$ yields $S \simeq b_o \delta T$ where $\delta = 2\beta - 4\alpha/3$. For large S and T, the leading term of V is

$$V \simeq 3(m_{3/2})^2[-1 + 2n/ST]; \quad m_{3/2} = |\, W\,|^2 /(8 S_R T_R^3)\,. \tag{4.4}$$

One therefore finds a minimum for V with

$$< S > \simeq (2n\delta b_0)^{\frac{1}{2}}; \quad < T > = (2n/\delta b_0)^{\frac{1}{2}} \tag{4.5}$$

yielding a gravitino mass of [15]

$$m_{3/2} \approx h \exp[-3 < S > /2b_0)]; \quad h \approx M_P\,. \tag{4.6}$$

For the Tian-Yau manifold there are n = 29(9) chiral field and the exponential is† $\approx 10^{-16}$ for $(\delta/b_0)^{\frac{1}{2}} \simeq 1$.

One is of course dealing in this argument with large numbers in an exponent which cannot be too accurately calculated. One sees, however, how a hierarchy might occur as a consequence of the large number of massless chiral fields at compactification (n $\gg$ 1) as occurs in all manifolds of this type.

4.2 *Renormalization Group Equations for SUSY m^2*

In order to calculate the scale M_I where the SUSY soft breaking (mass)^2 turn negative, one needs to know all the Yukawa couplings of the renormalizable $(27)^3$ and $(\overline{27})^3$ interactions. These, are not known, of course, for the general manifold. A detailed calculation, however, has been carried out[17] for the symmetric Tian-Yau model[8]. Here the topological approach[11] yields the unnormalized coupling

† Note, however, that the mass in Eq. (4.3) is not necessarily small, i.e. using Eq. (4.5) one has $\exp(-\alpha T/M_P) \approx 10^{-3}$ for $\alpha \approx \delta/4$.

constants. For reasonable sized normalizations ($10^{-1} \lesssim N \lesssim 10$), the intermediate scale breaking occurs very rapidly i.e.[17]

$$M_I/M_c \simeq 10^{-1} - 10^{-2} . \tag{4.7}$$

The reason for this is again the large number of light chiral fields (n $\gg$ 1) which drives the R.G. equations rapidly to spontaneous breaking. Thus barring exceptional circumstances, one expects $M_I \approx 10^{16}$ GeV for string models with n $\gg$ 1.

4.3 *Complex Structure*

At compactification, manifolds with different moduli $a_i = < A_i >$, $b_i = < B_i >$ etc. represent physically distinct theories giving rise to different Yukawa couplings. They are, however, a priori all degenerate. Supersymmetry breaking can remove this degeneracy, so that a unique complex structure can be energetically picked out[18]. This can come about because (as discussed in Sec. 3) supersymmetry breaking gives rise to non-vanishing VEVs $< N >$ and $< \nu^c >$ at $\mu = M_I$, and the moduli fields A_i, B_i etc. have *non-renormalizable couplings* to the 27 and $\overline{27}$ massless generations. Thus minimizing a total effective potential including the moduli interactions, the VEV growth of N and ν^c can be transmitted to the moduli determining $a_i = < A_i >$ etc.

To see how this might come about, we consider a toy model[19] involving a single moduli field A and one 27 and one $\overline{27}$ generation. We take for the non-renormalizable superpotential the form

$$W_{nr} = \frac{\lambda}{6}(27\ \overline{27})^3/M_c^3 + \frac{\lambda'}{6}\ A\ (27\ \overline{27})^3/M_c^4 \tag{4.8}$$

where λ, λ' are coupling constants and we have used the compactification mass M_c to scale these non-renormalizable terms. Keeping only the N and $\bar{N}$ components of the 27 and $\overline{27}$ for simplicity and setting $< N > = < \bar{N} >$ (for D flatness), the effective potential reads

$$V = (-m^2)N^2 + (\lambda' A/M_c + \lambda)^2 N^{10}/M_c^6 + \lambda'^2 N^{12}/M_c^8 \tag{4.9}$$

where the first term is the supersymmetry soft breaking mass term [m = O (1 TeV)] and we have assumed that the R.G. equations have turned this

$(\text{mass})^2$ negative. Minimizing V by $\partial V/\partial N = 0 = \partial V/\partial A$ yields two extrema:

$$\begin{aligned}
&\text{(i)}\ <N> = 0, \quad <A> = \text{arbitrary}; \quad <V> = 0 \\
&\text{(ii)}\ <N> = [mM_c^4/\sqrt{6}\lambda']^{\frac{1}{5}} \simeq 5 \times 10^{15}\ \text{GeV} \\
&\qquad a \equiv <A> = -(\lambda/\lambda')M_c \equiv O\,(M_P) \\
&\qquad <V> = -\frac{5}{6}\,(m^{12}M_c^8/6\lambda'^2)^{\frac{1}{5}} < 0\,.
\end{aligned} \tag{4.10}$$

We see that the absolute minimum of V determines *both* $<N>$ and the moduli a. (Solution (i) with no intermediate scale breaking leaves the moduli arbitrary and the different manifolds still degenerate.) Note also that the moduli a (in units of M_P) can differ significantly from zero, showing that the correct complex structure may be quite distant from the symmetric manifold (a=0).

5. Conditions for Achieving the Standard Model

We turn now to a detailed discussion of the conditions the complex structure and Yukawa couplings must obey in order that the model correctly reduce to the Standard Model. We assume that flux breaking of E_6 to $[SU(3)]^3$ has occurred at M_c, leaving massless the generations discussed in Eq. (3.4).

5.1 *Proton Stability*

A basic problem in all grand unified models is to prevent too rapid proton decay. In string models, this can be achieved in a natural fashion by requiring matter parity, M_2, invariance[20]. For the Tian-Yau manifold, M_2 is defined by $M_2 = C\ U_z$ where C acts on the Calabi-Yau coordinates by $C\ (x_2, y_2) \leftrightarrow (x_3, y_3)$ and U_z is an element of $[SU(3)]^3$ i.e. diag $U_z = (1,1,1) \times (\alpha,\alpha,\alpha) \times (\alpha,\alpha,\alpha)$ with $\alpha^3 = 1$. Since $[SU(3)]^3$ is conserved at M_c, any manifold that is C invariant preserves M_2. From Eq. (1.1) one sees that the general form of the Tian-Yau manifold for M_2 invariance requires $a_1 = a_2$, $c_2 = c_3$, $c_4 = c_5$, $b_1 = b_2$ i.e.

$$\begin{aligned}
P_1 &= \Sigma\ x_\alpha^3 + a_1(x_0x_1x_2 + x_0x_1x_3) = 0 \\
P_2 &= x_0y_0 + c_1x_1y_1 + c_2(x_2y_2 + x_3y_3) + c_4(x_2y_3 + x_3y_2) = 0 \\
P_3 &= \Sigma\ y_\alpha^3 + b_1(y_0y_1y_2 + y_0y_1y_3) = 0\,.
\end{aligned} \tag{5.1}$$

This is still a fairly general manifold. M_2 can also be defined for the Gepner model[9], though it is not yet clear that it can be defined for the general Schimmrigk model.

Matter parity invariance guarantees adequate proton stability provided[21] (i) the D and D^c quarks of Eq. (3.3) are superheavy ($M_D \gtrsim 10^{15}$ GeV) and (ii) the "exotic" generations (those generations other than the three light Standard Model generations) are sufficiently heavy ($\gtrsim 10^8$ GeV). We will see below how these conditions arise.

5.2 *Intermediate Scale Breaking at M_I*

As discussed in Sec. 3, intermediate scale breaking of $[SU(3)]^3$ to the Standard Model requires that $< N_i >$ and $< \nu_i^c >$ be non-zero [where N_i, ν_i^c are defined in Eq. (3.3)]. To determine whether this occurs one must minimize the effective potential $V = V_m + V_F + V_D$ where V_m is the SUSY soft breaking mass term, V_F is the F term arising from $W_{nr} = \Sigma\ \lambda_i(27_i \times \overline{27}_i)^n / M_c^{2n-3}$ and V_D is the D term. We assume that C invariance holds at $\mu = M_c$ and hence M_2 invariance holds for $\mu \geq M_I$. The general form of V is then[22]

$$
\begin{aligned}
V = & \sum_i (-m_i^2)(x_i + y_i + z_i + w_i) + \sum_i (-\bar{m}_i^2)(\bar{x}_i + \bar{y}_i + \bar{z}_i + \bar{w}_i) \\
& + \Sigma\ \frac{n^2 \lambda_i^4}{M_c^{4n-6}} (f_i f_i^\dagger)^{n-1} [x_i + \bar{x}_i + y_i + \bar{y}_i + z_i + \bar{z}_i + w_i + \bar{w}_i] \\
& + \frac{g_L^2}{6} [\{\Sigma[x_i - \bar{x}_i + y_i - \bar{y}_i - \tfrac{1}{2}(z_i - \bar{z}_i + w_i - \bar{w}_i)]\}^2 \\
& + 3 \mid \Sigma(H_i^{'\dagger} \nu_i^c - \bar{\nu}_i^{c\dagger} \bar{H}_i') \mid^2] \\
& + \frac{g_R^2}{8} [\{\Sigma(-z_i + \bar{z}_i + w_i - \bar{w}_i)\}^2 \\
& + \frac{4}{3} \{\Sigma[x_i - \bar{x}_i - \tfrac{1}{2}(y_i - \bar{y}_i + z_i - \bar{z}_i + w_i - \bar{w}_i)]\}^2 \\
& + \{\Sigma(y_i - \bar{y}_i - z_i + \bar{z}_i + w_i - \bar{w}_i)\}^2 + 4 \mid \Sigma(N_i^\dagger \nu_i^c - \bar{\nu}_i^{c\dagger} \bar{N}_i) \mid^2]
\end{aligned} \tag{5.2}
$$

where $x_i = N_i N_i^\dagger$, $y_i = \nu_i^c \nu_i^{c\dagger}$, $z_i = H_i H_i^\dagger$, $w_i = H_i' H_i^{'\dagger}$, etc., m_i, $\bar{m}_i = O\,(1\text{ TeV})$ are the SUSY soft breaking masses, $f_i = N_i \bar{N}_i + \nu_i^c \bar{\nu}_i^c + H_i \bar{H}_i + H_i' \bar{H}_i'$, and $g_{L,R}$ are the $SU(3)_{L,R}$ gauge coupling constants. (We have assumed here for simplicity a flat Kähler potential. The results, however, do not change qualitatively for the general case.)

Eq. (5.2) is a rather complicated potential. One can, however, prove the following result[22]: if $SU(2) \times U(1)$ is preserved at M_I, i.e. $< H_i >= 0 = < H_i' >$ (as is phenomenologically necessary) then the lowest lying extrema is the one that preserves M_2 invariance. Thus proton stability is maintained by the intermediate scale breaking. Under these circumstances one can always make a linear transformation so that the only non-zero $< N_i >$ lies in one C-even generation which we label $i = 1$, and the only non-zero $< \nu_i^c >$ lies in one C-odd generation labeled $i = 2$. One finds then for the non-zero VEVs the expressions[22]:

$$< N_1 > \cong \left[\frac{\Sigma_1^2 M_c^{4n-6}}{2n^2(2n-1)\lambda_1^4}\right]^{\frac{1}{4n-4}} \quad ; \quad < \nu_2^c > \cong \left[\frac{\Sigma_2^2 M_c^{4n-6}}{2n^2(2n-1)\lambda_2^4}\right]^{\frac{1}{4n-4}} \tag{5.3}$$

$$\begin{aligned} < N_1 >^2 - < \bar{N}_1 >^2 &\cong \frac{1}{4}\, g_R^2 (g_L^2 + \frac{1}{4}\, g_R^2)^{-1} \\ &\quad [2(g_L^2 + g_R^2)\Delta_1^2 + (g_R^2 - 2g_L^2)\Delta_2^2] \\ < \nu_2^c >^2 - < \bar{\nu}_2^3 >^2 &\cong \frac{1}{4}\, g_R^2 (g_L^2 + \frac{1}{4}\, g_R^2)^{-1} \\ &\quad [2(g_L^2 + g_R^2)\Delta_2^2 + (g_R^2 - 2g_L^2)\Delta_1^2] \end{aligned} \tag{5.4}$$

where $\Sigma_i^2 = m_i^2 + \bar{m}_i^2 > 0$ and $\Delta_i^2 = m_i^2 - \bar{m}_i^2$. Eq. (5.4) shows that there are deviations from exact D-flatness.

One may estimate the size of the VEVs from Eq. (5.3). One finds[22]

$$< N_1 > \gtrsim O\ (10^{15}\ \text{GeV}) \text{ for } n \gtrsim 3, 4 \tag{5.5}$$

where n is the power in $(27\ \overline{27})^n$ in W_{nr}. Since the D-quark mass is O $(\lambda^3 < N >)$ (where λ^3 is one of the $(27)^3$ coupling constants) we see that the condition for adequate proton stability in Sec. (5.2) requires that the *complex structure must exclude the* $n = 2$ *term in* W_{nr} *for the intermediate scale symmetry breaking* $i = 1, 2$ *generations.* Along with C invariance at compactification, this represents one of the basic constraints on the complex structure to achieve an acceptable low energy phenomenology.

An important parameter we will need in future discussion is

$$\epsilon \equiv < \nu_2^c > / < N_1 > = \left[\left(\frac{\Sigma_2}{\Sigma_1}\right)^{\frac{1}{2}} \frac{\lambda_1}{\lambda_2}\right]^{\frac{1}{n-1}} \quad ; \quad n \gtrsim 3, 4\ . \tag{5.6}$$

In general, Calabi-Yau coupling constants at the symmetric point in moduli space are comparable, and from calculations of couplings away from the symmetric point for the Tian-Yau[8] or Schimmrigk manifold[23] one would not expect them to differ by more than a factor of 10^3. Thus unless a coupling or mass is unexpectedly suppressed one roughly estimates

$$30 \gtrsim \epsilon \gtrsim 0.03\ . \tag{5.7}$$

5.3 *Exotic Generations*

Aside from the three light generations of the Standard Model (guaranteed by the index theorem) there are the additional "exotic" generations and mirror generations of Eq. (3.4) remaining after flux breaking. One expects the extra $6L + 6\bar{L}$ generations (and $4Q + 4\bar{Q}$ and $4Q^c + 4\bar{Q}^c$ for the Tian-Yau case) to pair up and become superheavy. Unless this happens, one will not get a correct value for $\sin^2\theta_W$, and as discussed in Sec. (5.2) this is required for matter parity invariance to control proton decay.

Whether the exotic generations become superheavy is a dynamical phenomena, as their mass growth arises from W_{nr} after intermediate scale breaking when $< N_1 >, < \nu_2^c >$ become non-zero[24]. The general form of W_{nr} is

$$W_{nr} = \sum_{i,j=1,2} \lambda_{ij}(27_i\overline{27}_j)^n/M_c^{2n-3} + \sum_{i,j}{}' \tilde{\lambda}_{ij}(27_i\overline{27}_j)^{n_e}/M_c^{2n_e-3} \tag{5.8}$$

where Σ' means that terms linear in $i,j = 1$ or 2 (where $< N_1 >, < \nu_2^c >$ are non-zero) are excluded. The mass matrix of the exotic generations come from the terms quadratic in them with the remaining factors in the $i,j = 1,2$ generations and hence replaced by VEVs $< N_1 >$ or $< \nu_2^c >$. One finds that the *exotic generations become superheavy provided*[24] $n_e < n$ e.g. $M_{\rm exotic} \gtrsim 10^{15}$ GeV for $n_e = 2$, $n \gtrsim 3, 4$. Thus the condition of heavy exotic generations does not put a stringent constraint on the complex structure, once Eq. (5.5) holds since the $n_e = 2$ term is normally expected to be present.

6. Light Exotic Particles

While the condition on W_{nr} of Sec. (5.3) makes all the exotic generations superheavy, there are four additional states that remain light[24] (i.e. of electroweak size mass). These lie in the $i = 1,2$ symmetry breaking generations and are $SU(3) \times SU(2) \times U(1)$ neutral. One may locate them by explicitly carrying out the Goldstone/Higgs analysis of intermediate scale breaking, and examining the mass matrix.

We first note that when $[SU(3)]^3$ breaks to $SU(3) \times SU(2) \times U(1)$, there are 12 massless Goldstone bosons in the mass spectrum which are absorbed by the 12 vector bosons which become superheavy. Second while Eq. (5.4) shows that $< V_D > = O\ (m^4)$ and hence small, the mass matrix from V_D i.e. $M^D_{ab} = < \partial^2 V_D / \partial\phi_a \partial\phi_b >$ has superheavy elements. This leads to 12 superheavy bosons. These are just the 12 spin zero components of the (broken) massive SUSY vector multiplet. Further, one may show that the mass matrix of the V_F part of Eq. (5.2), $M^F_{ab} = < \partial^2 V_F / \partial\phi_a \partial\phi_b >$ has only elements of O (m^2) and hence are light. Hence any states that are orthogonal to the 12 Goldstone bosons and 12 superheavy SUSY Higgs bosons, are candidates for light exotic states. The renormalizable $(27)^3$ and $(\overline{27})^3$ interactions, make some of these states superheavy. However, these interactions do not give mass to the N_i and ν^c_i fields. Direct calculation shows that there are precisely four combinations of N_i, ν^c_i, $i = 1, 2$ orthogonal to the 12 Goldstone and 12 superheavy Higgs states[24].

One is lead to the following theorem:

Consider a three generation Calabi-Yau model for which (i) E_6 has flux breaking to $[SU(3)]^3$ at M_c and (ii) $[SU(3)]^3$ breaks to the Standard Model at M_I in a fashion that maintains matter parity invariance. If we denote $i = 1, 2$ as the C-even, odd states with non-vanishing VEVs $< N_1 >$, $< \nu^c_2 >$, then this model always has four extra "exotic" light states of electroweak size mass (in addition to the three light Standard Model generations) given by

$$n_1 = \frac{1}{\sqrt{2}}\,(N_1 + \bar{N}_1); \quad n_2 = \cos\theta\ N_2 + \sin\theta\ \bar{\nu}^c_1$$
$$\hat{\nu}^c_2 = \frac{1}{\sqrt{2}}\,(\nu^c_2 + \bar{\nu}^c_2); \quad \bar{n}_2 = \cos\theta\ \bar{N}_2 + \sin\theta\ \nu^c_1 \tag{6.1}$$

where $\epsilon \equiv \tan\theta = < \nu^c_2 > / < N_1 >$.

The above theorem is quite general in that hypotheses (i) and (ii) are very weak conditions to obtain a theory consistent with low energy phenomena and a sufficiently stable proton. The theorem guarantees the existence of the four light states of Eq. (6.1). Any additional light exotic states can occur only if the complex structure produces "accidental" zeros in the mass matrix.

The additional light states do not effect the value of $\sin^2\theta_W$. If we assume that the only light states below M_I are the three Standard Model generations (in accord with Sec. (5.3)) and one pair of light Higgs, $\sin^2\theta_W$ is determined by

M_I. One finds in the one loop approximation[25]

$$\sin^2\theta_W = 0.230 \text{ for } M_I = 1 \times 10^{16} \text{ GeV} \tag{6.2}$$

in agreement with the experimental value of 0.230 ± 0.005. This is in accord with the expected size of M_I discussed in Sec. (4.3).

7. Conditions for Light Higgs

Any phenomenologically acceptable model must have one pair of light Higgs doublets, H and H', to complete the $SU(2) \times U(1)$ breaking at the electroweak scale. (Two pairs of Higgs doublets could produce unwanted flavor changing neutral interactions, as well as making the fit for $\sin^2\theta_W$ more difficult.) We consider next the conditions needed to achieve a light pair of Higgs.

Superheavy Higgs masses arise from two sources: the renormalizable $(27)^3$ and $(\overline{27})^3$ and the non-renormalizable interactions. The former give rise to the mass terms:

$$W_{(27)^3} = -\lambda^3_{ijk}[H_i H'_j < N_k > + H_i < \nu^c_j > \ell_k] \tag{7.1}$$

where $\ell_k = (\nu_k, e_k)$ and λ^3_{ijk} is a $(27)^3$ coupling constant. (A similar form holds for $W_{(\overline{27})^3}$.) From Eq. (5.8), the non-renormalizable mass terms are

$$W_{nr} = \Sigma\, \lambda_{ijk\ell}(< 27_i \overline{27}_j >^{n_e - 1} / M_c^{2n_e - 3}) L_k \bar{L}_\ell\,. \tag{7.2}$$

Since matter parity is conserved, we can divide the mass matrix into M_2-even and M_2-odd parts. We also introduce the notation $n = (1, m)$ for the C-even generations and $r = (2, s)$ for the C-odd generations. (With this convention only $< N_1 >$ and $< \nu^c_2 >$ are non-zero.) The Higgs states are contained in the M_2-even mass matrix which has the form:

$$\begin{array}{c|cccccc|}
 & \bar\ell_2 & \bar\ell_{s'} & \bar H'_1 & \bar H'_{m'} & H_{m'} & H_1 \\
\hline
\ell_2 & 0 & 0 & 0 & 0 & M^3_{2m'} & M^3_{21} \\
\ell_s & 0 & M^1_{ss'} & 0 & 0 & M^3_{sm'} & M^3_{s1} \\
\bar H_1 & \bar M^3_{12} & \bar M^3_{1s'} & \bar M^2_{11} & \bar M^2_{1m'} & 0 & 0 \\
\bar H_m & \bar M^3_{m2} & \bar M^3_{ms'} & \bar M^2_{m1} & \bar M^2_{mm'} & M^1_{mm'} & 0 \\
H'_m & 0 & 0 & 0 & M^1_{mm'} & M^2_{mm'} & M^2_{m1} \\
H'_1 & 0 & 0 & 0 & 0 & M^2_{1m'} & M^2_{11} \\
\hline
\end{array} \tag{7.3}$$

In Eq. (7.3) M^1_{ij} comes from W_{nr} while $M^2_{ij} = -\lambda^3_{ij1} < N_1 >$, $M^3_{ij} = -\lambda^3_{ij2} < \nu^c_2 >$, $\bar{M}^2_{ij} = -\bar{\lambda}^3_{ij1} < \bar{N}_1 >$ and $\bar{M}^3_{ij} = -\bar{\lambda}^3_{ij2} < \bar{\nu}^c_2 >$. All these masses are superheavy. In general, one may show that all the eigenvalues of Eq. (7.3) are superheavy (and there would then be no light Higgs) unless some of the entries vanish (a phenomena that does indeed happen for symmetric manifolds). If we assume no fine tuning cancellations between the renormalizable $(27)^3$, $(\overline{27})^3$ and the non-renormalizable $(27\ \overline{27})^n$ couplings, then it turns out that there is only one phenomenologically acceptable way of getting a light Higgs i.e.

$$\lambda^3_{11n} = 0 \tag{7.4}$$

which implies $M^2_{1n} = 0$.

Eq. (7.4) is an additional constraint on the Yukawa couplings needed to achieve the Standard Model. One finds from Eq. (7.3) that the light H' Higgs then is $H' = H_1$ and the H Higgs is a linear combination of H_n, $\bar{H}'_n$, $\bar{\ell}_r$, but is mostly H_1. Initially the light Higgs are massless. They grow electroweak size masses after electroweak breaking.

8. Low Energy Interactions

Below the intermediate scale M_I, one may integrate out the superheavy fields with mass $M = O\ (M_I) \gtrsim 10^{15}$ GeV. Terms with three light fields gives rise to the low energy superpotential. Terms with neutrinos, Higgs and one heavy field give rise to seesaw neutrino masses. These are the interactions predicted by string theory that are accessible to experimental test at current and future accelerators, i.e. this is the physical output of the theory.

One finds for the low energy effective superpotential the result

$$\begin{aligned} W = &\{\lambda^{(\ell)}_{pp'} H' e^c_p \ell_{p'} + \lambda^{(u)}_{pp'} H q_p u^c_{p'} + \lambda^{(d)}_{pp'} H' q_p d^c_{p'}\} \\ &+ [(\lambda_p H \ell_p n_2 + \bar{\lambda}_p H \ell_p \bar{n}_2) + (m_1 n_2 \bar{n}_2 + m_2 n_2 n_2 + m_3 \bar{n}_2 \bar{n}_2) \\ &+ (m_4 n_1 \hat{\nu}^c_2 + m_5 n_1 n_1 + m_6 \hat{\nu}^c_2 \hat{\nu}^c_2)] + W_{\text{seesaw}} \; . \end{aligned} \tag{8.1}$$

In Eq. (8.1) $p = 1, 2, 3$ runs over the light generations, $\lambda^{(\ell)}_{pp'}$, $\lambda^{(u)}_{pp'}$, $\lambda^{(d)}_{pp'}$, λ_p, $\bar{\lambda}_p$ are coupling constant [which can be related to the fundamental constants in $(27)^3$, $(\overline{27})^3$ and $(27\overline{27})^n$] and the $m_i = O$ (1 TeV) are the masses of the new exotic states of Eq. (6.1). In addition to the above, the $SU(3)_L$ gauginos λ^L_a, $a = 4 \dots 7$ mix with the light leptons to give a Lagrangian

$$\mathcal{L}_{\text{gaugino}} = (g_L\ U_p \sin\theta) \ell_p \gamma^0 [\bar{n}_2 H'^{\dagger} + 2^{-\frac{1}{2}} n_1 \tilde{\ell}^{\dagger}_1] + h.c. \tag{8.2}$$

U_p is the component of λ_a^L in the light lepton sector.

The brace of Eq. (8.1) is just the Standard Model. The bracket of Eq. (8.1) and Eq. (8.2) represents the *new physics* of string theory in the low energy domain induced by the four new light exotic states.

9. Neutrino Masses

The existence of additional superstring neutral fields coupling to the leptons gives rise to neutrino masses after $SU(2) \times U(1)$ breaking occurs and Higgs bosons grow VEVs. Contributions to neutrino masses come from two sources: W_{seesaw} and the new light exotic interactions. The mass contribution to the seesaw interactions reads

$$\begin{aligned} W_{\text{seesaw}} = & [\lambda_{ps} < H > \nu_p N_s + \bar{\lambda}_{ps} < H > \nu_p \bar{N}_s + \lambda_{pm} < H > \nu_p \nu_m^c \\ & + \bar{\lambda}_{pm} < H > \nu_p \bar{\nu}_m^c] + \frac{1}{2}[M_s N_s^2 + \bar{M}_s \bar{N}_s^2 \\ & + M_m (\nu_m^c)^2 + \bar{M}_m (\bar{\nu}_m^c)^2] \end{aligned} \tag{9.1}$$

where M_s, $\bar{M}_s$ etc. are superheavy i.e. $\gtrsim 10^{15}$ GeV. One may integrate out the superheavy $N_s, \bar{N}_s$ $\nu_m^c, \bar{\nu}_m^c$ fields to obtain a Majorana mass matrix for the neutrinos:

$$\frac{1}{2}\nu_p \mu_{pp'} \nu_{p'}; \quad \mu_{pp'} = -\lambda_{ps} \frac{1}{M_s} \lambda_{p's} < H >^2 + \cdots . \tag{9.2}$$

Comparing with the charged lepton mass matrix, one finds

$$\mu_{pp'} \approx \frac{m_\ell^2}{M_s} \approx 10^{-6} \text{ eV}; \quad \text{for } m_\ell = m_\tau . \tag{9.3}$$

The new exotic interactions of Eqs. (8.1), (8.2) give a mass contribution of

$$W_{\text{exotic}} = \mu_p \nu_p n_2 + \bar{\mu}_p \nu_p \bar{n}_2 + [m_1 n_2 \bar{n}_2 + m_2 n_2 n_2 + m_3 \bar{n}_2 \bar{n}_2] . \tag{9.4}$$

Using again the charged lepton mass matrix one finds

$$\mu_p = \lambda_p < H > \simeq g\, (\frac{3m_e}{m_\tau})\, \epsilon < H > \tag{9.5}$$

where $\epsilon =< \nu_2^c > / < N_1 >$ and g is a coupling constant. Thus the set $(\nu_p, n_2, \bar{n}_2)$

couple in a 5 × 5 mass matrix:

$$
\begin{array}{c|ccccc|}
 & \nu_1 & \nu_2 & \nu_3 & n_2 & \bar{n}_2 \\
\nu_1 & \mu_{11} & \mu_{12} & \mu_{13} & \mu_1 & \bar{\mu}_1 \\
\nu_2 & \mu_{21} & \mu_{22} & \mu_{23} & \mu_2 & \bar{\mu}_2 \\
\nu_3 & \mu_{31} & \mu_{32} & \mu_{33} & \mu_3 & \bar{\mu}_3 \\
n_2 & \mu_1 & \mu_2 & \mu_3 & m_2 & m_1 \\
\bar{n}_2 & \bar{\mu}_1 & \bar{\mu}_2 & \bar{\mu}_3 & m_1 & m_3
\end{array}
\tag{9.6}
$$

It is easy to show that a matrix of this type has the following eigenvalues:

$$m_{\nu_1} = O\ (\mu_{pp'}) \approx 10^{-6}\ \text{eV} \tag{9.7}$$

$$m_{\nu_2}, m_{\nu_3} = O\ (\frac{\mu_p^2}{m_i}) \approx g^2(\frac{3m_e}{m_\tau})^2\epsilon^2\frac{< H >^2}{m_i} \approx 1\ \text{eV} \tag{9.8}$$

upon using the lower bound $\epsilon = 0.03$ of Eq. (5.6), and $g^2 = 10^{-2}$ and $m_i = 1$ TeV. In addition there are two large eigenvalues

$$m_4, m_5 = O\ (m_i) \approx 1\ \text{TeV}\ . \tag{9.9}$$

Thus two seesaws operate on the neutrino masses: the conventional one, m_ℓ^2/M_I, for ν_1 and the additional one, μ_p^2/m_i for ν_2 and ν_3, from the new exotic interactions predicted from intermediate scale breaking. These latter two neutrinos are expected to have a mass $\approx$ 1 eV unless g and ϵ are anomalously suppressed. The current experimental bound on neutrino masses is $m_\nu \lesssim 13$ eV and so a neutrino in the eV range may be accessible to future experiments. The theory also predicts neutrino oscillations which may be accessible to experimental tests.

10. Conclusions

One of the features that has emerged from this analysis, is that the non-renormalizable $(27\overline{27})^n$ couplings play as important a role in Calabi-Yau string theory in determining the low energy predictions as the renormalizable $(27)^3$ and $(\overline{27})^3$ interactions. Thus the role of the Planck mass states is more than just to make quantum gravity finite, and string theory appears to be a theory where high energy and low energy properties are intimately connected.

It is not clear yet whether there is any three generation Calabi-Yau manifold or conformal field theory model which is consistent with the Standard Model at low energy. However, we have determined the conditions any such model must obey if it is to correctly reproduce known low energy physics. One may use these constraints to limit the search for viable Calabi-Yau manifolds. In addition one may use these constraints to make model independent low energy predictions of Calabi-Yau string models. Two such predictions were discussed here, i.e. the existence of four new light $SU(3) \times SU(2) \times U(1)$ singlet states which couple to the Higgs and leptons, and qualitative estimates of neutrino masses.

Acknowledgements

This research was supported in part under National Science Foundation Grant Nos. PHY-8907887 and PHY-8706873 and the Texas Advanced Research Program No. 3043.

References

1. A.H. Chamseddine, R. Arnowitt and P. Nath, *Phys. Rev. Lett.* **49** (1982) 970. For a review of supergravity GUTs see P. Nath, R. Arnowitt and A.H. Chamseddine, *Applied N=1 Supergravity* (World Scientific, Singapore, 1983) and H.P. Nilles, *Phys. Reports* **110** (1984) 1.
2. D.V. Nanopoulos and D. Ross, *Phys. Lett.* **108B** (1982) 351.
3. D. Gross, J. Harvey, E. Martinec and R. Rohm, *Phys. Rev. Lett.* **55** (1985) 502; *Nucl. Phys.* **B258** (1985) 253.
4. G. Tian and S.T. Yau, Proc. of Argonne Symposium, *Anomalies, Geometry and Topology*, ed A. White (World Scientific, Singapore, 1985).
5. R. Schimmrigk, *Phys. Lett.* **193B** (1987) 75; D. Gepner, *Nucl. Phys.* **B296** (1988) 757.
6. L.E. Ibañez, H.P. Nilles and F. Quevedo, *Phys. Lett.* **192B** (1987) 332; L.E. Ibañez, J. Mas, H.P. Nilles and F. Quevedo, *Nucl. Phys.* **B301** (1988) 157.
7. I. Antoniadis, J. Ellis, J.S. Hagelin and D.V. Nanopoulos, *Phys. Lett.* **205B** (1988) 459.
8. B. Greene, K.H. Kirklin, P.J. Miron and G.G. Ross, *Phys. Lett.* **180B** (1986) 69; *Nucl. Phys.* **B278** (1986) 667; **B292** (1987) 606; S. Kalara and

R.N. Mohapatra, *Phys. Rev.* **D36** (1987) 3474.

9. G.G. Ross, in *Strings '89* ed. R. Arnowitt, R. Bryan, M. Duff and C. Pope (World Scientific, Singapore, 1989).
10. C. Vafa and N. Warner, *Phys. Lett.* **218B** (1989) 51 and D. Gepner, *Nucl. Phys.* **B296** (1988) 757.
11. A. Strominger and E. Witten, *Comm. Math. Phys.* **101** (1985) 341; A. Strominger, *Phys. Rev. Lett.* **55** (1985) 2545; P. Candelas, *Nucl. Phys.* **B298** (1988) 458.
12. J. Distler and B.R. Green, *Nucl. Phys.* **B309** (1988) 295.
13. P. Candelas, G.T. Horowitz, A. Strominger and E. Witten, *Nucl. Phys.* **B258** (1985) 46.
14. M. Dine, R. Rohm, N. Seiberg and E. Witten, *Phys. Lett.* **156B** (1985) 55.
15. G.G. Ross, *Phys. Lett.* **211B** (1988) 315.
16. M. Dine, N. Seiberg, X.-G. Wen and E. Witten, *Nucl. Phys.* **B278** (1986) 769; **B289** (1987) 319.
17. F. del Aguila and C.D. Coughlan, *Phys. Lett.* **215B** (1988) 93.
18. G.G. Ross, *Phys. Lett.* **200B** (1985) 441.
19. R. Arnowitt and P. Nath, unpublished.
20. M.C. Bento, L. Hall, and G.G. Ross, *Nucl. Phys.* **B292** (1987) 400.
21. R. Arnowitt and P. Nath, *Phys. Rev. Lett.* **62** (1989) 222.
22. P. Nath and R. Arnowitt, *Phys. Rev.* **D39** (1989) 2006.
23. J. Wu, private communication.
24. R. Arnowitt and P. Nath, *Phys. Rev.* **D40** (1989) 191.
25. P. Nath and R. Arnowitt, *Phys. Lett.* **62** (1989) 1437.

List of Participants

R. Acharya
Arizona State Univ.
Dept. of Physics
Tempe AZ 85281

R. Arnowitt
Texas A&M Univ.
Dept. of Physics
Coll. Station TX 77843

I. Batalin
Lebedev Phys. Inst.
Lenin Prospect 53
Moscow 117924 USSR

Z. Bern
Los Alamos National Lab.
Physics P.O. Box 1663
Los Alamos NM 87545

R. Bluhm
Indiana Univ.
Dept. of Physics
Bloomington IN 47405

D. Boyanovsky
Univ. of Pittsburgh
Dept. of Physics
Pittsburgh PA 15260

S. Bruce
Univ. of Notre Dame
Dept. of Physics
Notre Dame IN 46556

D. Caldi
Univ. of Connecticut
Dept. of Physics
Storrs CT 06268

L. Clavelli
Univ. of Alabama
Dept. of Phys. & Astr.
Tuscaloosa AL 35487

P. Coulter
Univ. of Alabama
Dept. of Phys. & Astr.
Tuscaloosa AL 35487

P. Cox
Texas A&I Univ.
Dept. of Physics
Kingsville TX 78363

M. Cvetic
Univ. of Pennsylvania
Dept. of Physics
Philadelphia PA 19104

D. Depireux
Univ. of Maryland
Dept. of Physics
College Park MD 20742

J. Dijkstra
Univ. of Alabama
Dept. of Math
Tuscaloosa AL 35487

P. Di Vecchia
Nordita
Blegdamsvej 17
Copenhagen 2100 Denmark

L. Dolan
Rockefeller Univ.
Dept. of Physics
New York NY 10021

P. Frampton
Univ. of North Carolina
Dept. of Physics
Chapel Hill NC 27599

P. Freund
Enrico Fermi Inst.
5640 Ellis Ave.
Chicago IL 60637

T. Gabriel
Oak Ridge Nat'l Lab.
Physics P.O. Box 2208
Oak Ridge TN 37831-6369

S. Gates
Univ. of Maryland
Dept. of Physics
College Park MD 20742

J.-L. Gervais
Lab. de Phys. Theorique
Ecole Normale Sup.
F-75231 Paris France

J. Hadley
Dept. of Phys. & Astr.
Georgia State Univ.
Atlanta GA 30303

P. Hall
Univ. of Alabama
Dept. of Math
Tuscaloosa AL 35487

B. Harms
Univ. of Alabama
Dept. of Phys. & Astr.
Tuscaloosa AL 35487

E. Harris
Univ. of Tennessee
Dept. of Physics
Knoxville TN 37996-1200

R. Holman
Carnegie-Mellon Univ.
Dept. of Physics
Pittsburgh PA 15213

S. Jones
Univ. of Alabama
Dept. of Phys. & Astr.
Tuscaloosa AL 35487

T. Kephart
Vanderbilt Univ.
Phys. & Astr. Box 1807 B
Nashville TN 37235

B. Keszthelyi
Univ. of Florida
Dept. of Physics
Gainesville FL 32611

G. Kleppe
Univ. of Florida
Dept. of Physics
Gainesville FL 32611

H. Konno
Univ. of Alabama
Dept. of Phys. & Astr.
Tuscaloosa AL 35487

A. Kostelecky
Indiana Univ.
Dept. of Physics
Bloomington IN 47405

B.-H. Lee
Univ. of Minnesota
School of Physics
Minneapolis MN 55455

T. McCarty
Univ. of Miami
Physics PO Box 248-046
Coral Gables FL 33124

J. Mogilski
Univ. of Alabama
Dept. of Math
Tuscaloosa AL 35487

A. Morozov
ITEP
B. Cheremushkinskaya ul., 25
117259 Moscow USSR

R. Nepomechie
Univ. of Miami
Physics PO Box 248-046
Coral Gables FL 33124

H. Nishino
Univ. of North Carolina
Dept. of Physics
Chapel Hill NC 27599

M. Pal'chik
Lebedev Inst.
Lenin Pr. 53
117924 Moscow USSR

J. Polchinski
Univ. of Texas
Dept. of Physics
Austin TX 78712

C. Pope
Texas A&M Univ.
Dept. of Physics
Coll. Station TX 77843

B. Radak
Univ. of Maryland
Dept. of Physics
College Park MD 20742

P. Ramond
Univ. of Florida
Dept. of Physics
Gainesville FL 32611

S.-J. Rey
Univ. of California
Dept. of Physics
Santa Barbara CA 93106

A. Schwarz
Inst. for Adv. Study
Dept. of Physics
Princeton NJ 08540

J. Shapiro
Rutgers Univ.
Dept. of Physics
New Brunswick NJ 08903

S. Shatashvili
Steklov Math. Inst.
Vavilov St. 42
117333 Moscow USSR

S. Shen
Texas A&M Univ.
Dept. of Physics
Coll. Station TX 77843

B.-S. Skagerstam
Dept. of Physics
Chalmers Univ. of Tech.
S-40220 Goteborg 5 Sweden

B. Spokoiny
Landau Inst.
GSP-1 Kosygin St. 2
117940 Moscow USSR

A. Stern
Univ. of Alabama
Dept. of Phys. & Astr.
Tuscaloosa AL 35487

R. Svoboda
Louisiana State Univ.
Physics & Astronomy
Baton Rouge LA 70803

C. Taylor
Case Western Reserve Univ.
Dept. of Physics
Cleveland OH 44106

C. Thorn
Univ. of Florida
Dept. of Physics
Gainesville FL 32611

A. Tseytlin
Lebedev Inst.
Lenin Prospect 53
117924 Moscow USSR

A. Tsuchiya
Tokyo Inst. of Tech.
Oh-okayama, Meguro-ku
Tokyo 152 Japan

S.-H. Tye
Cornell Univ.
Lab. of Nuclear Studies
Ithaca NY 14853

M. Walton
Dept. of Physics
Laval University
Quebec Canada G1K7P4

O. Yasuda
Univ. of North Carolina
Dept. of Physics
Chapel Hill NC 27599

S. Yost
Univ. of Florida
Dept. of Physics
Gainesville FL 32611